MW01154729

The Magnificent Machines of Milwaukee
and the Engineers who Created Them

Cover Illustrations

Top

One of the five Port Washington steam turbine-generators of the Wisconsin Electric Power Company, designed and constructed by Allis-Chalmers of West Allis. The plant was the most thermally efficient electric generating station in the world, starting from its initial operation in 1935 and continuing for approximately thirteen years. The five 80-megawatt turbine-generators served the company for almost seventy years, before being replaced with combined-cycle generating units. Photograph courtesy of We Energies. Further details are provided on page 147.

Bottom

Hamilton Aero Manufacturing Company of Milwaukee designed an all-metal airplane, which was dubbed the Hamilton Metalplane H-18 and christened *Maiden Milwaukee*. The principal designer of the aircraft was James McDonnel, who had previously worked for Stout and Ford, and incorporated similar features from those companies' airplanes into the H-18, along with some new ideas. The aircraft used a tubular frame with corrugated skin and a thick single wing projecting out of the fuselage underneath the open cockpit. It was powered by a 200-horsepower 'J-4' Wright radial engine and a Hamilton metal propeller. *Maiden Milwaukee* received the first US air certificate for an all-metal airplane. The airplane pictured is likely a Hamilton H-47—a later model which was better able to accommodate passengers and freight. It was photographed in St. Paul, Minnesota while freight was being transferred from a Rock Island railroad train to the airplane. Photo used with the permission of the Minnesota Historical Society. See page 135 for additional information.

Back

Several drawings of the Triple Expansion steam engine designed by Irving Reynolds of the E.P. Allis Company for installation at Milwaukee's North Point Pumping Station, superimposed over an artist's drawing of the facility. This water-pumping engine established a world record and proved so incredibly economical that it led to orders from major cities all over the country. The pumping station is long gone—replaced by a parking lot on Milwaukee's lakefront. However, the iconic water tower on the bluff continues to serve as a testament to the original pumping station. See page 44.

The Magnificent Machines of Milwaukee

and the Engineers who Created Them

Thomas H. Fehring, P.E.

ISBN-13: 978-1542549165
ISBN-10: 1542549167
Library of Congress Control Number: 2017901481

CreateSpace Independent Publishing Platform, North Charleston, SC

Produced by NorCENergy Consultants, LLC. While every effort was made to be both accurate and comprehensive in this publication, no guarantee is provided that the information contained is accurate either at the time of publishing or at any time in the future. If you spot a mistake or wish to cite additional Milwaukee-area innovations, please send a note to the author at tfehring@norcenergy.com. Because of the dynamic nature of the internet, web addresses or links contained in this book may have changed since publication and may no longer be valid.

Sponsored by the Milwaukee County Historical Society
910 North Old World Third Street
Milwaukee, WI 53203-1591
414.273.8288

April 2017

Dedicated to Milwaukee's future engineers and innovators—
perhaps including Bryce, Dylan, Sydney, Max, Sara, Zoe, Olivia and Beckett.

May these stories inspire them to create the future.

ABOUT THIS BOOK

This book tells the stories of the engineers who developed the incredible machines that were manufactured in the greater Milwaukee area. In doing so, it also attempts to show how Milwaukee became known as "The Machine Shop of the World." It has been suggested that the greater Milwaukee area contributed to the country's "age of largeness much as Silicon Valley, California, now pioneers in microelectronics, is presently contributing to the age of smallness."[1]

Since the late 1970s I have been exploring the Milwaukee area's rich industrial heritage, concentrating on the numerous engineering accomplishments and innovations that have occurred over the years. I found over one hundred significant engineering accomplishments, and recently started pulling together these discrete individual achievements to help tell the larger story of the incredible innovation that has occurred in the area. The individual accomplishments are like pieces of a puzzle that when assembled together in the correct pattern reveal a larger picture.

The book is filled with information and illustrations that highlight the machinery produced during this era. In each case, the book focuses on the engineering accomplishments these machines represent and on the individuals that are credited with their design and manufacture. Finally the book illustrates the "industrial commons" of engineering innovation that existed in Milwaukee and discusses the attributes that contributed to this innovation.

The commerce that resulted from the innovation of these industrial companies was essential to the development of the City and to the livelihood of many thousands of its citizens. Many of these companies survive in some form, and several have grown to become major international firms. I believe their stories reveal important characteristics that help to point the way toward enhanced innovation and commerce in the future.

Genesis

I started exploring Milwaukee's industrial heritage by happenstance. Early in my career, one of the managers I worked for brought a number of us engineers together and challenged us to "give back" to our professions by volunteering to work for our respective engineering societies. I took the advice to heart and showed up at a meeting of the local section of the American Society of Mechanical Engineers (ASME). When I asked how I could help, the executive committee offered to assign me to chair their history and heritage committee.

I said "yes" before I realized that the position had never been filled before, and that the Society was gearing up to celebrate its Centennial year by recognizing its history. However, since I knew I had the support of my employer, I started to explore ways of celebrating Milwaukee's engineering heritage. It occurred to me that many of Milwaukee's industrial companies were established because of some significant engineering innovations. So I set out to prepare a catalog of those engineering advances.

With the support of my department's secretaries, I sent out letters to the chief executive officers of dozens of area companies and asked them to nominate engineering accomplishments to this catalog. I was somewhat overwhelmed with the reaction—I began getting periodic packages in the mail, nominating achievements for consideration. Many of the submitted nominations included supporting materials such as books, photographs, and further historical documentation.

Armed with this information, I began to publish articles about Milwaukee's significant mechanical engineering achievements in ASME-Milwaukee's newsletter. The articles were well received and helped to generate additional nominations. Then, in 1980, I gathered articles from about fifty of these engineering accomplishments into a small book entitled *Mechanical Engineering: A Century of Progress*. It was a modest, self-published book, but well over a thousand copies were distributed and sold. It also resulted in a story in the Milwaukee Journal, which also helped to generate additional nominations.

Over the years, I have continued to document additional engineering achievements from Milwaukee and the surrounding area. For many years, however, I didn't have the time to do much with the information. Upon retirement I pledged to do something about it. My initial thought was to publish a book about one hundred local mechanical engineering accomplishments, and I set about to do that. But as I started the process I remembered some advice I received years ago about providing some context—my 'elephant epiphany.'

Describing the Elephant

Shortly after publishing the *Century of Progress* book, I was invited to present a paper at a meeting of the Society for the History of Technology, which was holding its annual meeting in Milwaukee that year. I wrote a paper entitled, *Technological Contributions of Milwaukee's Menomonee Valley Industries.*[2] It was presented as part of a panel that was discussing various aspects of Milwaukee's industrial heritage. It was a peer-reviewed session and one of the reviewers likened the various papers to a group of blind men describing an elephant. He was referring, of course, to an ancient, well-known Indian parable. We were all discussing various aspects of the elephant's anatomy, but collaboratively we weren't fully describing the animal. As one example of an aspect not described, he asked: "What became of all the arsenic?" One of the papers discussed Milwaukee's tanning industry. The reviewer noted that arsenic sulfide was often used for removing hair from animal hides, and that it may well have been disposed of in the Milwaukee River. The question was thought provoking, but I must admit that I have never fully investigated the disposition of any arsenic from Milwaukee's tanneries. However, over the years I have challenged myself to develop a better sense for the "elephant" that represents Milwaukee's industrial heritage.

As a result, I wanted this book to be more than a compilation of engineering innovations. I have tried to put Milwaukee's engineering innovations into context, and to explore questions like, "Why did these innovations occur here?" and "Was Milwaukee unique?" and "What can we learn from all of this that might help us in the future?"

To help accomplish this, I considered several different approaches to tell the story of Industrial Milwaukee before I ultimately decided to organize this book around the individual neighborhoods where manufacturing companies were located. By describing the important industrial companies in each neighborhood and the significant machines these companies manufactured, I could illustrate how their proximity may have been essential to the innovation that developed. I believe the approach also helps to provide some insights into the characteristics that were important to support the incredible innovation that occurred in these Milwaukee neighborhoods during this era.

As an aside, many, if not most, of the companies included in this book moved as they needed more space for growth. For example, E.P. Allis's 'Reliance Works' was initially located on the west bank of the Milwaukee River near the site of today's Marquette Park. As it needed room for expansion it moved to Walker's Point, and as it merged with other companies and required a much larger facility, it relocated operations to West Allis. The book discusses the companies in the neighborhoods where important engineering innovations occurred. In the case of E.P. Allis, the company is discussed in all three neighborhoods. For some other companies, I concentrated upon the location where a company built its most significant machinery.

I decided to include short bios of some of the principal engineers that were involved in this innovation. Again, the thought was that this would provide insight into what may have led to the engineering accomplishments they achieved.

I recognize that this effort comes far short of fully *describing the elephant*. Many additional aspects could have been included. For example, one of Milwaukee's most significant innovations was the successful adoption of pulverized coal for the generation of electricity. While the book discusses the significant reduction in the cost of electricity attributable to this innovation, it doesn't deal with the environmental impacts—the impact of the additional emissions of fly ash from power plant chimneys. The list of additional avenues for exploration into Milwaukee's industrial heritage is almost unlimited. That, perhaps, is one of the most exciting aspects of researching the past—there is always more of the story to be told. I hope to be able to continue to find other ways to help describe the elephant.

The Task

The decision to place Milwaukee's engineering innovations into historical context added a good deal of complexity, and time, to the process of writing this book. This I anticipated. What I didn't fully anticipate, however, was the ripe additional information that would be uncovered in the process. Frankly, almost every time I opened a different research avenue I stumbled across additional information about innovative Milwaukee companies and the machines they built. As a result, the information covered in the book expanded significantly.

At the same time, the process of writing the book became a good deal more interesting. I found myself thoroughly enjoying the process of discovery and of assembling the material into an orderly fashion. I found some of the stories fascinating—like that of Niels Christensen and his eventual perfection of the O-ring, of Joseph Merkel and his innovations in motorcycles, of Warren Johnson's pneumatic clocks, of Thomas Hamilton's *Maiden Milwaukee* airplane which received the first US air certificate for an all-metal aircraft, of Sig Haugdahl's *Wisconsin Special* speed-car, of Otto Widera's counterblow hammer built for Ladish Drop Forge, and of Reuben Stanley Smith's automated auto frame factory referred to as the "mechanical marvel."

I have attempted to focus on the companies that built truly innovative machines, and upon engineers who have made significant engineering innovations. Time did not permit inclusion of all of Milwaukee's industrial companies, and I am certain that I have overlooked some important firms and their innovations. While I spent a good deal of effort to identify the companies that created the most significant machines of the area during this period, I undoubtedly missed many—I apologize for any omissions.

A Word About Engineers

As noted, this book in an effort to explore and document the engineering creativity that has occurred in the greater Milwaukee area.

This effort takes a broad view of the engineering profession. The great scientist-engineer Theodore von Kármán once said, *"Scientists investigate that which already is; engineers create that which has never been."*[a] Most of the individuals that developed the machines illustrated in this book are referred to as 'engineers.' While a good number of the inventors had engineering degrees, many of them did not. In the 1800s, in particular, there were two tracks for obtaining and demonstrating engineering expertise—the 'shop culture' and the 'school culture.'[3] Some of the most famous engineers, such as Edwin Reynolds who served for a time as the president of the American Society of Mechanical Engineers, received their training under the shop culture, serving initially as apprentices working for other engineers who may or may not have attended an engineering college. Any of the innovators that demonstrated engineering expertise are considered engineers in this book.

This book also focuses on mechanical engineers, as opposed to practitioners of the other engineering professions, since mechanical engineers are the ones that design and build machines. An exception to this was made regarding the electrical engineers who designed motor controls, because of their importance in building devices to control machines, as well as Milwaukee's prominent role as a supplier of controls to operate the machines of the world.

Additional Notes

There are many engineers, draftsmen, and designers that were responsible for the achievements cited herein whose names are not included. During this period in history, it was "common precedent for an employer to take credit for work done by an employee."[4] As a result, many true innovators were never given credit for their work. This book is dedicated to the engineers, designers and builders whose work has gone unrecognized.

I would like to thank the many individuals that contributed to this effort and provided help, guidance and encouragement. They include (in no particular order) Ken Wirth and Charles Kempker of Johnson Controls, Marv Klowak from Briggs & Stratton, Charles Wright of A.O. Smith, Barry McNulty of We Energies, John Hoylman of Hoylmedia, Scott Kramer from Milwaukee Electric Tool, Karl Schneider of GE-Waukesha Motors, Thomas Pelt formerly from Bradley Corporation, Brian Eskra of Power Engineers Collaborative and Juhl Energy, Mark Frank, Otto Widera formerly assistant dean of the Marquette University college of engineering, John Widera of California Box, Theodore Wilinski from the Milwaukee Area Technical College, William O'Brien of Marquette University, Maripat Blankenheim from Harley-Davidson, Angela Hersil of Rexnord, John Favill formerly of Harley-Davidson, James Kieselburg director of the MSOE-Grohmann Museum, Megan Sharp from the Stuhr Museum of the Prairie Pioneer, Dennis Tollefson of the Waukesha Engine Historical Society, Douglas Robbins from the Microsoft Office Community, Bruce Erickson and William Finke formerly of Wisconsin Electric, Walt Smith of SmithPumps, Jeff Anthony of the Midwest Energy Research Consortium (M-WERC), Caley Clinton of Joy Global, Mark Sklar a descendant of Harry Soref of Master Lock, Bruce Boczkiewicz of Nordco Inc., Nick Wichert of Northwestern Mutual Life Insurance, Greg Meier of Cardinal Stritch, Joseph Zimmermann III, Sue Gast of Emerson-Vilter, Jean Davidson of the Davidson Family (H-D), Kevin Keefe formerly of Kalmbach Publishing, Bill Graffin of the Metropolitan Milwaukee Sewerage Distract, Steve Schaffer of the Milwaukee County Historical Society, Al Muchka of the Milwaukee Public Museum, Michael Schultz photographer of the Ladish forge, Thomas Bentley of Bentley World Packaging, Douglas Armstrong formerly of the Milwaukee Journal, John Bernaden formerly of Rockwell, Anne Bingham who

---

[a] This quotation is often credited to Albert Einstein. However, it appears that the statement was first made by Theodore von Kármán, a Hungarian-American mathematician, aerospace engineer and physicist. He is regarded as the outstanding aerodynamic theoretician of the twentieth century

provided editing advice, and the officers of ASME-Milwaukee including Allen Perkins their current president. I'm certain that I have missed many individuals that should have been listed, for which I apologize.

The staff of the Milwaukee Central Library's Frank P. Zeidler Humanities Department is amazing. They provided invaluable help—without their assistance, this book would never have come together. The staff answered my numerous questions quickly and they often went well beyond my inquiry to provide additional insights that were invaluable. Yet this team goes about their business with little fanfare and they often work anonymously. While on a family vacation in Singapore, I attempted to continue my progress on the book. The task at the time involved determining the original locations of a number of early Milwaukee companies. The library team usually had answers to my inquiries available the next morning, which allowed me to continue to do my writing while on the other side of the world! I suspect that few folks recognize what an incredible resource the City provides by funding this department. I not only thank them but also thank the City of Milwaukee for continuing to make this service available.

Finally, I would like to acknowledge the support of my family and especially my wife Suzan. Sue has put up with my strange hobby for years, all the while being dragged along to look at factories, hydroelectric plants, and numerous industrial artifacts. She has allowed me to spend countless hours in my attempt to record Milwaukee's industrial history. I am very grateful for her support and encouragement.

Thomas H. Fehring, P.E.

FOREWORD

Before Steve Jobs, there was Edward Allis. Before Silicon Valley, there was the Menomonee Valley. Before the army of innovators who ushered in the Digital Age, there were Milwaukeeans like Henry Harnischfeger, Lynde Bradley, Bruno Nordberg, A.O. Smith, Ole Evinrude, and dozens of others who helped lead America into the Machine Age. Some were immigrants and most lacked engineering degrees, but all had bright ideas and the talent, vision, and pluck to turn them into realities. Together they made their hometown the self-proclaimed Machine Shop of the World. From the QWERTY keyboard to the pressed-steel automobile frame to the rubber-tired tractor, it all started in the city that made beer famous.

Until *The Magnificent Machines of Milwaukee*, the stories of these innovations and the men behind them had been told largely in fragmentary fashion—an article here, a scholarly reference there. Tom Fehring has assembled the entire cast of characters in a single book that is a testament to talent, an ode to ingenuity, and a singular contribution to the history of American industry.

John Gurda

John Gurda
Milwaukee Writer and Historian

Table of Contents

INTRODUCTION TO MILWAUKEE'S INDUSTRIAL ACCOMPLISHMENTS

This book concentrates on a century of engineering innovation that occurred in Milwaukee from the 1860s through the mid-1900s. This was an era of incredible innovation, not only in Milwaukee but also throughout the United States. In the decades after the Civil War, Americans in great number turned their attention to inventive activity. The United States Patent Office issued four times as many patents in the 1860s as it had issued in its entire previous seventy years of existence. That number of patents doubled in the 1870s, and again in the 1880s.[5,b] Arguably, the most critical phase in the evolution of American industry occurred in these one hundred years following the Civil War. The period was truly a *Century of Progress.*

As we'll discover in this book, the Milwaukee area was one of the principal centers of industrial innovation in the United States. William Bruce, former secretary of Milwaukee's merchants' and manufacturers' organization, stated it well when he said,

> ***There is a great romance in Milwaukee industry.*** *It has its inception in the lowly condition which existed when many of our present industries were founded. It lies in the transformation of back yard shacks into great factories—from the tumble down huts where some gritty man with real vision started fashioning with his own hand some article which would be useful to mankind....Then came the step to quantity production, scientific organization and distribution on a systematic scale. Today those little enterprises have forged ahead and into huge industrial units of the city. Romance? I'll tell the world it is!*[6]

This book features the incredible machines built in the greater Milwaukee area during this period. In the process, it highlights the engineers who created these machines and summarizes the history of the numerous companies that helped the greater Milwaukee area achieve prominence in industrial design and manufacturing.

The stories of Industrial Milwaukee are not just of historical curiosity. The engineering innovation that occurred during this period resulted in commerce that was essential to the development of the City and to the livelihood of its many thousands of citizens. Many of these companies survive in some form, and several have grown to become major international firms. Their stories reveal important characteristics that may help to point the way toward enhanced innovation and commerce in the future.

THE INDUSTRIAL NEIGHBORHOODS OF MILWAUKEE

Today, we rightly think of Milwaukee as a city of neighborhoods. Each neighborhood has distinguishing characteristics. Some are largely residential or commercial. Some historically were highly industrialized. This book explores early Milwaukee's history in several discrete industrial neighborhoods.

As in most cities, manufacturing in the Milwaukee area was concentrated—even before restricted by zoning laws. Manufacturing companies tend to need similar things—transportation, access to affordable metals and other materials, availability of a productive and capable workforce, access to water and an available energy source to power the machinery. These attributes caused industrial companies to locate in discrete Milwaukee-area neighborhoods.

---

[b] Note: References are designated numerically and are located at the back of the book. Comments are designated alphabetically and are footnoted.

As illustrated in the next section, manufacturing in Milwaukee started on the banks of the Milwaukee River where waterpower was available to run the needed machinery and where the river could be used to transport manufactured goods. Soon, however, the steam engine became affordable and generally available, allowing manufacturing activities to move away from congested areas along the river to other areas of the City. Initially, companies relocated from the Milwaukee River into the Walker's Point area, which provided a rich environment for incubation of small manufacturing firms. The concentration of industrial commerce led to the Walker's Point neighborhood becoming an effective *industrial commons,* which helped to seed and incubate other companies. As these companies grew, they eventually spread out into the balance of the Menomonee Valley. The Walker's Point/Menomonee Valley location provided industrial companies with access to Lake Michigan for transportation of goods by ship, and access to rail transportation afforded by the Chicago, Milwaukee and St. Paul Railway. Suitable housing was also nearby for its productive workforce.

Bay View also became an early industrial center—largely because it was the home of the Milwaukee Rolling Mill. The mill was a critical asset to Milwaukee's industrial development because it provided an affordable source of iron and eventually steel—both of which were essential to most area companies. Other industries were attracted to the Bay View area, some to provide services to the mill and others to capitalize on the neighborhood's industrial site attributes.

As companies in Milwaukee's Walker's Point and Menomonee Valley continued to grow, they needed to expand their manufacturing capabilities further. Companies generally followed the rail lines to locate suitable factory space located in what was then the "outskirts" of the city. Access to reliable rail transportation was a critical requirement for any manufacturer of large machinery. As a result, it was inevitable that industry would move into areas along the rail lines that also provided room for expansion, access to water, and an available workforce.

Some companies moved west along the tracks of the Chicago, Milwaukee and St. Paul Railway into the communities of West Milwaukee, into the newly formed city of West Allis and to the city of Waukesha. Others expanded to the south, also generally aligned along the railway corridors. And still others built their manufacturing plants to the north, along what was once called the *Beer Line* railway on the west bank of the Milwaukee River, or along the tracks of the Chicago and Northwestern Railway on the east side of the river. Finally, several notable companies expanded northwest, alongside the tracks of the Chicago, Milwaukee and St. Paul Railway—an area that is now referred to as Milwaukee's 30th Street Industrial Corridor.[c]

This book explores each of the important industrial neighborhoods and attempts to show how the companies were interconnected—and how this helped to foster Milwaukee's industrial growth.

[c] Complicating the factory neighborhood picture, some companies expanded into more than one neighborhood—others moved several times. This book attempts to highlight the locations where significant engineering accomplishments occurred and often ignores intermediate company sites.

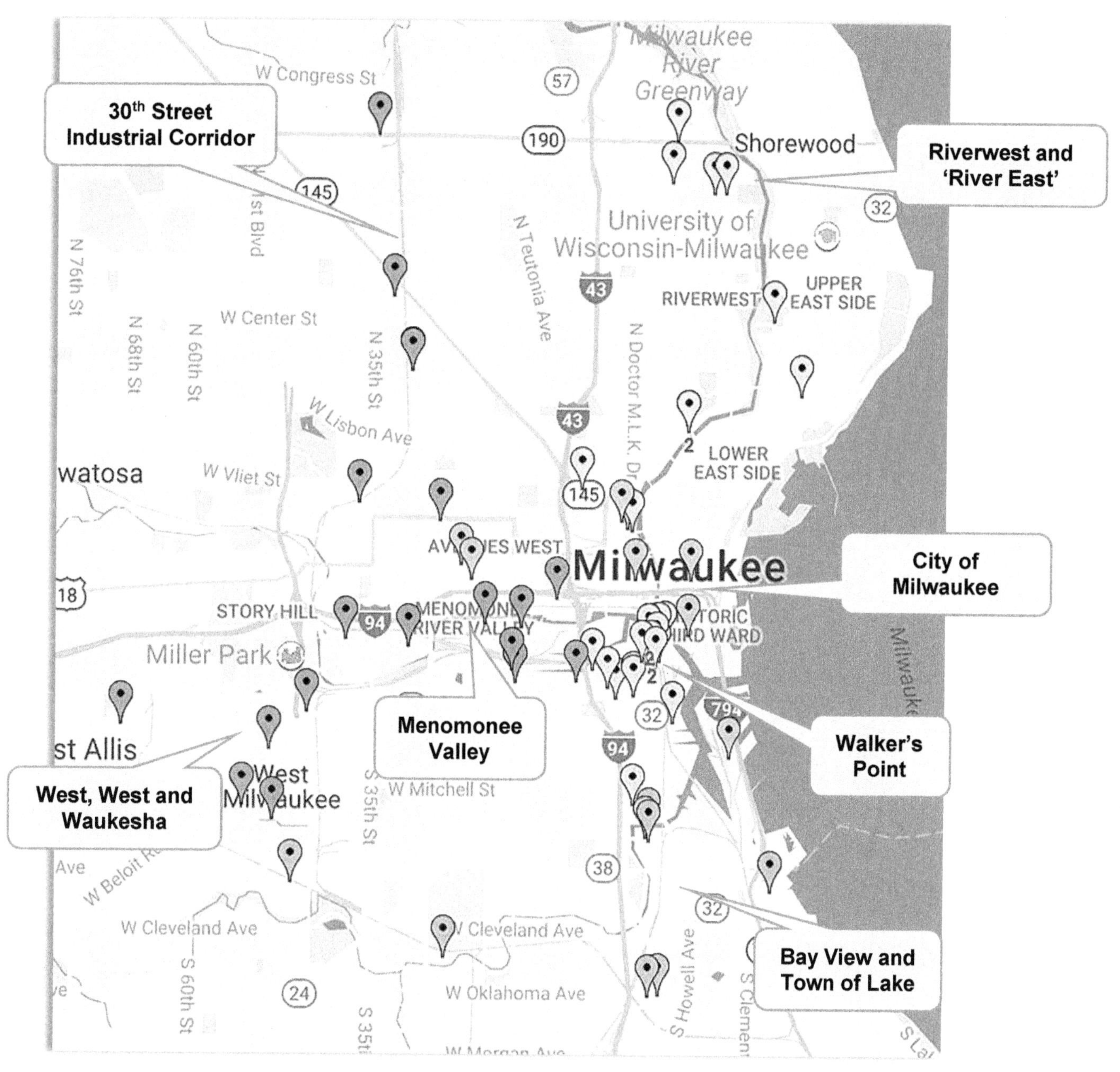

This map shows the Milwaukee's industrial neighborhoods. Over seventy industrial sites are identified in this book and are grouped in their industrial neighborhoods. Notice the high concentration of sites in the Walker's Point neighborhood—an early "industrial commons." Not all sites are individually shown on the map, due to scaling issues. Base map courtesy of Google, using its BatchGEO feature.

CHAPTER 1: INDUSTRY NEAR THE BANKS OF THE MILWAUKEE RIVER

Milwaukeeans often describe their home as "a great place on a great lake." While the lake was an essential attribute, Milwaukee actually grew up around its rivers. The confluence of the Milwaukee, Menomonee and Kinnickinnic rivers as they discharge into Lake Michigan created an excellent location for a port and an ideal location for a major city.[d]

As early settlers began settling in the area, they generally built their homes and businesses near the banks of the Milwaukee River. Other sites were not as favorable for settlement: there were high bluffs along Lake Michigan, the Menomonee River bed was largely marshland, and only a short section of the Kinnickinnic was navigable. For many businesses, it was important to have direct access to the Milwaukee River for boat navigation, as well as for waterpower. As a result, the Milwaukee River's banks became lined with numerous factories and storage yards for lumber and coal.

THE MILWAUKEE AND ROCK RIVER CANAL

In the early 1800s, water navigation was the most cost-effective way to ship bulk goods, as well as for the transportation of people. The early success of the Erie Canal, which opened in New York in 1825, caused one of Milwaukee's founding fathers, Byron Kilbourn, to consider the construction of a canal that would allow boats to travel from the Milwaukee harbor to the lead mining region in southwestern Wisconsin, through a series of locks. Kilbourn, who was himself a surveyor and had worked for a time as an engineer on some canal building projects in Ohio, brought renowned scientist and engineer Increase Lapham to Milwaukee to oversee the canal's development. Lapham had experience working on the Erie Canal, as well as other canals and locks. The two men laid out the Milwaukee and Rock River Canal, with a proposed route that would have gone from the Milwaukee River on a route through Menomonee Falls, Pewaukee, Delafield, and Fort Atkinson, where it would have joined the Rock River. The Rock River eventually discharges into the Mississippi River. The route was surveyed and work was started on the eastern end in 1840 with the construction of a timber dam on the Milwaukee River, just south of North Avenue. However, only a little over a mile of the canal was built.

Had the canal been attempted fifteen years earlier, it might have successfully met Kilbourn's vision of his canal as the "last connecting link between the Hudson and the Mississippi."[7] However, by the 1840s, cost efficient rail transportation was taking hold in the United States. Kilbourn abandoned the canal project and in 1847 invested in the Milwaukee & Waukesha Railroad Company. He was elected its president in 1849 and the name was changed the following year to the Milwaukee & Mississippi Railroad Company—which continued to expand and eventually became the Chicago, Milwaukee, St. Paul and Pacific Railroad Company, and was known to most as *The Milwaukee Road.*[8]

While the effort to build the canal was not successful, the dam and its short canal segment built adjacent to the Milwaukee River provided an unintended benefit. The canal provided a convenient 'head race' for the production of hydraulic power. The year following completion of the short canal, Samuel Brown and

---

[d] The word 'Milwaukee' is believed to have originated from either the Potawatomi word 'minwaking', or the Ojibwe word 'ominowakiing', meaning "Gathering place by the water" – an apt description for Milwaukee's location. See: *A Short History of Milwaukee, Wisconsin*, William George Bruce, The Bruce Publishing Company, 1936, p. 15–16. LCCN 36010193.

Benjamin Moffat built a sawmill and a man named Rathbone put up a large grist mill, both powered by water from the canal. The waterpower from the canal was looked upon as the "most important element of the present and future prosperity of Milwaukee."[9] By mid-century, no fewer than twenty-five Milwaukee industries would be powered by water from the canal and its dam.[10]

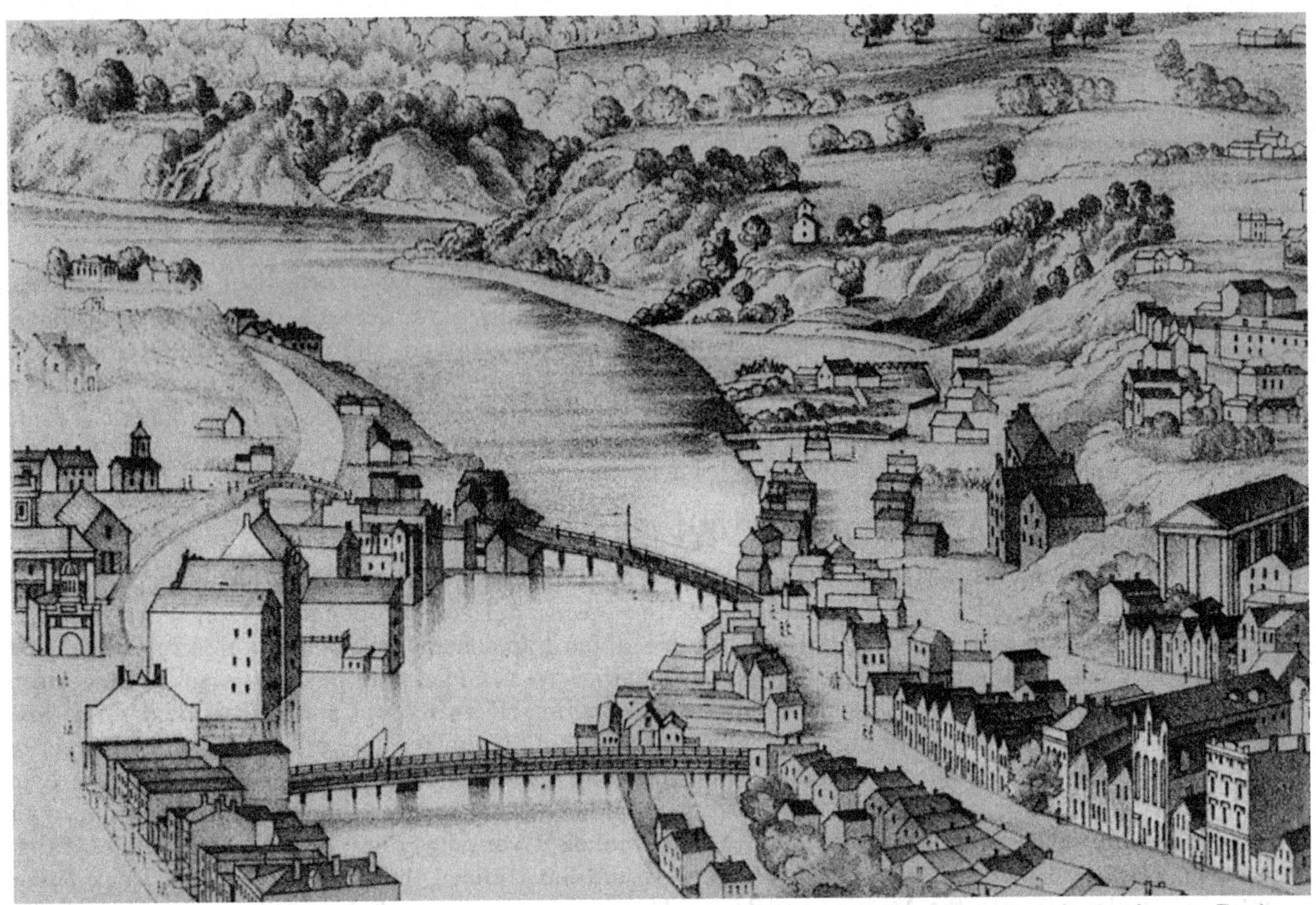

A portion of an 1858 illustrated map of Milwaukee showing the Milwaukee River and, to its left, the former Rock River Canal. Factories located between the canal and the river were able to use the water from the canal, pass it through a water wheel and discharge it into the Milwaukee River—thus powering their factories.

It was in this serendipitous manner that many machine shops and manufacturers prospered in Milwaukee, each paying the established rent for the water of $75 a year for one hundred cubic feet per minute.[11]

While little is known about many of the mill powered factories from this era, it is known that one of these mills was the factory where a breech-loading musket was manufactured for use by the US Army during the Civil War. This same mill was also the location where the Milwaukee-built version of the first commercially successful typewriter—the Sholes & Glidden 'Type Writer'—was produced.

An early photograph of the North Avenue dam, which impounded water for the planned Rock River Canal. Photograph from the Historic Photo Collection of the Milwaukee Public Library

BYRON KILBOURN was one of the three founders of the City of Milwaukee. Born in Granby, Connecticut in 1801, he moved with his family to Worthington, Ohio while a young child. Kilbourn worked in Ohio as a surveyor and as a state engineer. Among his jobs was working on canal projects.

He first visited Wisconsin in 1834, working as a government surveyor. This task provided him with a familiarity of the Milwaukee area, which he considered so promising that he purchased land on the west banks of the Milwaukee River. In 1837 he founded Kilbourntown, which was one of the three communities that eventually merged in 1846 to form the City of Milwaukee—the other two communities being George Walker's 'Walker's Point' and Solomon Juneau's 'Juneautown.'

Kilbourn was elected mayor of Milwaukee in 1848, and again in 1854. He is also credited with founding the City of West Bend (in 1845) and Wisconsin Dells (initially known as Kilbourn City)

He was active in the early railroad industry, helping to form the Milwaukee and Mississippi Railroad Company, serving as its first president. The company, which eventually became known as The Milwaukee Road, is more formally known as the Chicago, Milwaukee, St. Paul and Pacific Railway.

Near the end of his life, Kilbourn moved to Florida. He died in 1870 at the age of 69, and was buried in Jacksonville. In 1998, his remains were transported to Milwaukee for permanent interment at Forest Home Cemetery.

INCREASE A. LAPHAM was born in Palmyra, New York in 1811. His family eventually settled in Ohio where his father, Seneca Lapham, worked on various canal systems. Increase Lapham displayed an early talent for scientific observation, and shared his father's interest in canals. He worked with Bryon Kilbourn on the construction of the Miami and Erie Canal in Ohio and, in 1836, he joined Kilbourn in Milwaukee and worked with him on various business ventures.

Lapham continued his interest in exploring geography and nature. In 1836, he published the *Catalog of Plants and Shells, Found in the Vicinity of Milwaukee, on the West Side of Lake Michigan*, perhaps the first scientific work published west of the Great Lakes.

Throughout the course of his career, Lapham published numerous papers and books dealing with geology, archaeology and history, and flora and fauna of Wisconsin. In 1844, he published a book on the geography of the Wisconsin Territory and in 1846 he published his first map of Wisconsin. Lapham also published many other early maps that were used for civil projects for the railroads and canals.

Increase Lapham is considered "Wisconsin's first great scientist." He has also been called the "Father of the U.S Weather Service," because of his use of weather data to forecast storms on Lake Michigan as well as his efforts to lobby Congress to create an agency to forecast storms . In 1848, Lapham founded the Wisconsin Natural History Association, a predecessor of the Wisconsin Academy of Sciences, Arts, and Letters of which he also was a charter member. Lapham was buried at Forest Home Cemetery in Milwaukee.

THE SMITH AND BIRGE MILL

Henry Smith and Charles S. Birge's mill was located at 454 Canal, on a narrow strip of land between the Milwaukee River and the Milwaukee and Rock River Canal. It was described as a two-and-a-half story building of masonry made up of large cut stones (referred to as ashlar). It used hydropower to turn its line-shafts, which in turn ran its machinery.

Smith was a wheelwright and likely used the mill to turn line-shafts for various tools for his trade. While little more is known about the early years of this mill, early directories show that it was used by James Paris Lee in 1864 and 1865 to manufacture his carbine for the US Army. Amazingly, this same mill was also used periodically by Christopher Latham Sholes, Carlos Gilmore and Samuel Soule during the development of their early typewriters, and was rented in 1872 by James Densmore and used by Mathias Schwalbach to manufacture the Milwaukee-built Sholes & Glidden 'Type Writers' prior to outsourcing production to Remington.

The Smith & Birge mill was also used for a time by Carlos Glidden, who was developing a spader that he believed would be better than plows on the market at that time.[12]

The next sections explore two important innovations that occurred at the Smith & Birge mill—manufacture of James Paris Lee's carbine, and the development of the Sholes & Glidden 'Type Writer.' In Chapter 11, the innovations that occurred at the Smith & Birge mill are presented as an illustration of the importance of collaboration and of 'collisions' of individuals with different skills, toward fostering innovation.

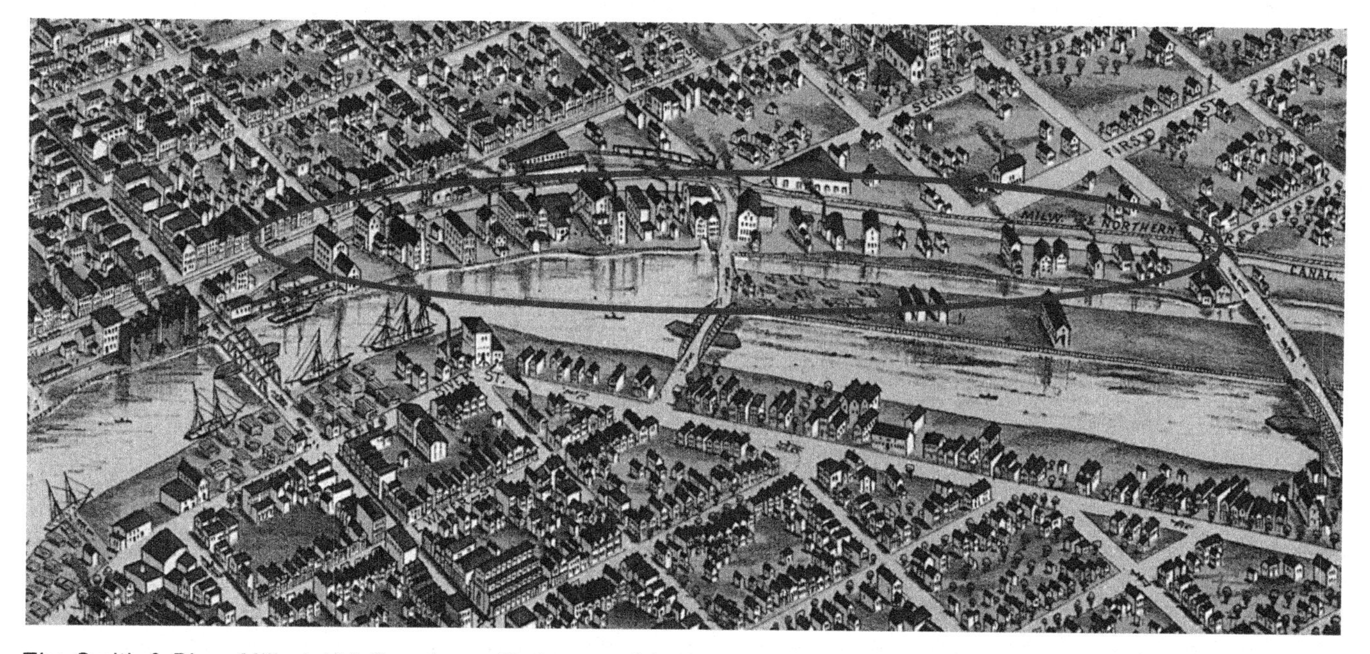

The Smith & Birge Mill at 454 Canal was likely one of the buildings pictured in this 1872 Illustrated Map of the City of Milwaukee, located on a narrow strip of land between the Milwaukee and Rock River Canal and the Milwaukee River. Their mill was able to use the hydropower created by the difference in elevation between these two bodies of water. Smith was a wheelwright. The mill was eventually used by Lee's Fire Arms Company, to produce rifles for the Civil War, and later by Mathias Schwalbach to manufacture the Sholes & Glidden Type Writer.

LEE'S FIREARMS COMPANY

James Paris Lee was a watchmaker and firearms designer, born in Scotland in 1831. His family immigrated to Ontario when he was four or five. At the age of 12, Lee built his first firearm, which failed to function properly and injured him. He apprenticed at his father's watch shop when he was 17. He married Caroline Chrysler in 1852 and moved with his family to Janesville, Wisconsin in 1859. He soon relocated to Stevens Point.

Shortly after arriving in Wisconsin, he continued his pursuit of firearm design. With the onset of the Civil War, Lee developed a breech-loading cartridge conversion for the Springfield Model 1861 Rifled Musket, obtaining a patent for his innovation in July 1862. The US Army was impressed with the design and in 1864 ordered a thousand rifles for use during the Civil War. Lee immediately relocated to Milwaukee to manufacture firearms, establishing Lee's Firearms Company.

DEVELOPMENT OF THE SLANT BREECH-LOADING RIFLE

Lee set up a Milwaukee factory in the former Smith & Birge Mill located at 454 Canal to manufacture his breech-loading musket. While Lee apparently could cast, forge and machine the barrel and other parts of his carbine, he lacked the capability of accurately boring the barrel. As a result, he subcontracted boring of the rifle barrels to E. Remington & Sons of Ilion, New York.

Lee's Firearms Company initially delivered 200 rifles, but the entire lot was rejected by the government because the barrels had the incorrect bore. A misunderstanding of some sort caused Remington to bore the rifles to the incorrect caliber. As a result, the weapon did not see use during the Civil War.

Photograph of Lee's Slant Breech-Loading Rifle, courtesy of the Rock Island Auction Company

James Paris Lee's slant breech-loaded carbine was one of the very few firearms manufactured in the West during the Civil War. Lee eventually completed two hundred and fifty-five carbines, of which just over one hundred were sold. When he ceased their manufacture, approximately two hundred more were near completion and a few hundred more in various stages of production. It is not known whether these additional rifles were ever completed. In any case, these carbines are rare and highly collectible.[e]

[e] Lee's Slant Breech-Loaded Carbine can be readily identified – the left side of the barrel is marked in capital lettering "Lee's Firearms Co., Milwaukee, Wis./Pat'd July 22, 1862."

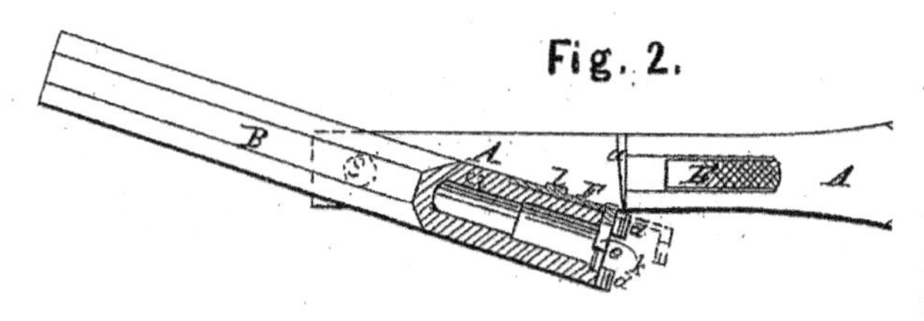

This detail is from Lee's 1862 Patent Drawing (US35941A granted July 22, 1862). The view looks down at the top of the breech of the rifle barrel, showing how it pivots and swings to the side to allow a cartridge to be loaded.

Lee also manufactured a 'sporting rifle' in Milwaukee. It was also a breech loading rifled. Its octagonal barrel was substantially longer than the military carbine—up to 28 inches. Overall, an estimated thousand such rifles were made. The Milwaukee Public Museum has both a Lee carbine and a Lee sporting rifle in its collection.

Lee soon relocated to the New York area and continued his collaboration with Remington. That company ultimately was the manufacturer of most of Lee's firearm designs.

Lee eventually developed a vertical box magazine for rifles—solving a problem of carbine detonation within tube magazines. His magazine was revolutionary. It is reported that virtually all existing bolt-action rifles were influenced by it. The Lee Model 1879 rifle, a landmark rifle design, incorporated a turn-bolt action and Lee spring-loaded column-feed magazine system. It was adopted by the US Navy, as well as by the Chinese military.

Interest in Lee's magazine system prompted James Lee and his wife Caroline to travel to Britain and continental Europe through the 1880s marketing guns, and these trips ultimately led to the British adoption of the Lee rifle in 1889, after extensive trials. It formed the basis for the standard British service arm for many decades.[13]

JAMES PARIS LEE was a firearms designer best known for the rifles that bear his name—the Remington-Lee, the 1895 Lee Navy, and the Lee-Enfield. He also invented the detachable box magazine, used first in the Lee-Metford rifle which combined Lee's rear-locking bolt system and ten-round magazine with an innovative seven groove rifled barrel designed by William Ellis Metford.

James Lee was born on August 9, 1831, in Hawick, Scotland. He emigrated with his family to Galt, Ontario in Canada in 1836 when he was five.

In 1858, James Lee and his wife Caroline Lee (née Chrysler, of the later automotive family) moved to Wisconsin, where they had two sons: William, (born in 1859) and George (1860). The Lees moved to Milwaukee in 1864, where James established Lee's Firearms Company to manufacture an order of rifles for the US Army during the American Civil War.

He later moved to Ilion, New York to work with Remington—an association that lasted for many years.

Lee's patent (221,328 on November 4, 1879) for the vertical box magazine is considered a significant advance in firearms because it solved a serious problem associated with cartridge detonation in tube magazines.

The Lee Model 1879 rifle, a landmark rifle design, incorporated turn-bolt action, the Lee spring-loaded column-feed magazine system and was his first successful magazine-fed rifle. The Model 1879 was adopted by the US Navy. Two later designs—the Remington-Lee M1885 and the Winchester-Lee or Lee Navy M1895—were also adopted militarily and sold commercially.

Following extensive trials, Lee's bolt and magazine design soon interested British ordnance authorities. In 1889, the British military adopted his design and it was the basis for the standard British service arm for many decades and in official service for nearly a century.

James Paris Lee is described as "one of the 19th Century's greatest gun designers."[14]

Lee died in Short Beach, Connecticut in 1904.

SHOLES & GLIDDEN

It was commonplace in the mid-1860s for small shops to spring up to support larger factories. Known as *job shops,* these businesses provided specialty services such as patternmaking and machining. In Milwaukee, a number of job shops were located near the banks of the Milwaukee River in order to support adjacent industry. One of these early job shops was Charles F. Kleinsteuber's machine shop, located a short distance from the river on State Street.

Kleinsteuber's shop must have been an interesting place. In addition to providing machining and foundry services, the shop served as an informal incubator for would-be inventors. Carlos Glidden spent time there working on designs for a steam-driven rotary plow and a mechanical spade. Machinist Mattias Schwalbach was working on a design for sewing machines, which he patented in 1866, and various designs for clocks. Christopher Latham Sholes used the shop to develop a page numbering device and a newspaper addressing machine.[f]

It is reported that in 1867 Glidden, while observing Sholes's work on perfecting his page numbering device, encouraged him to develop a mechanical writing machine. With the aid of Glidden, Schwalbach and fellow printer and inventor Samuel Soulé, Sholes produced a functioning machine by the fall of that year.[15]

The machine shop of Charles F. Kleinsteuber, as it appeared in 1867. Photograph from the Historic Photo Collection of the Milwaukee Public Museum

Sholes and various associates toiled for nearly seven more years before his model for what would become the world's first practical typewriter was introduced for mass production in 1874. The innovations and refinements that occurred during this seven-year period are what distinguished the Sholes & Glidden typewriter from that of the numerous other innovators who had previously attempted to create a mechanical typing machine. They converted a crudely built model into a device that typed reasonably well, plus being reliable, rugged and able to be manufactured in large numbers.

The person who was most responsible for taking the Sholes & Glidden typewriter from a rough model to a final product was James Densmore. Densmore, a former newspaper associate of Sholes, provided financing to assist in moving the development into manufacture. Of equal importance, Densmore continuously prodded Sholes to tweak the design in order to improve reliability and ease of use.

---

[f] Sholes, Glidden, Soule and Schwalbach also used the second floor of a mill located adjacent to the Milwaukee River and owned by Henry Smith, for their collaboration on the typewriter.

Historian Richard N. Current provided the most extensive record of the development work during this time.[16,17] He relied on extensive correspondence between Densmore and Sholes and others involved in the development. According to these accounts, Densmore agreed to provide financing in exchange for an ownership share, prior to actually seeing the device. By the time he saw the typewriter in March 1868, there were two versions: the original relied on long wires to connect the type bars and key levers, and a refinement developed by Samuel Soulé involved a simplified arrangement for striking the keys onto paper. In the summer of that year, Densmore attempted to manufacture the refined machines in Chicago. After making fifteen typewriters and observing them in use—some of which were used in a school for telegraphers in Chicago, he concluded that design was not yet suitable for the market.

Richard Current reports that this prompted Sholes, for the first time, to look into the record of what previous inventors had done. He concluded that all had failed because they had not satisfied one or more of the "fundamental ideas" that he and Densmore considered "essential to success." These ideas were that "the machine must be simple and not liable to get out of order," that "it must work easily and be susceptible of being worked rapidly," and that "it be made with reasonable cheapness." Additionally, Densmore insisted that a successful typewriter be capable of writing on paper of ordinary thickness—as opposed to the early designs by Sholes that only printed satisfactorily on paper that was tissue-thin.

To satisfy this last requirement, Sholes abandoned the flat platen design and devised a revolving cylindrical platen to serve as the paper carrier. Sholes employed the cylinder in a novel way to avoid infringing on an existing patent. The cylinder rotated to space the letters and indexed along its axis to change the lines—essentially perpendicular to what is now known as the conventional typewriter layout. While this permitted thick paper to be used, the page was limited to the width of the cylinder—roughly three inches.

In September 1869, Sholes declared that he had perfected all the necessary principles, writing to Densmore on the machine, "I am satisfied the machine is now done."

Densmore continued to press for additional improvements, much to the annoyance of Sholes. Somewhat reluctantly, Sholes continued to work on refinements. He next adopted a refined keyboard, devised by Schwalbach, which involved four rows of metal key levers and buttons set in ascending banks. At the urging of a customer who tried this design, a space bar was added underneath the four rows.

In the summer of 1871, Densmore manufactured in Milwaukee a sufficient number of typewriters to "supply the present demand, pay up the debts and have one or two over to sell." These machines were far from perfect. In addition to durability issues, the type bars wouldn't stay in line. Sholes, Glidden, Schwalbach and Densmore's stepson, Walter J. Barron, worked together to solve this problem. The design was using short, stiff wires, which directly connected the key levers and type bars and pulled at an angle. Glidden initially suggested a system of intermediate levers. Despite Sholes's disapproval, Densmore pressed on with this approach—which ultimately failed. Barron meanwhile suggested an alternative method that reduced the angle. Sholes and Schwalbach redesigned the machine using this approach.

That year, Sholes also tackled an issue for another customer. D.N. Craig, of the Automatic Telegraph Company, told him that his typewriter would be much more useful if it could accommodate a continuous roll of paper. To meet this request, Sholes redesigned the cylindrical platen to move lengthwise to space the typed letters, and to rotate to index to the next line. Since letters were typed on the underside of the cylinder, Sholes also hinged the mechanism so that it could be swung up to allow the typist to review typed print. While that was a significant improvement, it infringed on a patent that had recently been issued to Charles A. Washburn of San Francisco, requiring the payment of a license fee to Washburn.

Finally, to reduce the nuisance of type-bar collisions, which were frequent, Sholes and Densmore worked out a non-alphabetical arrangement for the keys, resulting in the QWERTY layout that became and remains the standard for keyboards everywhere.

With these improvements in hand, Densmore began his third attempt to manufacture typewriters for sale. He rented a former wheelwright's mill between the Milwaukee River and the Rock River Canal.[g] With Schwalbach's assistance, he equipped the shop, using water from the canal to power the machinery. The typewriters were produced individually, which allowed Schwalbach to continue to incorporate changes in design as the manufacturing process progressed.

By the end of 1872, the appearance and function of the Sholes & Glidden typewriter had assumed the form that would become standard in the industry and remain largely unchanged for the next century.

The resulting Sholes and Glidden typewriter incorporated several components adapted from existing devices, such as escapement (a geared mechanism governing carriage movement) adapted from clock works, keys adapted from telegraph machines and type hammers adapted from the piano.[18] However, this was the first device that put all of these components together in a functional typewriter that was commercially available.

Picture of a Milwaukee-built Sholes & Glidden 'Type Writer,' courtesy of the Buffalo Museum.

Outsourcing Manufacturing

While Schwalbach and his workmen were turning out typewriters in their improvised factory, Densmore calculated that the machines were costing more to build than the selling price. For advice, he turned to his friend and former business associate, George Washington Yost, who was then managing a farm implement factory at Corry, Pennsylvania. Yost visited Densmore in Milwaukee to observe his operations and suggested that he contact E. Remington & Sons. Remington manufactured guns, farm implements, and sewing machines, in Ilion, New York. Yost made the introductions and arranged to visit their factory in Ilion with Densmore.

On March 1, 1873, Densmore signed a contract under which Remington agreed to have their lead mechanics, William Jenne and Jefferson Clough, rework the machine and to produce a minimum of a thousand machines. Under the contract, Densmore agreed to pay them $10,000 for manufacturing the typewriters, plus agreed to pay a royalty for the services of Remington's lead mechanics. Jenne and Clough prepared the design for mass production, encasing the production version in metal instead of wood, and somewhat reducing the overall size. In principle, however, the final Remington-built Sholes and Glidden typewriters were the same in form and function as the last typewriters built in Milwaukee.[h]

[g] The building rented was located at 454 Canal and owned by Henry Smith and Charles S. Birge, listed as Wheelwrights and Machinists.

[h] Under the contract, Remington agreed to produce as many as 24,000 typewriters, at its discretion. Remington did not initially acquire the patent rights for the device, but acted as a contract manufacturer. Densmore had to borrow the funds for the advance payment.

The Sholes & Glidden *Type Writer*,[i] manufactured by Remington in Ilion, New York under contract for James Densmore and partners. The typewriter pictured is in the collection of the Milwaukee Public Museum and was designated a landmark by the American Society of Mechanical Engineers as the world's first commercially successful typewriter. Photograph by the author at the permission of the Milwaukee Public Museum.

Significance of the Typewriter

The Sholes & Glidden typewriter is particularly noteworthy in that it represents the first commercially successful typewriter to be manufactured in quantity for sale to the public. It was also the first typewriter that enabled operators to write significantly faster than a person could write by hand.

The Sholes and Glidden typewriter set off a revolution in the conduct of commerce and business, as well as in communications. The ability of a skilled operator to type uniform, easily read text at high speed, and to employ the use of carbon paper to make multiple copies, created significant increases in efficiency and economy in the workplace.

One measure of the significance of the Sholes & Glidden 'Type Writer' was the competition it attracted. Several other makes of typewriters were developed and marketed by 1885, including the Hall, the Caligraph, the Crandall, and the Hammond. While each looked different from the Sholes & Glidden/Remington, they were all clearly inspired by the Sholes machine, leading to patent litigation. *The Stenographer*, a professional magazine, counted no fewer than forty-seven makes of typewriters on the market by 1891. By that time, it is reported that all sizable offices had at least one resident typist. By 1910, there were at least eighty-nine typewriter manufacturers,[19] all founded in part on the Sholes & Glidden design, albeit with modifications and improvements.

By the early 1900s, clerical typing pools and stenographers became universally employed in all modern offices, and typing courses were offered in most secondary schools. As the price came down, families

---

i Sholes used two words to describe his device. The term was eventually combined to form the more familiar term—'typewriter.'

also purchased a typewriter for home correspondence and student use. By the mid-1900s, the typewriter had become commonplace throughout modern society.

It is clear that millions of typewriters were manufactured over the years by United States manufacturing firms such as Remington, Smith Corona Corp., Royal, Oliver, International Business Machines, and Underwood, as well as by a number of foreign companies.[j]

The use of the typewriter was also an important tool for writers. Mark Twain claimed in his autobiography that he was the first important writer to present a publisher with a typewritten manuscript, for *The Adventures of Tom Sawyer* (1876).[k] While he may have been the first, the typewriter eventually became a standard tool for most writers, poets, and reporters, as well as anyone that wrote for his or her profession.

WOMEN AND THE TYPEWRITER[20]

One historian has commented that "perhaps one of the greatest or even the greatest achievement of the typewriter is the transformation it wrought in the social order. A strong prejudice existed ... against the employment of women in business. Then the typewriter came, soon to be followed by the girl typist, who blazed the way for other women to enter every department of business life."[21]

This association of women with the typewriter can be traced to the earliest advertising for the machine. Before Remington acquired the design rights, Sholes's daughter was employed to demonstrate the device and to appear in promotional images. It is reported that Remington's marketing included the use of attractive women to demonstrate their typewriter.

In 1874, less than four percent of clerical workers in the United States were women. By 1900, the number of women clerical employees had increased to approximately seventy-five percent.

Before his death, Sholes remarked of the typewriter, "I do feel that I have done something for the women who have always had to work so hard. This will enable them more easily to earn a living."

A typist operating a Sholes & Glidden Type Writer, as depicted in an 1872 article in Scientific American.

A group of women typists with early Remington typewriters.

[j] It is reported that Underwood alone produced over 5 million typewriters by 1939.

[k] Typewriter collector and historian Darryl Rehr challenged this claim, stating that Twain's memory was faulty and that the first novel submitted in typed form was *Life on the Mississippi* (1883).

THE QWERTY KEYBOARD

The Sholes & Glidden typewriter had one entirely original feature—the arrangement of the keyboard. Schwalbach came up with the four-row configuration. However, it has been reported that Sholes arranged the keys in the now familiar QWERTY layout to minimize the possibility of jamming the typebars. The acronym QWERTY was adopted from the first six letters of the top row of alphabetic keys. The layout was commonly adopted by other typewriter manufacturers and, as a result, it became commonplace. While other arrangements have been attempted, QWERTY became the standard. As a result, the world will likely have to live with this letter configuration forever, even on computers and electronic mobile devices.

Above: QWERTY keyboard arrangement on the Remington-built Sholes & Glidden 'Type Writer'

Practice QWERTY keyboard template by Remington Sewing Machine Company

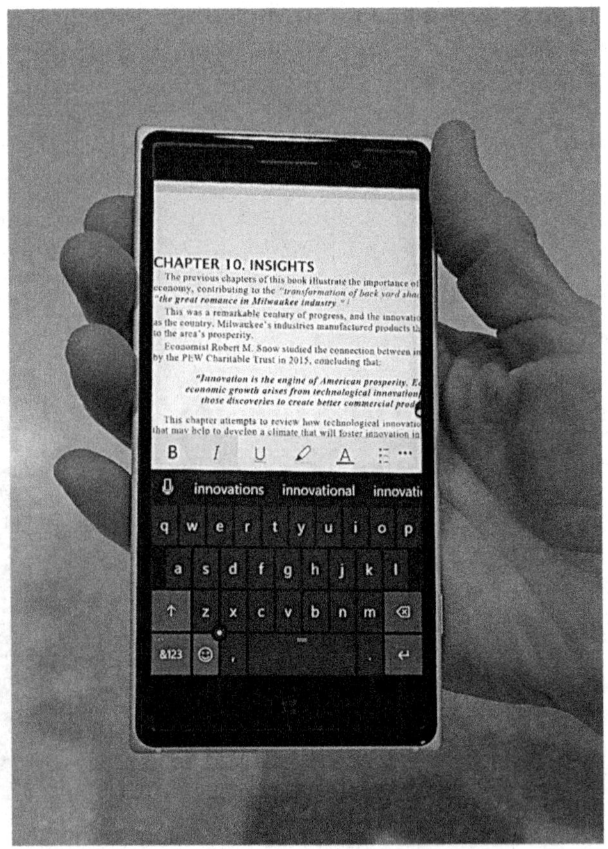

Picture of the QWERTY simulated keyboard on a Lumia 830.

THE SHOLES 'VISIBLE' TYPEWRITER

The Sholes & Glidden 'Type Writer' was an 'upstrike' typewriter. Upstrikes were blind-writers—they printed on the underside of the platen and operators could not see their work while they were typing. The basic mechanism for striking the typebar onto paper on the Sholes & Glidden Type Writer is illustrated in the adjacent figure. In order to assist typists in correcting errors, Sholes designed the typewriter with a hinged carriage.

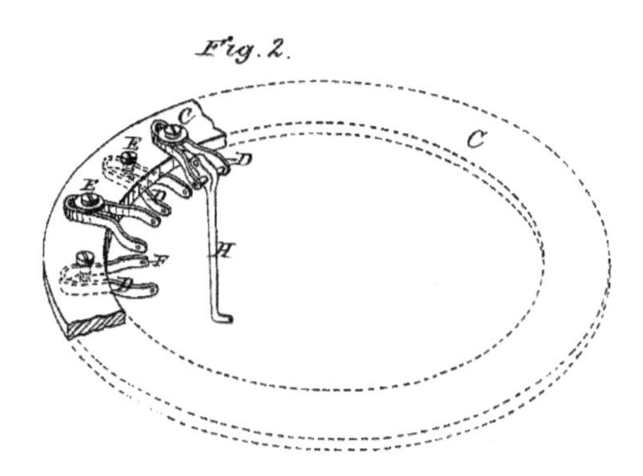

Above: The typebars were attached to the circumference of a metal ring such that they would strike a common center.

Left: This photograph shows how the hinged carriage could be swung up to allow the operator to check what had been typed. Photograph by the author at the permission of the Milwaukee Public Museum.

Sholes recognized that the hinged carriage was a less-than-ideal solution and he spent much of the balance of his career designing a 'visible' typewriter. In 1889, he filed for a patent for a typewriter that solved the problem (Patent No. 464,902). Others were also working on various approaches that would allow the typist to see the text as it was typed on the page.

Sholes died before his 'visible' typewriter went into production. His sons, Louis C. and Frederick Sholes, carried his plans forward and eventually began producing the Sholes Visible under the C. Latham Typewriter Company name. While their venture was not commercially successful, they sold the rights to the design to August D. Meiselbach of Milwaukee in 1900, who manufactured the machine in Kenosha for several years.

We will discuss this innovative typewriter later in the book in the section covering the Meiselbach Machine Company.

CHRISTOPHER LATHAM SHOLES was born in 1819 in Mooresburge, Pennsylvania and as a youth worked there as an apprentice to a printer. He moved to Wisconsin in 1837, along with two of his brothers. He became a newspaper publisher and politician, serving in both the Wisconsin State Senate (1848-1849 and 1856-1857) and the Wisconsin State Assembly (1852-1853). Sholes was instrumental in the successful movement to abolish capital punishment in Wisconsin.

Sholes lived in Kenosha (known as Southport at the time) for several years and published the *Southport Telegraph.* He later moved to Milwaukee and served as editor of the *Milwaukee Sentinel.* It was while he was employed in Milwaukee that he began work on the development of a typewriter—eventually producing what has been described as the world's first commercially successful device.

Sholes died in 1890 and is buried in Milwaukee's Forest Home Cemetery.

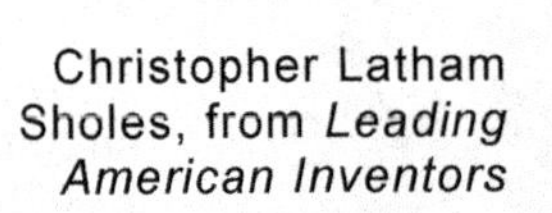

Christopher Latham Sholes, from *Leading American Inventors*

CARLOS GLIDDEN, MATTHIAS SCHWALBACH AND JAMES DENSMORE all made important contributions toward the development of the first commercially practical typewriter. Schwalbach, in particular, should be recognized for turning Sholes' ideas into working prototypes. He was a gifted clock-maker and continued to manufacture and repair large clock mechanisms of his own design.

Densmore, however, was the moving force behind the team. He provided funding, encouragement, and criticism, and actively marketed the product. He also established the relationship with Remington whereby that company took over the manufacture of the device—refining it further along the way.

Pictures from The Story of the Typewriter: 1873-1923, published by the Herkimer County Historical Society, NY, 1923.

SCHWALBACH'S STAR TOWER CLOCK COMPANY

In 1872, Matthias Schwalbach established his own business, the Star Tower Clock Company at what is now 1336 North Ninth Street, just south of Vliet Street. His company manufactured church and tower clocks of his own design. The company also produced various kinds of small machinery. Schwalbach was also a dealer in accordions and sewing machines.

Schwalbach obtained patents for clock escapements in 1874, 1880 and 1890, all of which involved his version of the '*remontoire*' (constant force) mechanism.

He also obtained a patent for a sewing machine mechanism in 1866, which was witnessed by Samuel Soulé and Charles Kleinsteuber. However, it appears that Schwalbach put his sewing machine design aside for fifteen years while working on the Sholes & Glidden Typewriter, but then returned to the mechanism in 1881. A patent for a slightly modified design was issued in that year. It is not known if any sewing machines were produced.

Schwalbach's clock company installed at least seventy tower clocks during its forty-two years of existence.[1] In Mathias Schwalbach's lifetime, his company installed fifty-five clocks in eleven different states. His son Robert continued to manufacture his clock designs after Mathias's death.

Picture of Schwalbach's Star Tower Clock Company, photograph from the Historic Photo Collection of the Milwaukee Public Library

Cover of an Illustrated Catalogue of Schwalbach's Church and Tower Clocks

[1] Some Schwalbach clocks have serial numbers over 100, indicating that more than one hundred may have been installed.

Schwalbach 'Tower' Clocks

Mathias Schwalbach's Star Clock Company manufactured many of the most significant clock towers in Milwaukee, as well as throughout the Midwest and as far east as New York, and as far south as Texas. In Milwaukee, Schwalbach design and built clocks were installed in at least twenty churches, starting with the oldest known installation in 1875 at St. Stanislaus Church—a well-known Milwaukee landmark with its twin, four-sided clock towers.

The scale of Schwalbach's tower clock mechanisms is not immediately evident by viewing the photographs. They are quite large; a typical clock mechanism is seven feet tall, four feet long and three and a half feet deep. The striking mechanism pulls a hammer that can weigh as much as forty pounds, and the pendulum would normally weigh one hundred and twenty-five pounds.

As noted earlier, Schwalbach tower clocks featured *remontoire* (constant force) mechanisms. Remontoire is derived from the French word *remonter,* which means 'to wind.' Remontoire mechanisms are generally associated with higher quality clock movements. It supplies a smooth, constant source of power to the escapement.[22]

Most tower clocks use a dead beat escapement that is typically powered by falling weights on pulleys. Since the falling weight force drives the clock gearing, it is susceptible to variance caused by external factors such as weather conditions. The difficult environment in which clock towers typically operate, where they are exposed to large seasonal and daily temperature differences, and the clock hands are periodically impacted by wind, ice and snow, all work against the efficiency of the mechanism to deliver a constant force to the escapement. These factors deliver varying amounts of force to the dead beat escapement, affecting its oscillations, or timing accuracy. The remontoire alleviates this variance by isolating the escapement from the rest of the clock trains. The remontoire mechanism operates off of the falling weights, which periodically rewinds the remontoire spring.

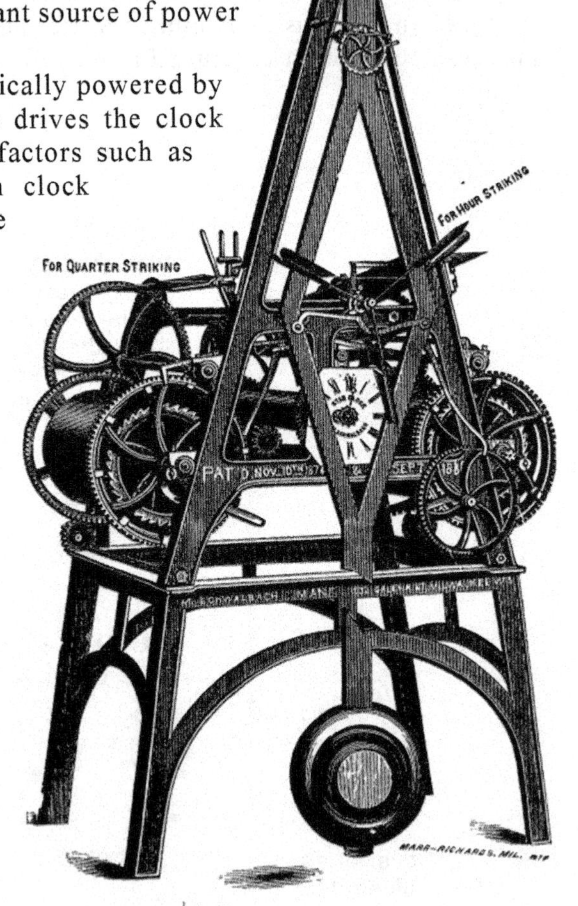

This illustration accompanied an article about Schwalbach's tower clocks from the 1886 book entitled, *Industrial History of Milwaukee.*

The remontoire escapement provides an elegant solution to timing inaccuracies by detaching the escapement from the clock trains. The Schwalbach remontoire uses a separate internal spring that provides a constant force to the pinwheel escapement. This system allows the force to be transferred to the pendulum with greater accuracy. The remontoire began to fall out of favor in tower clocks with Denison's invention of the double three-legged gravity escapement in the mid-1800s.

Remontoire clocks are interesting to watch in action. Typically, every thirty seconds or so the remontoire mechanism goes into action, usually employing a fan that acts as an airbrake to help regulate speed, to rewind the escapement.

In referring to his mechanism in one of his patents, Schwalbach states, "The objects of the improvement are to transmit the force or impetus from the motive power to the pendulum uniformly without regard to the heft of the weight or tension of the spring which moves the clock; also, to decrease the force required to produce a given number of vibrations of the pendulum and to increase the accuracy of the clock in keeping time."[23]

It is likely that Schwalbach had been exposed to such mechanisms during his apprenticeship in Germany.

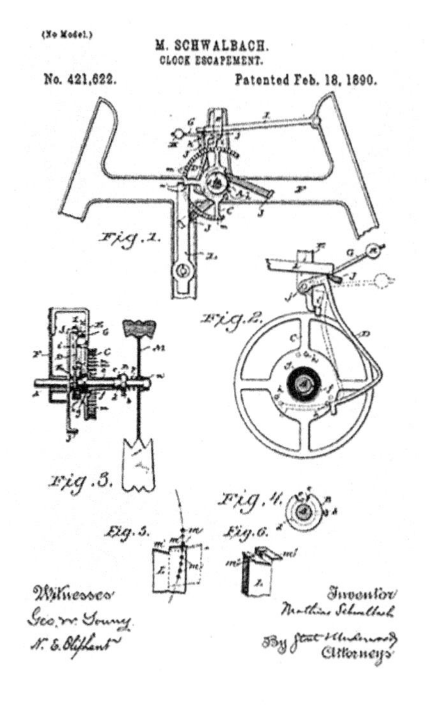

Patent drawing from Schwalbach's 1880 patent for a refinement to his remontoire mechanism. Notice the 'pinwheel' type fan illustrated in Figure 3.

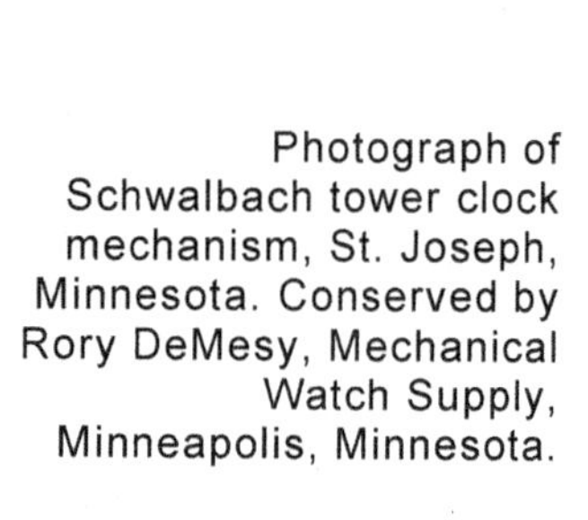

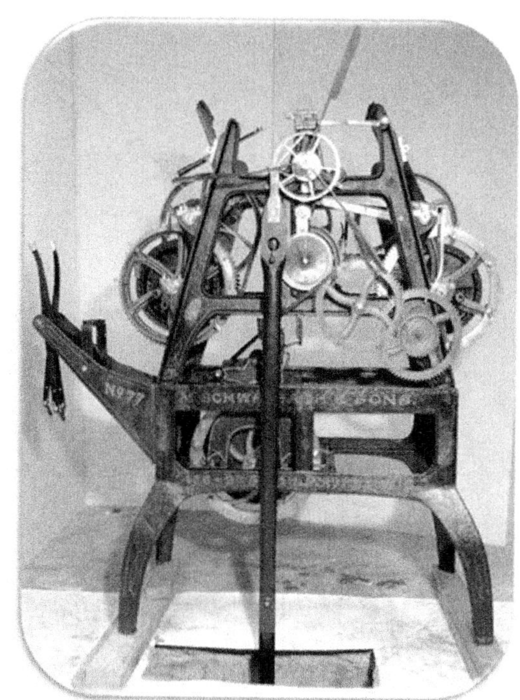

Photograph of Schwalbach tower clock mechanism, St. Joseph, Minnesota. Conserved by Rory DeMesy, Mechanical Watch Supply, Minneapolis, Minnesota.

Schwalbach tower clocks typically were thirty-hour models, needing to be rewound essentially once a day, although it should be noted that a few eight-day models were also produced. Because of the prominence of clock towers on churches and other buildings, and the numerous people that relied upon early clock towers for keeping track of the time, building owners wanted to make sure they were properly wound and maintained. Given the difficulty of accessing the mechanisms in towers only accessed with narrow, winding stairways, during often adverse weather conditions, this could be a difficult task. Many have been subsequently electrified to eliminate the need to manually wind the clocks.

One of the churches that continue to incorporate its original Schwalbach clock is the Church of St. Joseph in central Minnesota. The clock not only displays the time, it also regulates the striking of the tower bell on the hour and quarter hour, and the playing of the Angelus. After more than one hundred years of service, it underwent conservation in 2007. It is expected to run for another hundred years with proper care and maintenance.

An architect's drawing of St. Stanislaus Catholic Church in Milwaukee, located at 524 West Historic Mitchell Street. Note the prominent twin four-sided clock towers, each originally supplied with a clock mechanism from Schwalbach's Star Tower Clock Company.

MATHIAS SCHWALBACH[m] was born the son of Mathias and Gertrudis Simon Schwalbach in Malberg, a municipality in western Germany, on December 17, 1834. He arrived in the United States in 1857, at the age of twenty-three, and found a job working as a machine journeyman in Albany, New York. After two years, Schwalbach moved to Syracuse, New York, before eventually relocating to Milwaukee in 1853, where he spent the rest of his life.

In Milwaukee, Schwalbach initially worked for mechanic and engraver Charles Kleinsteuber for nine years. While with Kleinsteuber, he also assisted Christopher Latham Sholes and his associates and made significant contributions to the development of the first commercial typewriter, as noted in the article about the Sholes & Glidden 'Type Writer.'

It is generally believed that Schwalbach's contributions to the Sholes & Glidden 'Type Writer' were never properly recognized, and that without his skills it is unlikely that it would have been achieved. He is credited with the design of the four-row typewriter keyboard used on the Sholes typewriter, which found its way onto almost all subsequent typewriters.

Following his association with the typewriter developers, Schwalbach went on to establish his own business, the Star Tower Clock Company. The company, founded in 1872, manufactured church and tower clocks, and produced various other small mechanisms. His clocks were patented designs and involved constant force escapement mechanisms similar to a German design called '*remontoire*' escapements.

Schwalbach also designed a sewing machine, for which he received two patents.

Mathias Schwalbach outlived three wives. He fathered twenty-four children, although not all survived early childhood.

In November 1918 guardianship proceedings were initiated by nine of his children: Louis, Robert, Gertrude (Boehlein), Mathias, Theodore, Katherine (Walters), Felix, Elizabeth (Lohrer) and Helen (Fishang). His son Robert's family cared for him in his final years of life. He died in Milwaukee on February 29, 1920, at the age of eighty-six. He is buried in the Calvary Cemetery under a wrought-iron grave monument.

[m] Mathias Schwalbach's first name is occasionally recorded as 'Matthias.'

THE 'RELIANCE WORKS'

One of the many companies located along the banks of the Milwaukee River in the mid-1800s was Dexter and Seville's Reliance Works, a manufacturer of milling equipment and cast-iron stoves. The Reliance Works complex stood at a site known today as Pere Marquette Park, just a few blocks downstream from the old canal.

The Reliance Works was located downstream from the canal, alongside the Milwaukee River. In this portion of an 1858 illustrated map of Milwaukee, it is one of the factories on the left of the picture with smoke discharging from its chimney. Kleinsteuber's machine shop was located about a block west of the Reliance Works—and would likely have provided parts and machinery services to the Reliance Works, as well as other factories in this area.

The Reliance Works was one of a dozen or so small factories in the area and took advantage of the increasingly affordable steam-powered shop engines to drive its equipment. With the increasing affordability of the steam engine, factories no longer needed to rely on waterpower. A typical factory had a common steam engine that provided power to an overhead shaft via a wide belt. This common shaft then was used to drive other overhead shafts (called 'line shafts), with take-offs for individual machinery—all driven by leather belts.

The Financial Panic of 1857 hit the Reliance Works hard and, in 1861, the struggling company was acquired by a young industrialist named Edward Phelps Allis. Allis had moved to Milwaukee in 1846 at 21 years of age. His family was in the tanning business in New York and, with the financial support of his father, Edward entered into the business in Milwaukee, which had a thriving tanning enterprise. After a decade of successful business, Allis was interested in other ventures and acquired the Reliance Works in a sheriff's sale.

It is not by accident that the achievements of the Edward P. Allis Company (which eventually became the Allis-Chalmers Corporation) occupy a good portion of this book. The accomplishments of the E.P. Allis Company during the late 1800s clearly stand out because of their diversity and engineering innovation. This company, more than any other, seems to exemplify the remarkable advances made by Milwaukee industries.

Under Edward Allis's stewardship, the business expanded rapidly. However, the Reliance Works was landlocked, inhibiting his ability to expand his new company. So in 1867 Allis floated his buildings down the Milwaukee River to a new location with easy access to Lake Michigan shipping—right next to the bustling Milwaukee Harbor.

The E.P. Allis story continues in Chapter 2 at page 31, and as the Allis-Chalmers Corporation in Chapter 8 at page 307.

CHAPTER 2: WALKER'S POINT

Walker's Point was initially aptly described as a *point*—it was a peninsula that surrounded by marshy lowlands. Early developers began filling in the lowlands to create buildable space adjacent to the Menomonee River and the lower portion of the Milwaukee River.

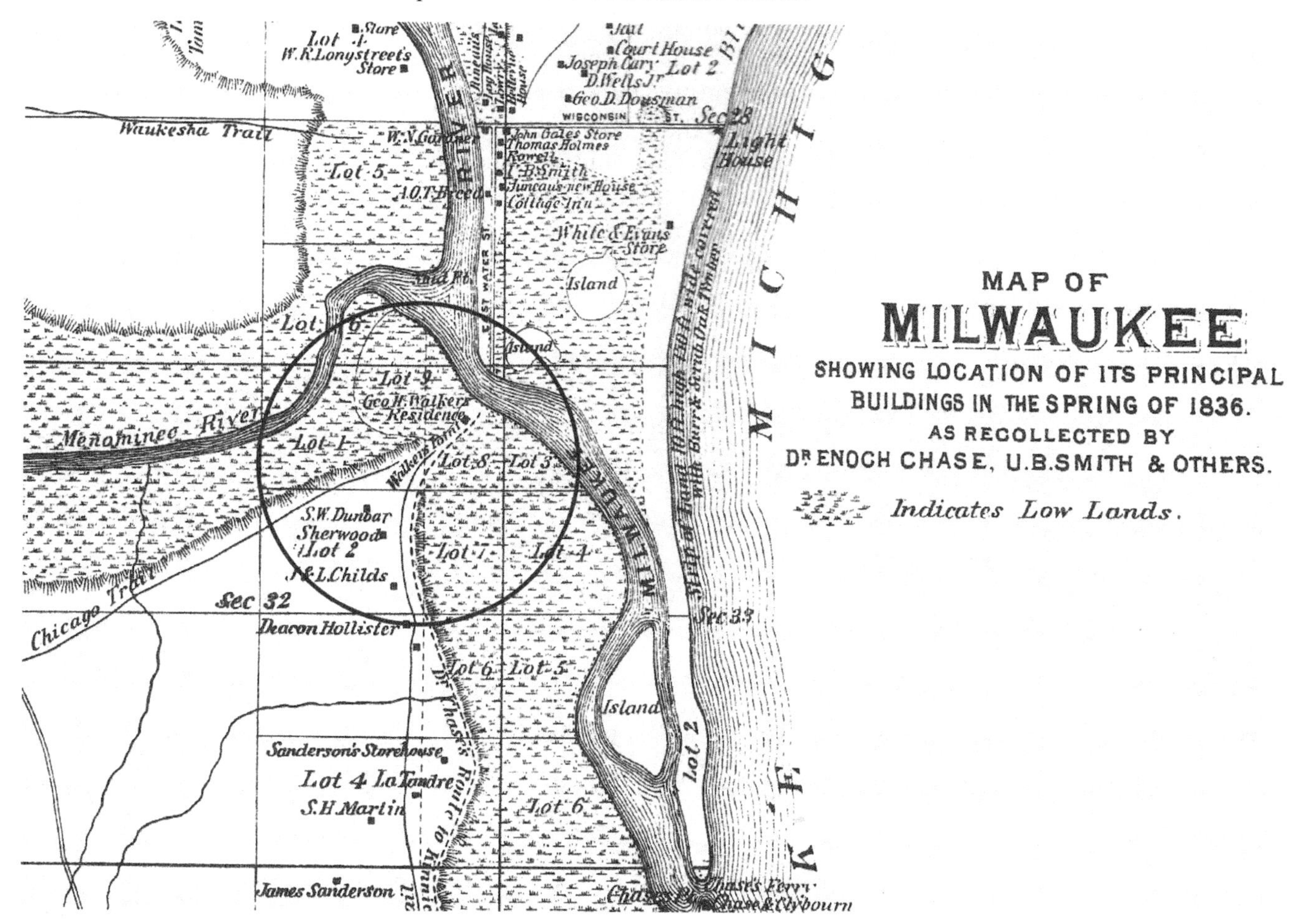

Early map showing the Walker's Point area (encircled) in the mid-1830s

As noted in Chapter One, E.P. Allis moved his factory into the Walker's Point neighborhood. It quickly attracted other manufacturers.

With open land readily available, small businesses flocked to this neighborhood, drawn in part by the giant industrial magnet that was E.P. Allis. They were interconnected, sharing ideas, feeding off each other, driven by a common passion for progress. The engineering achievements of the firms that eventually located within this river valley are remarkable.

Milwaukee's Walker's Point and its Menomonee Valley became the local epicenter of innovation and enterprise. Today the neighborhood surrounding the Reliance Works of E.P. Allis would be described as an *industrial commons*[24]—a breeding ground for the best and brightest engineers, inventors, designers and manufacturers.

The term *industrial commons* was coined by economists Gary Pisano and Willy Shih in an article entitled *Restoring American Competitiveness.* They draw their concept of "the commons" from the historical example of shared lands in a town where animals belonging to people in the community could graze. This shared property, or *commons,* was a benefit shared by all in the community. They describe an *industrial commons* as providing "a foundation of knowledge and capabilities (technical, design and operational) that is shared within an industry sector, such as research and development know-how, advanced process development and engineering skills, and manufacturing competencies related to a specific technology." That description all applies to the Menomonee Valley of the late 1800s.

In fact, Milwaukee's Menomonee Valley might be one of the best early examples of an industrial commons in the United States. The closely located companies benefited from their relationships with one-another. People moved from firm to firm. Technical knowledge moved from company to company. In essence, the area may be considered a 19th century, industrial version of today's Silicon Valley.

This chapter explores the companies that got their start in the Menomonee Valley, and shows how they were often linked by proximity, goods and services, and by employees—and how they benefited from interacting with one-another.

Many of these companies eventually relocated to other areas of the City, to expand their manufacturing facilities further. When this was the case, the early history of these companies is provided in this chapter, with later company history provided in other chapters of this book.

This illustrated map is part of an 1872 map showing the Walker's Point area. The industrial buildings can be clearly distinguished from the residential areas. E.P. Allis Company is shown between Clinton and Barclay streets and south of Florida Street. The map was created by Milwaukee Lithographing & Engraving Company and published by Holzapfel & Eskuche, Stationers and Book Sellers. Map courtesy of the Library of Congress.

E. P. ALLIS COMPANY

E.P. Allis moved his factory to Walker's Point in 1867. The new Reliance Works campus occupied several city blocks in Walker's Point, south of Florida Street along what was then Clinton, and today is South First Street. For decades, some of the greatest machinery in the world was produced at the plant located near the confluence of the Milwaukee and the Menomonee rivers.

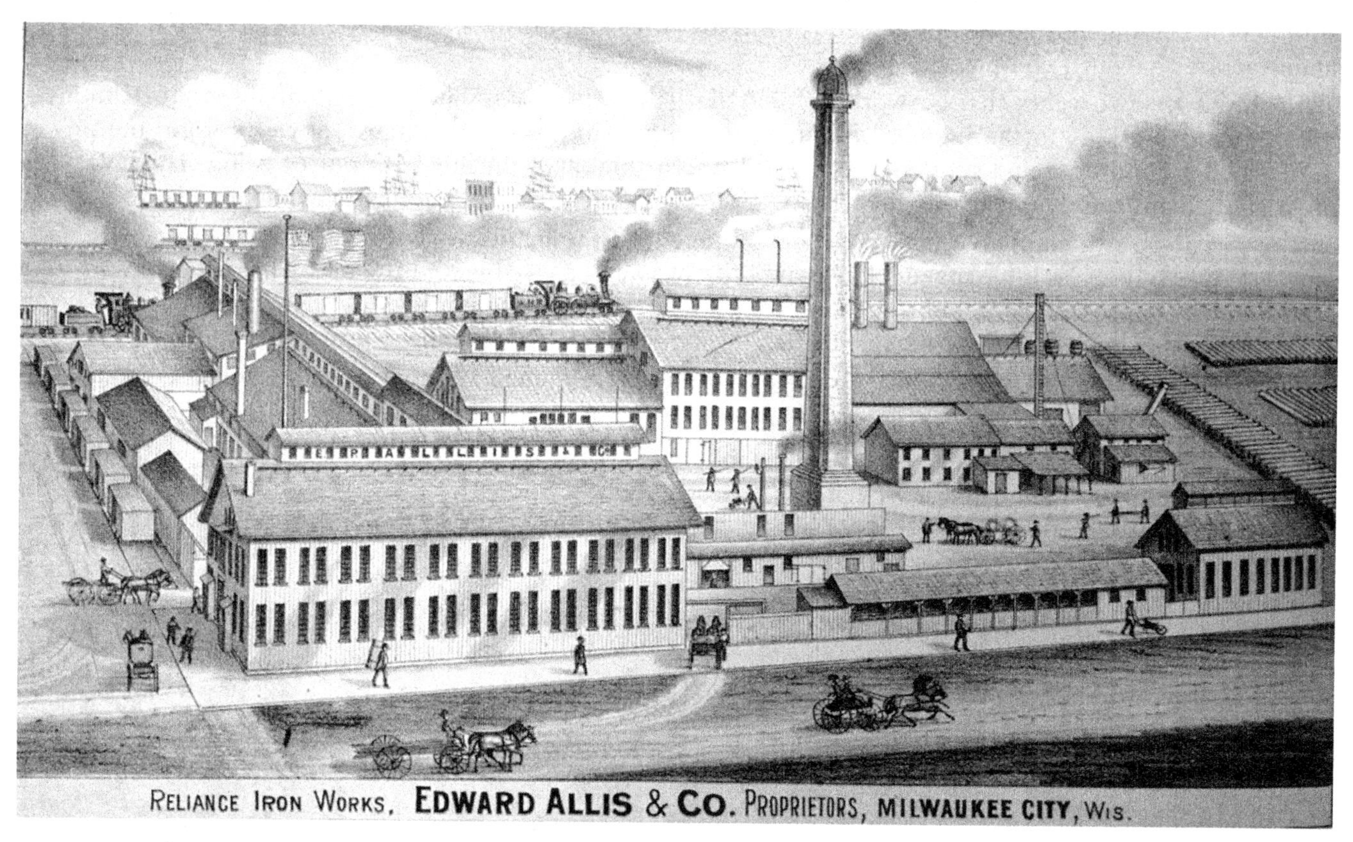

E.P. Allis Reliance Works, as illustrated in its Walker's Point location at Clinton (now known as 1st Street) and Florida Streets. Illustration from ~1875.

By the late 1880s, the E.P. Allis Company was Milwaukee's largest industrial employer. It had built a world reputation for its production of heavy machinery for mines, power plants, public utilities and steel mills.

Allis achieved success by pursuing a policy of acquiring the services of the best engineers in their respective fields to supervise the design and manufacture of his basic product lines—flour milling equipment, sawmills, and heavy-duty steam engines and other large equipment.

Allis built a solid team of engineering talent by granting lucrative and almost irresistible benefits. He allowed his primary engineers to retain part, or all, of their patents—and then paid them for their use. He also gave his engineers a great deal of fame by carrying their names on the equipment they designed.[25]

In his book *An Industrial Heritage*, Walter F. Peterson refers to three of these men—Hinkley, Gray, and Reynolds—as the "engineering triumvirate which (led) the Reliance Works to international fame and financial success."[26]

Sawmill Innovations

The first of these engineers, George Madison Hinkley, was persuaded to join Allis in 1873 as the head of its fledgling sawmill department. Hinkley had previously built and operated a shingle mill in Milwaukee before going out on his own to design sawmill equipment. For several years, he designed and sold patent equipment that was produced for him by Filer and Stowell, an early Milwaukee firm that still manufactures sawmill equipment.[n]

In the thirty-two years that he worked for Allis, Hinkley accumulated 35 patents. Among Hinkley's notable contributions to the industry included improvements to the automatic power swage, the power guide, and a hand-operated setworks to control board thickness during each cut of a log. However, his most significant achievement was the design and development of the practical sawmill bandsaw.

The circular saws employed before the development of the band saw cut a half inch of *kerf* with each pass. With Hinkley's high-speed band saw, the waste was reduced by half. Throughput was also significantly improved. This improvement, along with Hinkley's numerous other achievements, caused E.P. Allis to become one of the most significant manufacturers of sawmill equipment.

Upon Hinkley's death in 1905, the American Lumberman likened Hinkley's "improvements of sawmill machinery to that of Edison to electrical development."[27]

An 1890 catalog of E.P. Allis Co. illustrates Hinkley's band saw. The company noted that the design was patented and any infringement would be prosecuted vigorously.

[n] Filer and Stowell is discussed later in this chapter.

The Trout Power Set Works

William Henry Trout joined George Hinkley's sawmill department at E.P. Allis in 1884, having worked at Filer & Stowell for a short period, and before that for the Hamilton Manufacturing Company of Peterborough, Ontario. Trout was recognized for having an intuitive understanding of the principles of mechanics and for his innovative design skills. His first job was to assist Hinkley with the design of his band saw.

Trout participated in numerous design improvements, while most often not receiving patent credit for his innovations.[28] During the course of his long tenure with E.P. Allis, this occasionally led to confrontations when Trout insisted upon receiving patent recognition, as well as potential remittance under the company's policies. As noted earlier in this book, it was "common precedent for an employer to take credit for work done by an employee."[29] In the case of Allis's company, it was common for the chief engineer of each department to take credit for his employees.

In 1897, Trout designed a band saw with teeth on both sides of the blade, along with the necessary apparatus to allow saw cuts to be made. With appropriate changes to the related carriage works and other apparatus, this permitted a saw-cut to be made in both directors, significantly reducing the time for sawing logs into boards. Trout argued that he should be named on the patent application and was able to sustain his appeal for the first time.[30] However, this was the exception rather than the rule.

The most notable innovations by Trout had to deal with the log 'set works.' To understand this better, it is helpful to describe the mechanism used to move logs through the band saw. Trout described the mechanism as follows: "A log carriage is made to run back and forth along this side of the band mill; and the log to be sawn is mounted and secured on suitable blocks on the carriage. A set works, which may be operated by a hand lever or by power, is also mounted and connected with the blocks, so that the log may be set forward by it. When the log is thus set forward sidewise beyond the sawing line, the forward movement of the carriage brings the end of the log against the saw, and a slab is taken off the log; then after the backward motion of carriage takes place, the log is again set forward, and a board or plank sawed off; and this is continued till the log is all sawn."[31]

This photograph from an A-C catalog shows a log on a carriage being moved toward the band saw. The set works is the mechanism that moves the log forward to establish the depth of the board being cut.

During his career at E.P. Allis, William Trout introduced several refinements designed to make the setting of board thickness accurate and uniform. This is not a trivial task, considering that sawmills were commonly designed to handle massive logs weighing several tons initially and as much as sixteen feet in length. Uniform thickness for the entire length of the log is important.

The initial setworks designs were for manual operation. Trout eventually designed a power mechanism that was operated from a dial mechanism, allowing thickness to be regulated to 1/32nd of an inch. It could be either powered by a rope mechanism from underneath, or from an electric motor mounted on the log carriage.

The Power Setworks controlled the thickness of lumber cuts with much greater precision than previously possible. Illustrations from Allis Chalmers 1919 Log Machinery Catalog.

Trout reported did all of his design work on his final setworks mechanism from home. When he was satisfied with the design, he applied for patent protection. In the meantime, he convinced his employer, now Allis-Chalmers, to begin offering the power setworks design. When the patent was approved in 1905, Trout negotiated a royalty to allow the company to continue to use the mechanism, which by that time had become a popular offering.

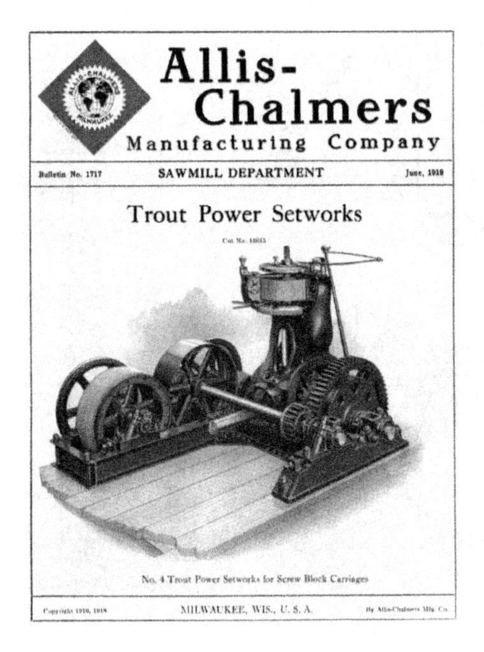

PRACTICAL ROLLER FLOUR MILL

The second of the three prominent engineers hired by E.P. Allis was William Dixon Gray. Gray joined Allis on January 8, 1877, to head the flour mill department. Gray was a millwright, draftsman, and mill engineer. Shortly after joining the company, he perfected a belt driven roller mill that eventually revolutionized the flour milling process in the United States.

Prior to this innovation, the grinding mechanism of mills consisted of two grooved or dressed circular stones, between which the kernels of wheat passed and were cut and crushed. In the process the outer husk, kernel, and germ were all crushed into one mass.

The development of the roller mill allowed better control over the operation, which allowed the efficient production of a higher grade of flour by breaking the grain open between corrugated, accurately spaced rollers, allowing the grain to remain in granular form. The introduction of the roller mill also eliminated the need for periodic dressing, which was required by millstones.[32]

Spring wheat could also be economically milled with Gray's mill, which itself led to an agricultural revolution by allowing two crops to be planted yearly.[33]

Gray's Noiseless Patent Roller Mill. By 1894, over 85% of flour milled in the Midwest was reduced with this roller mill. Illustration courtesy of Allis-Chalmers.

Steam Engines, Pumps and Blowers

The most innovative and best known of the three Allis engineers was Edwin Reynolds, who joined the firm in July of 1877. Reynolds left a better paying job as general superintendent of the Corliss Steam Engine Company to join the Reliance Works, which speaks well of Edward Allis's ability to persuade good men to join his firm.

Edwin Reynolds grew up at a time when formal engineering education in America was almost unavailable, but through his genius and his experience working under such masters as George H. Corliss and William Wright, he developed into one of the best mechanical engineers in the country.[34]

The Reynolds-Corliss Engine

Immediately upon coming to Allis, Reynolds began designing and building steam engines of a modified Corliss design. Corliss had patented a valve design that allowed steam engines to be much more efficient than others of its day. His engines employed separate inlet, exhaust valves, and varied the steam inlet time to regulate engine speed. These features resulted in much lower thermal losses than were experienced in common slide-valve engines which utilized common inlet and exhaust valves and varied inlet pressure to regulate engine speed.[35]

Since the basic Corliss patent had expired in 1870, Reynolds was free to copy and improve upon it.[15] The "Reynolds-Corliss" engine became a "synonym for simplicity, economy and reliability."[16] The first of Reynolds' engines was relatively straightforward in design and well suited for production in the Reliance shops, which at that time lacked the tooling and equipment to handle more sophisticated designs.

By early 1878, the new engine line was so successful that the firm had a six-month backlog of orders.[17]

Reynolds was responsible for numerous improvements to the basic Corliss design, notably in valve-gear and governor configurations. Under Reynolds' leadership, Allis became the world's premiere supplier of steam engines and related equipment. Reynolds engine and pump designs would forever alter the machines that helped power industries like mining, steel production, water supply and rapid transit.

Illustration of an early Reynolds-Corliss steam engine, showing the valve gearing on the steam chest. Illustration courtesy of Allis-Chalmers.

Blowing Engines

In 1880, Reynolds contracted to build two huge steam engines coupled to pumps to provide air for the steel-making process. These "blowing engines" differed markedly from those of conventional design. The Reynolds air pumps had cast steel air valves instead of the common leather valves and achieved far greater efficiency than their predecessors achieved.

Their success led to many orders and helped America achieve a six-fold increase in blast furnace capacity—which allowed the United States to surpass Britain in steel production for the first time.[36]

In 1894, Niels Christensen joined the company to assist Reynolds with the design and construction of blowing engines. It is reported that Christensen developed a new type of blowing engine that was adopted in "all the great steel mills of the United States and many of them abroad."[37] Christensen went on to develop numerous accomplishments related to pneumatic systems—many of which are covered later in this book.

An illustration of a large blowing engine built by the E. P. Allis Co. of Milwaukee. The steam cylinder is 42" diameter, the air cylinder 84" and the stroke 60". Valve gear on the steam engine is of the Reynolds-Corliss type. Illustration courtesy of Allis-Chalmers.

CENTRIFUGAL PUMPS

In 1884, Edwin Reynolds devised a centrifugal sewage pump, largest in the United States at that time. It had a capacity of seventy million gallons per day. Its twelve-foot impeller was driven by a tandem-compound Corliss steam engine directly connected to the vertical pump shaft. Installed at the northeastern edge of Jones Island, this successful unit was in continuous use for thirty years. This was the first centrifugal pump built by E. P. Allis & Co.

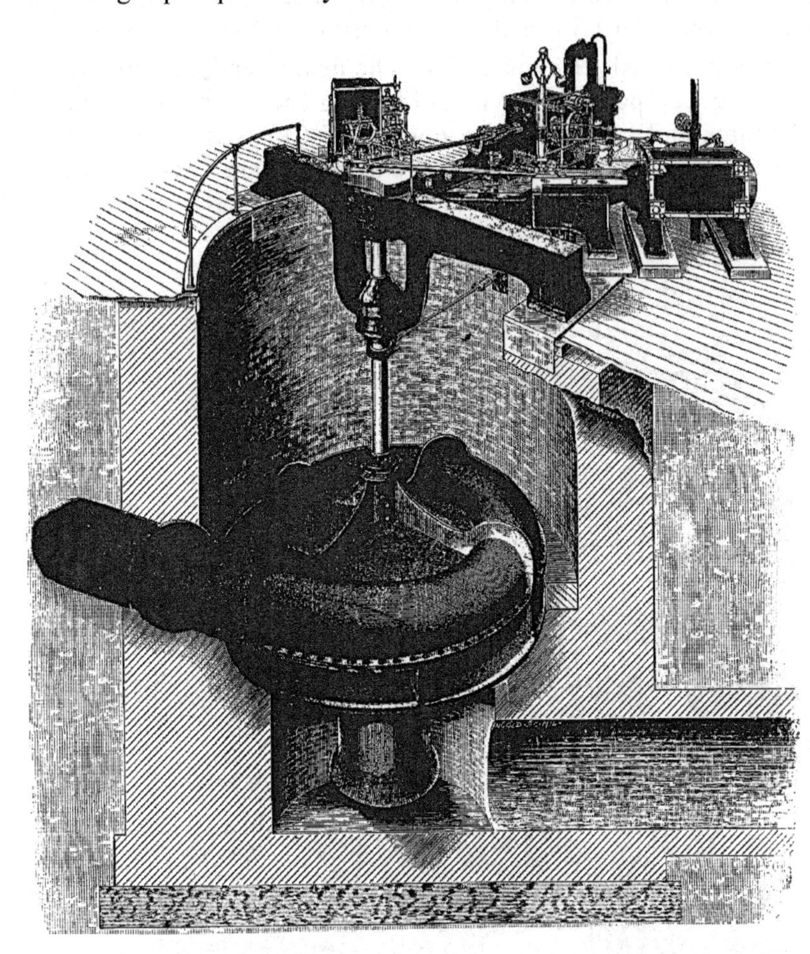

The first centrifugal pump built by E. P. Allis & Co. was rated at a capacity of 70 million gallons per day. It was built for pumping sewage at Jones Island, Milwaukee. Illustration courtesy of Allis-Chalmers.

Reynolds' centrifugal pump played an important role in helping to deal with part of Milwaukee's sewage problems of the time. Toward the end of the decade of the 1880s, Milwaukee's population was rapidly approaching 200,000. The city had nearly 165 miles of sewers—most of which discharged untreated into the Milwaukee, Menomonee, and Kinnickinnic Rivers. The rivers were virtually stagnant at some times of the year and produced unpleasant odors referred to delicately as the *river nuisance*. Resentment increased as more people recognized the public health danger of sewage flowing through the most populous section of a great city. For more than two years resentment continued to develop as the belief grew that it was harmful to the public health to have the sewage of the great city of Milwaukee pouring into its main streams.

At that time, the sewerage system was under the control of the Board of Public Works of the City of Milwaukee. Under its oversight, the first regular system of sewer construction was begun in 1869. It was in the form of a combined system, which disposes of both sewage and surface drainage storm water. In addition, another 247 miles of sewers had been constructed by 1897, not including the additional four miles of the so-called Menomonee special sewer.

This Menomonee special sewer intercepted all sewage emptying into the Menomonee Valley from the north and south. It also received all the sewage from the slaughterhouses, factories, and ships located in the valley. In addition, the Menomonee sewer line also conveyed about 70,000,000 gallons of river water from the ends of various canals, slips, and rivers in the valley to the sewage pumping station at the lakeshore south of the harbor. This sewer was built of brick along the river. An inverted siphon was built and laid in the dredged river to carry the sewage to Jones Island.

The Allis pump was installed at the northeastern edge of Jones Island and used continuously for thirty years. The sewage was ejected into the lake without treatment during the next several decades. However, the same pumping station was used for experimental work in 1914 that resulted in developing the new activated sludge process for the treatment of sewage.

The pump was removed from service when completion of a new sewer system rendered its use unnecessary.[19]

Milwaukee River Flushing Station

As noted previously, the direct discharge of raw sewerage into the Milwaukee, Menomonee and Kinnickinnic Rivers caused extremely unpleasant odors, especially during the summer months. Resentment increased as Milwaukeeans recognized the public health danger of sewage flowing through the most populous section of the city. In addition to the installation of a sewerage pump to discharge much of the sewerage flow into Lake Michigan, as discussed above, the City proposed that lake water be pumped into the Milwaukee River to increase the flow rate and flush the river. While these practices seem amazing today, in the late 1800s the adage commonly used at the time was, "the solution to pollution is dilution."

To accomplish this task, Reynolds proposed a "screw" pump with an impeller fourteen feet in diameter and a hub of six feet. It was driven by a vertical compound engine at fifty-five revolutions per minute, with a capacity at that speed of over 40,000 cubic feet of water per minute. The idea of the screw pump had originally been developed by Reynolds many years earlier when he built a very small pump for raising a wrecked steamer on the Ohio River. The Milwaukee screw pump was almost as efficient as plunger or piston pumps, performing about double the amount of work with the same fuel and costing half as much as centrifugal pumps adapted for the same service.

When placed in service it exceeded the contract capacity of five hundred million gallons in twenty-four hours, pumping a greater quantity of water than any machine in the world. The new pump eased the situation in Milwaukee until the City built an elaborate intercepting system to pump sewage one thousand feet into the lake.

In 1912, the original steam engine was replaced with an electric motor and the four horizontal tubular boilers that had supplied the steam were removed. In 1955, a large tree stump found its way into the impeller chamber, badly damaging the pump. Two blades were damaged beyond repair and were removed. After repairs the remaining two blades were salvaged and the pump placed back into operation. It remains in this configuration to this day.

The pump house is a prominent feature on Milwaukee's lakefront at 1701 North Lincoln Memorial Drive and now hosts a popular coffee shop. The pump remains operable. It is used occasionally during the summer months for its original intended purpose—flushing the Milwaukee River.

In 1888, E. P. Allis & Company built and installed the highest capacity water pump in the world at that time. The pump was a horizontal screw pump of a unique design by Edwin Reynolds. The pump was recognized as a landmark by the American Society of Mechanical Engineers in 1992. Illustration courtesy of Allis-Chalmers.

World's Columbian Exposition 'World's Fair' Engine

During the 1890s, Edwin Reynolds continued to make significant design improvements to steam engines—gaining Reynolds worldwide recognition for outstanding achievements in engine building. E.P. Allis built an engine for the World's Columbian Exposition held in Chicago in 1893. The engine was dubbed "Pride of Machinery Hall." The horizontal, quadruple-expansion Reynolds-Corliss engine was rated at 3,000 horsepower.

One of the impressive attractions at the 1893 World's Columbian Exposition in Chicago was this 3,000 horsepower, quadruple-expansion Reynolds-Corliss steam engine.

It was a dramatic moment when United States president Grover Cleveland pulled a lever that started the huge engine. The mighty cylinders of the steam-powered giant throbbed with power. Reports say that the crowd was galvanized at the sight as the thirty-foot diameter flywheel began to rotate. The entire engine weighed 325 tons.

Driving two Westinghouse 750-kw alternators, the Allis steam engine supplied the electricity for twenty-thousand 16-candlepower incandescent lamps. Fairgoers from around the world took home memories of electricity used on a lavish scale for the first time—and of the engine that produced that power.

It appears that E.P. Allis may have first met Niels A. Christensen at the World's Columbian Exposition. Christensen was a leading draftsman at Fraser and Chalmers in Chicago at that time. He worked briefly on electrical systems for the Columbian Exposition, where he would likely have interfaced with Reynolds and others on E.P. Allis's staff. In any case, Christensen was hired by Allis shortly afterward. His innovations are discussed later in this book.

Cornish Pump for the Chapin Mine

Edwin Reynolds also designed a huge steam-powered mine pump for the Chapin Mining Company of Iron Mountain, Michigan. His *Cornish* style mine pumping engine was regarded as "the largest pumping engine in the world" when installed in 1892.

The Chapin mine was discovered in 1879 when Dr. Nelson Hulst sent a crew to investigate an area that he had prospected seven years earlier. The crew of eight was headed by John Wicks. At the site of what is now the City of Iron Mountain, on wilderness property owned by Henry Chapin, the crew started digging the shaft of what became the Chapin mine. They struck one of the richest ore deposits in the world, and turned the area into a boomtown. It was a deep and extensive seam, requiring dewatering pumps to allow the mineshafts to be dug to accommodate extraction of the ore.

The engine Reynolds designed was a vertical tandem-compound steam engine with a high-pressure cylinder of fifty inches in diameter, a low-pressure cylinder of one hundred inches in diameter, and a piston stroke of 120 inches. It stood 54 feet tall and had a flywheel forty feet in diameter. The engine weighed 160 tons; largest of its type ever constructed. Underground, there were ten pumps in a vertical shaft, eight set at intervals of 192 feet and two at intervals of 170 feet on a single pump rod—extending a total of 1,500 feet below the surface. The pump's capacity of 3,400 gallons per minute was sufficient to handle the mine's normal flow of 3,000 GPM.

The pump was moved from its initial site in 1896, when it was disassembled and moved to the "C" or "Ludington" shaft, its present location. The steam engine was replaced with electric motors in 1914, but the pump continued to operate until the mine was permanently closed in 1934.[38] The engine was donated to the City of Iron Mountain, where it has become a tourist attraction.[39]

This Reynolds steam engine was built in 1891 by the E. P. Allis Company to power a 'Cornish' pump. It has become a tourist attraction—the centerpiece of a mining museum in Iron Mountain, located at the corner of Carpenter and Kent streets. The engine/pump was designated an ASME landmark in 1987.

Triple Expansion Water Pump Engines

The rapid development of an urban, industrial society in the late 19th century placed enormous pressures on municipalities in their attempts to provide basic services. Among those services, the increasing demand for fresh water by industry and private citizens posed a continuing problem. The Allis Company had provided nearly all the pumps used in the Milwaukee water system and for a great many other cities. But while the pumps devised by Edwin Reynolds in 1880 and 1884 were a vast improvement over the Hamilton pumps of 1874, they were not efficient enough to revolutionize municipal pumping systems and thereby seize a large portion of the growing market. The man who did this was Reynolds' nephew, Irving H. Reynolds, whom the elder Reynolds brought to Milwaukee in 1884.

Working from experience with marine engines and the pumps that the Reliance Works had previously produced, Irving Reynolds set to work to design better water pumps. He sought to reduce pulsation in the water main, achieve fuel saving, and build greater durability in a pump that could be serviced more easily. Working almost every night and every Sunday for two years, he finally developed his own engine, a triple-expansion pumping engine.

In May 1886, Milwaukee planned a high capacity pumping station and advertised for a six million gallon compound engine driven pump. The E.P. Allis firm submitted three bids, the most expensive of which was a new design for a triple-expansion pumping engine. By virtue of its bidding rules, the City could choose any of the bids in a multiple bid group and selected the new engine.

Irving Reynolds proposed a vertical, triple-expansion engine with a guaranteed efficiency of 115 million foot-pounds of work per one-hundred pounds of anthracite coal; this rating compared to a guarantee of 100 million foot-pounds for a three-cylinder compound engine. When installed, the actual output of the new pumping engine was measured at 118,186,312 foot-pounds per thousand pounds of steam.

The triple-expansion pumping engine developed by Irving H. Reynolds was first purchased by the City of Milwaukee in 1886. Its success led to orders from cities all over the Country.

In 1891, Irving Reynolds also built a triple-expansion pumping engine for the North Point Station with a capacity of 18 million gallons per day, developing over 154 million foot-pounds of work per thousand pounds of steam. This engine proved so incredibly economical that some skeptical engineers challenged the company's integrity, accusing it of deliberately hoodwinking the municipalities of the nation. The argument was resolved when Professor John Carpenter of Cornell University headed a team that subjected the engine to extensive tests. The results of the tests appeared in a paper by Dr. Robert H. Thurston, Dean of the College of Engineering at Cornell, presented at a meeting of the American Society of Mechanical Engineers. The results vindicated the Allis claims of performance by this new machine. Consequently, the Company rapidly gained leadership in municipal pumping, which it maintained for thirty years until the triple-expansion pumping engine gave way to turbine-powered centrifugal pumps in the 1920s.

By 1893, E.P. Allis had completed one hundred of these huge engines in many sizes and configurations.

Architect's illustration of Milwaukee's historic North Point Station. The pumping station is long gone—replaced by a parking lot on Milwaukee's lakefront. However, the iconic water tower on the bluff continues to serve as a testament to the original plant. Picture from Library of Congress.

The North Point Water Tower is all that remains from the North Point Station. Designed by architect Charles A. Gombert and completed in 1874, it serves as one of Milwaukee's best-known landmarks. The tower houses a standpipe that was used to reduce pulsations from the pumping engines, thus lowering the peak pressure on Milwaukee's water mains—minimizing the potential for breaks and resulting leaks.

The buildings for the North Point Pumping Station, including its water tower, were built by the Bentley Construction Company. Bentley Construction was founded by John Bentley, who founded the company in 1848.

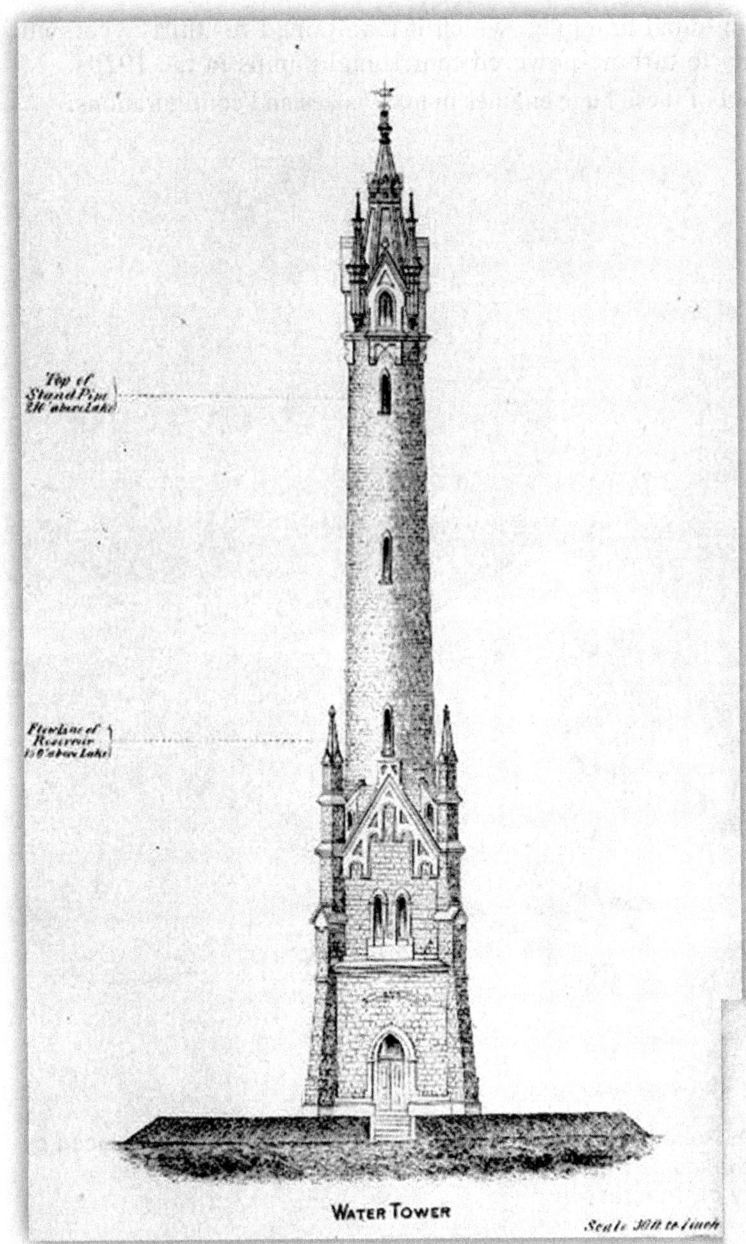

The North Point Water Tower, designed by architect Charles A. Gombert and completed in 1874, serves as one of Milwaukee's best-known landmarks. The tower houses a standpipe that was used to reduce pulsations from the pumping engines, thus lowering the peak pressure on Milwaukee's water mains and minimizing the potential for breaks.

Direct Shaft Mounting of Electric Generators

Another of the notable accomplishments that occurred under the direction of Edwin Reynolds was the direct shaft mounting of electric generators. Prior to 1892, electric generators were mounted on separate shafts and belt- or rope-driven from the steam engine's flywheel. It was believed to be impractical to mount generator rotors directly on the engine shafts because of the slow speed that such engines ran.

In March of 1892, Reynolds designed two Corliss-type cross-compound steam engines for the Narragansett Electric Light Company of Providence Rhode Island. The engines were the first in the United States to have "magnetic poles on the flywheel" for the generation of electricity.

Since the U.S. generator manufacturers were unwilling to risk such an unproven design, Reynolds began to plan their production at the Reliance Works.

When the Westinghouse and Thomson-Houston companies heard that E.P. Allis was about to produce its own generators, they agreed to each manufacture one generator if E.P. Allis would agree to design and build the rotors. This was apparently done in order to keep E.P. Allis from entering into the generator business.

These units were so successful that belt driving of generators was almost immediately abandoned; direct mounting of generators on shafts of large engines becoming standard practice almost overnight.[40]

Perhaps one of the most widely observed units of this type was the five-thousand horsepower Reynolds horizontal-vertical steam engine connected directly to a Bullock generator at the St. Louis World's Fair in 1904. All the electricity for use at the fair was supplied by this engine-generator set. Called "Big Reliable," the engine was awarded a grand prize, as was the Bullock generator.

During the very same year of 1904, Allis acquired the entire Bullock business, effectively entering the generator business.

An important contribution by Erwin Reynolds of E. P. Allis & Co. to the production of power was the shaft mounting of generator rotors directly to large engines, eliminating the need for long belts.

THE MANHATTAN HORIZONTAL-VERTICAL STEAM ENGINES

In 1899, E.P. Allis built the first of seventeen steam engines that were to become known collectively as the "Manhattan" engines. They were built for the Metropolitan Street Railway Company in New York and were "the largest stationary steam engines ever made."[41] They were four-cylinder, compound engines. In order to meet the capacity requirements in a relatively compact space, Edwin Reynolds designed each engine with its two 44-inch diameter high-pressure cylinders mounted horizontally and its 88-inch low-pressure cylinders vertically—both cylinders being connected to a common crank pin in what became known as the *horizontal-vertical* configuration. The E.P. Allis horizontal-vertical steam engine configuration provided compact engine settings, allowing steam engines to be built to tremendous sizes. The largest of the engines produced approximately ten-thousand horsepower each. They were used to electrify the rapid transit system.[42] Edwin Reynolds received a patent for this steam engine arrangement.

It is reported that the inspiration for the engine design occurred during a train trip that Edwin Reynolds took to discuss the new engines with the Metropolitan Street Railway Company.[43] He was faced with a dilemma. The railway company wanted to generate 96,000 horsepower on a small site. Reynolds realized that existing steam engine designs wouldn't be able to meet the small site's constraints. In addition to the engines, the boilers, chimneys, coal-handling equipment and other auxiliaries all needed to be crammed into the site—conventional steam engines just wouldn't provide enough power on such a small site.

Reynolds was also concerned about the required size of the flywheels, which would have been enormous.

He was aware of compound engine designs, where one cylinder was installed at 90 degrees to the other—the larger, low-pressure cylinder vertical and the smaller high-pressure cylinder horizontal. But the flywheel size would still pose a problem.

Reynolds' solution was to duplicate the compound engine design on both sides of the flywheel—effectively a double-compound engine configuration. As he thought it through, he realized that placing the crank pins at an angle of 135 degrees would spread the resulting eight power strokes evenly about each revolution of the crankshaft. By doing so, this smoothed out the torque from the engines, minimizing the flywheel requirements. In fact, he eventually determined that no flywheel would be required other than the revolving fields of the generator.

This configuration had the additional benefits of reducing the required size of the crankshaft and bearings and resulted in smaller foundations and, of course, a smaller footprint. Finally, it eliminated the need for a barring engine,[o] since these engines could never be stopped on dead center.

The first of these engines were placed in Manhattan at 74th Street, adjacent to the East River. Thus, this type of engine became known as the *Manhattan Engine*. Eventually, eight engines were installed in this powerhouse, which was completed in 1903. The powerhouse provided power to electrify the railway. Electricity from the plant powered electric locomotives that replaced 225 steam locomotives.

A near duplicate powerhouse was built at 58th Street and 11th Avenue, completed by 1905. The steam engines for the 58th Street powerhouse were built in West Allis by Allis-Chalmers. They were in use for many years and remained on standby duty as late as 1954.

---

[o] A *barring* engine is a small engine, usually a steam engine, that is used to turn the main engine to a favorable position from which it can be started. Smaller steam engines are usually moved to a favorable starting position manually, using iron bars inserted in the flywheel – thus the term *barring* became in common use.

In 1899, E.P. Allis built the first of 17 steam engines for the Metropolitan Street Railway Company in New York. Known as the "Manhattan" engines, they were the largest stationary steam engines ever made. The most powerful of the engines produced 10,000 horsepower. Photograph courtesy of Allis-Chalmers.

GEORGE MADISON HINKLEY was a prominent mechanic and inventor. He was born in the state of New York on May 24, 1832. His family moved to Ohio when he was an infant. After his parents died at a young age, George was raised by neighbors until he became apprenticed to a carpenter at age 12. Later he worked as an apprentice on building bridges in Michigan, before working on sawmills.[44]

In 1861, Hinkley enlisted as a private in the First Michigan cavalry and served for three years during the Civil War. Upon his return to Michigan, he became manager of a mill at Muskegon, but after a year he moved to Wisconsin to construct a sawmill. When the Wisconsin mill was complete, Hinkley managed the plant for three years. Resigning, he began designing various sawmill mechanisms.

In 1874 he was hired by E.P. Allis to head up that their sawmill department. He worked for E.P. Allis for thirty-two years, during which time he accumulated thirty-five patents. Among Hinkley's notable contributions to the industry included improvements to the automatic power swage, the power guide, and a hand-operated setworks to control board thickness during each cut of a log. However, his most significant achievement was the design and development of the practical sawmill bandsaw.

George Hinkley married Sarah Tubbs of Michigan in 1860. They had two children, George C. and Sarah. His wife died in 1865. Several years later Hinkley married Elizabeth Langdon, who was born near Janesville.

WILLIAM DIXON GRAY, JR. was born in Scotland on July 22, 1843. His family immigrated to Canada when he was a child, where they established a farm. As the oldest son, William attended public schools in Canada and worked on the farm. He also learned the carpenter's trade and worked in that occupation for several years. His work ultimately offered him an opportunity to work on a flourmill. He was attracted to milling machinery and he moved to Minneapolis to gain a better understanding of the practical, as well as scientific, aspects of milling.

Over a number of years, Gray worked as an apprentice, millwright and draftsman, before being promoted to mill engineer.

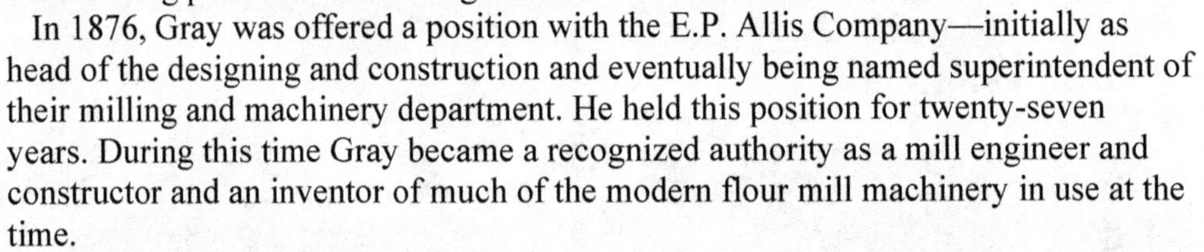

In 1876, Gray was offered a position with the E.P. Allis Company—initially as head of the designing and construction and eventually being named superintendent of their milling and machinery department. He held this position for twenty-seven years. During this time Gray became a recognized authority as a mill engineer and constructor and an inventor of much of the modern flour mill machinery in use at the time.

When E.P. Allis consolidated and formed Allis Chalmers, Gray decided to go into business for himself. He organized the Foster Construction Company in 1906 and served as its president. While there, he built a number of mills in different parts of the country, including a large mill in New York City. Gray was married to Kathrine E. Chipman—they had two daughters.

EDWIN REYNOLDS was born in Mansfield, Connecticut in 1931. He relocated to Milwaukee in 1877 to become superintendent of the Reliance Works of E.P. Allis Co., leaving a similar position at the famed Corliss Engine Works. Prior to working at Corliss, he was employed for 16 years at various jobs involving the machining and installation of steam engines. During that time, Reynolds helped John Ericsson build the ironclad Monitor, which, along with the Merrimac, ushered in the age of armored battleships.[45]

It is remarkable that E.P. Allis was able to entice Reynolds to leave a better paying job at Corliss Engine Works for the struggling Reliance Works that employed a smaller workforce in relatively crude facilities. It is likely that Reynolds was partly motived by the ability to retain the rights to his patents, as well as the privilege of having his name featured on the equipment he designed. He was also granted the freedom to explore improvements in the design of steam engines, pumps, blowers, and other equipment.

Since the patent for the Corliss trip-gear valve mechanism had run out, Reynolds were free to adopt it to his steam engines. His Reynolds-Corliss engine offered an improved valve design, which ran faster and allowed for better engine speed governance with less stress on the parts. Reynolds also adopted numerous other engine improvements that improved efficiency and power output. Thousands of steam engines were sold and Allis became one of the most prominent manufacturers in the world.

Edwin Reynolds had over forty patents for improvements to steam engines, valve gear, air compressors, mining machinery, and other devices.

Edwin Reynolds hired his nephew, Irving Reynolds in 1884 as his engineering assistant. Under his uncle's sponsorship, Irving developed the triple-expansion pumping engine, which was first installed for the Milwaukee Water Works in 1886. Its high efficiency and reliability led to installations across the country in major cities.

In 1901, when the Allis-Chalmers Company was formed through a merger of E.P. Allis, Fraser & Chalmers and the Gates Iron Works. Edwin Reynolds was appointed chief engineer. He oversaw the building of the company's new factory in West Allis, but eventually moved into a consulting role for the combined company.

Edwin Reynolds served as one of the first presidents of the American Society of Mechanical Engineers in 1902. Suffering from ill health, he retired in 1906—three years before his death.

Edwin's nephew, Irving Reynolds, continued to work for the company as manager of its engine department. Irving Reynolds retired in 1935 after working for the firm for 43 years.

William Henry Trout was born in Canada in 1834. His father was a millwright and carpenter, and William grew up exploring his flourmill in Norval, Ontario. The family moved around Ontario, and William began working with his father on various projects, constructing a sawmill and a flourmill. He also helped to build his father's mill. Using the mill for power, he and friends established Trout & Jay to manufacture agricultural hand tools.

During this time, Trout met naturalist John Muir, who lived with the family for over a year. Muir had left Wisconsin to live in Canada to avoid the draft during the Civil War. Muir eventually moved to Indianapolis, although the two remained lifetime friends and correspondents.

Trout attended Hiram, as well as Williamsville Academy, for a time, although he never received a degree.

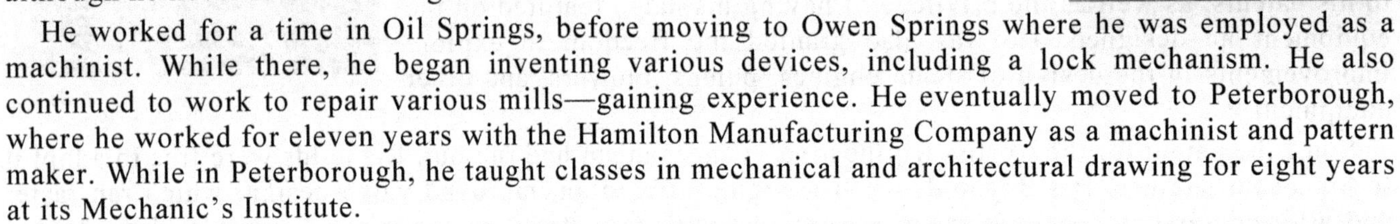

He worked for a time in Oil Springs, before moving to Owen Springs where he was employed as a machinist. While there, he began inventing various devices, including a lock mechanism. He also continued to work to repair various mills—gaining experience. He eventually moved to Peterborough, where he worked for eleven years with the Hamilton Manufacturing Company as a machinist and pattern maker. While in Peterborough, he taught classes in mechanical and architectural drawing for eight years at its Mechanic's Institute.

Because of declining Canadian business opportunities, Trout joined George Hinkley's sawmill department at E.P. Allis in 1884, having worked at Filer & Stowell for a short period. Trout was recognized for having an intuitive understanding of the principles of mechanics and for his innovative design skills. His first job was to assist Hinkley with the design of his band saw.

During his time at E.P. Allis, Trout was named on at least seventeen US patents. He also participated in many other design activities for which others were recognized. He is best known for his invention and refinement of the Trout Power Set Works.

Trout was also known for his written history of the Trout Family. He privately published his work in Milwaukee, Wisconsin in 1917, titling his remarkable compilation of 23 years of research as "*Trout Family History by W.H. Trout, Mechanical Engineer and Machinery Designer*." Complete with numerous family photos and family tree charts, this combination personal memoir and ancestral genealogy is a touchstone reference source. It is claimed that at least five subsequent books were largely based on his work.

Trout married Jane Barclay Knowles in 1867. They had eight children, including Walter Charles Trout (1874-1947), a sawmill and oilfield machinery designer/manufacturer of Lufkin, Angelina, Texas.

William Henry Trout died in 1917 in Milwaukee.

The story of E.P. Allis, its transition to the Milwaukee suburb of West Allis and the formation of Allis-Chalmers Corporation, is continued in Chapter 8.

MILWAUKEE BOILER WORKS

The Menomonee Valley company with closest ties to the E.P. Allis Company was the Milwaukee Boiler Company. Allis did not make boilers, but almost all of its customers needed them to provide the steam to run its steam engines. This need was often satisfied by the Milwaukee Boiler Works. Founded in 1890, it was located in Walker's Point at 222 East Oregon Street, just a short distance from the Reliance Works.

The Milwaukee Boiler Company was established by a group of businessmen with ties to E.P. Allis. Edwin Reynolds served as president of the company for a time. He is reported to have been a major partner in the firm. He contributed several improvements in boiler designs, including a vertical tubular boiler design that was useful when space was tight and quick steaming was a requirement.[46] It is likely that Allis salesmen routinely took orders for boilers from Milwaukee Boiler Works when selling steam engines. The largest boiler offered could provide the steam for a 170 horsepower Reynolds-Corliss engine.

After the demise of the steam engine, Milwaukee Boiler operated as a job shop, making parts tailored to customer's specifications. It specialized in the manufacture of steel and alloy weldments, ASME pressure and process vessels and systems, specialized machinery for the underground construction industry, and special vehicles for railroad locomotive servicing equipment.

This vertical boiler was designed by Edwin Reynolds and manufactured by the Milwaukee Boiler Works.

Milwaukee Boiler operated for well over one-hundred years, having relocated in 1921 to 1101 South 41st Street before going out of business in October 2013. All is its equipment was auctioned off at that time.

PAWLING & HARNISCHFEGER

Another business that located near E.P. Allis was Pawling & Harnischfeger. Alonzo Pawling, a Chicago-born patternmaker, and Henry Harnischfeger, a German immigrant and locksmith journeyman, had worked together for the Whitehill Sewing Machining Company in Milwaukee in 1882.[47] Two years later, they established their Pawling & Harnischfeger Machine and Pattern Shop. They moved into a poorly insulated, wood-frame building at 292 Florida Street—just north of the Reliance Works

They initially didn't have products of their own. The company essentially operated as a job shop for other companies. Among other products, they made wood patterns for area manufacturers, which included their neighbor, E.P. Allis Company. Wood patterns were important to the manufacture of castings. They were used to prepare a cavity in sand into which molten material would be poured during the casting process. The patterns had to be prepared to exacting standards. Sometimes the patterns were used repeatedly for an entire product line. At other times, they were made for one-off, specialized equipment and were shipped along with the manufactured products, to allow the owner to have replacement parts made in case of equipment failure or breakage.

Soon Pawling & Harnischfeger's machine shop developed a reputation for its expertise and began working for Milwaukee's growing knitting industry to design and repair knitting machinery. They also began to attract other manufacturers to manufacture products such as file-cutting machines, stamping presses, brick-making equipment and beer-brewing equipment.

The first breakthrough for Pawling & Harnischfeger came in the late 1880s through an association with Christopher Warren LeValley. LeValley, a former manager of the St. Paul Harvester Company, envisioned a replacement for the trouble-prone leather belting that was typically used to drive agricultural equipment. He devised a linked, flexible, metal chain belt and obtained patents for his design.

LeValley contracted with Pawling & Harnischfeger to help with the production of his metal chain belt. For five years, P&H made the patterns for the castings and handled the assembly of the chain.

Pawling and Harnischfeger's shop on Florida Street also made poppet valves for Bruno Nordberg for several years prior to the establishment of the Nordberg Manufacturing Company. These two early P&H clients—LeValley's Chain Belt and Nordberg—eventually joined forces to become the mining processing equipment manufacturing firm Rexnord Corporation with extensive operations around the world—more on that later.

In 1887, an overhead crane failure at E.P. Allis's Reliance Works, led Alton J. Shaw, an employee of Allis, to conceive of an improved crane design. Shaw oversaw the rebuilding of the rope-drive overhead traveling bridge crane at the Reliance Works, but the experience led him to design a crane with three electric motor drives—one for the bridge drive, one for the trolley drive and one for the hoist. Such a crane drive was revolutionary—having never previously been built.

E.P. Allis wasn't interested in the concept, so Shaw walked over to Pawling & Harnischfeger and presented the idea to them. P&H embraced the new technology. The design was far superior to anything else on the market, leading Pawling & Harnischfeger to hire Shaw and form a separate company to manufacture the new three-motor overhead crane.

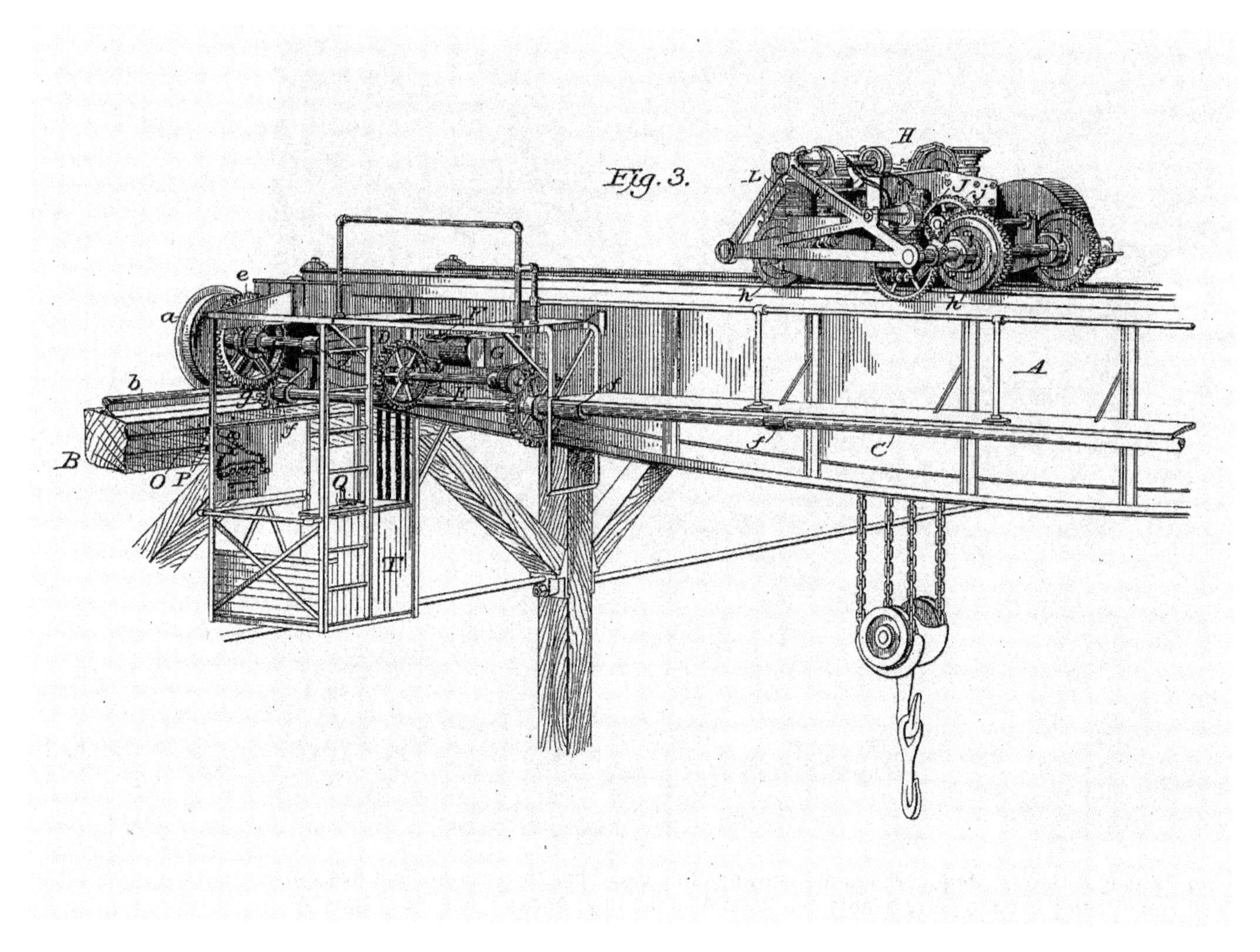

Shaw's electric crane design, from his 1890 patent.

Sales of the new crane got off to a good start. But in 1892 Shaw, apparently dissatisfied with the arrangement, left to form his own company—the Shaw Electric Crane Works. Since Shaw retained the patent rights to the design, Pawling & Harnischfeger lost its rights to manufacture the cranes. P&H had built a foundry and new factory to manufacture the crane and loss of the product line almost led to bankruptcy. A last minute order from Pabst Brewery for grain dryers kept them financially solvent, although they had to close the foundry.

Shaw lost the exclusive rights to manufacture his crane design when the patent expired, allowing Pawling and Harnischfeger to re-enter the business. P&H hired a chief engineer to design a complete line of cranes. Eventually, they added other hoisting machinery to the product line. Harnischfeger went on to become a world leader in the manufacture of overhead cranes.

Two Pawling & Harnischfeger cranes are shown in this view. The lower crane is a five-ton wall crane, being used to carry a ladle in a West Allis foundry. Above it is a P&W type "A" bridge crane. Both cranes are operated by crane operators, riding in their individual cabs.

P&H soon developed other products bearing its name, including steam-powered engines for sawmills, steam-powered steering gears for ships, and machines for the brewing industry. The company also developed its own electric motors, using production machinery purchased when the Gibbs Electric Company of Milwaukee was acquired by Westinghouse. The new line of electric motors improved the reliability of P&H overhead cranes and sales increased.

When fire destroyed their foundry building in 1903, P&H acquired a 26-acre manufacturing site located at 44th and National Avenue to accommodate their growing business. The company's story continues in Chapter 8.

HENRY HARNISCHFEGER was born in Germany, on July 10, 1855. He was the son of Konstantin and Christina (Adrian) Harnischfeger. His dad ran a tannery in Germany. Henry received his early education at German public schools. At age thirteen, he was indentured as an apprentice to a locksmith.

In 1872, he and six friends immigrated to New York. He initially secured employment with the Singer Sewing Machine Company, but left after a short time and took a job at the Rhode Island Locomotive Works in Providence. His next employment was with Browne & Sharpe—an early manufacturer of machine tools—but was laid off and returned to Singer. He worked there for nine years in their new plant in Elizabeth, New Jersey. He attended school in the evenings while employed in the New York area.

He moved to Milwaukee in 1881 to accept a job at the Whitehill Sewing Machine Company, where he obtained a job as foreman of their milling department.

With Alonzo Pawling, who he met at Whitehill, Harnischfeger founded the Pawling and Harnischfeger Company, which became one of the largest manufacturers in Milwaukee.

Harnischfeger was a member of Milwaukee's Deutscher Club, the German-English Academy, the Milwaukee Musical Society and the Milwaukee Turnverein. He married Marie Kauwertz and they had four children—two of which survived into adulthood—Frieda, born in 1894 and Water, born in 1895.

ALONZO PAWLING was born in Chicago in 1857. Moving to Wisconsin, he obtained positions at various manufacturers, including Filer & Stowell and in Racine at the J.I. Case Company before becoming a wood patternmaker at the Whitehill Sewing Machine Company. While at Whitehill, he met Henry Harnischfeger, as well as Maurice Weiss, who had come from New York with Harnischfeger.

In 1883, these three men went into business for themselves, starting the Milwaukee Tool and Pattern Shop. After Weiss left in 1884 to return to New York, Pawling and Harnischfeger formed the partnership of Pawling and Harnischfeger.

Pawling retired from active involvement with the company in 1912.

Alonzo was a member of St. John's Episcopal Church, of the Deutscher Club and the Old Settler's club. He died in 1914, after several months of poor health. In addition to his wife, he had six children—sons Harry, Robert, Joseph L, and Alonzo, Jr., and daughters Ether and Mrs. Walter Schmitz. He is buried in Forest Home Cemetery.

THE NORDBERG MANUFACTURING COMPANY

Bruno V. Nordberg joined the E. P. Allis Company in 1879 as a draftsman. He had come to Milwaukee from his native Finland where he received a degree in mechanical engineering from the Polytechnic School in Helsinki. A short time after starting employment with E.P. Allis he was promoted to the position of private designer for Irving Reynolds, Edwin Reynolds' nephew and successor as Chief Engineer. Nordberg assisted Irving Reynolds in the design of America's first triple-expansion pumping engine, calculating the cylinder sizes for this engine, which was installed at Milwaukee's North Avenue High Service Station.[48]

While at Allis, Nordberg also designed a steam-hoisting engine to work at one of the Upper Michigan copper mines to a depth of two-thousand feet. While designing this hoist, he was allowed to personally sign the design drawings. Today it would be commonplace—and expected—for the designer to sign any drawing he or she worked on. However, back in the 1800s, drawings were typically only signed by the chief engineer; others who worked on the design commonly worked in obscurity. The fact that Reynolds granted Nordberg the right to sign the drawings indicates the high regard he had for Nordberg's engineering ability. Reynolds may have soon regretted this decision, however. Before long, Nordberg left E.P. Allis and formed his own company and competed for business with his former company.

Nordberg initially rented space on the third floor of Pauling & Harnischfeger's facilities, just north of the Reliance Works. He started with six employees. In 1892, Nordberg rented an old brewery building at 8th and Virginia Street formerly occupied by the Pabst Brewing Company, where the firm initially manufactured a cut-off governor for slide valve engines. The cut-off governor was designed to retrofit slide-valve engines and improve their efficiency by eliminating the reduction in steam pressure that was commonly used to regulate engine load. Nordberg also began making his own simple poppet-valve steam engines.

Nordberg's first shop was located above Pawling & Harnischfeger's factory on Clinton (now 1st Street)

Within a short period of time, Nordberg Manufacturing also began the manufacture its own Corliss engines, and used them to drive the numerous machines that it designed and manufactured, including air compressors, pumping engines, stamp mills and, eventually, steam-driven hoisting engines for the mining industry.

Quadruple-Expansion Steam Engines with Feedwater Heaters

In the late 1890s, Bruno Nordberg began a quest to increase the efficiency of steam engines. Typical steam engines exhausting to atmosphere typically had thermodynamic efficiencies in the range of one to ten percent, while those with condensers and multiple expansion and high steam pressure/temperatures historically operated in the range of ten to twenty percent. Steam engine efficiency improved as the operating principles were discovered, which led to the development of the science of thermodynamics.

The efficiency of steam engines is primarily related to the steam temperature and pressure and the number of stages or expansions. Triple expansion engines with condensers had been used by Edwin and Irving Reynolds and others, but each successive cylinder added cost, complexity and size. Each additional cylinder was larger than the previous, and the third-stage cylinders were as large as 94 inches in diameter[49].

Bruno Nordberg undertook the design and construction of several quadruple-expansion steam engines that established world records for steam engine efficiency. One of these was extensively reviewed in the Transactions of the American Society of Mechanical Engineers—an engine for the Wildwood Pumping Station.

The Wildwood Pumping Station was located near Pittsburgh, Pennsylvania. Nordberg designed and built a four-cylinder, quadruple-expansion steam engine to power the water pump. It operated at two-hundred pounds of steam pressure and could deliver its rated capacity of six million gallons per day, pumping water from the Allegheny River for the Pennsylvania Water Company. The engine delivered 712 horsepower to the water pumps.

Not only did the use of four successive cylinders increase efficiency by utilizing as much energy as practical from the steam, but Bruno Nordberg employed a novel use of feedwater heaters. He was one of the first to recognize that the overall efficiency of the thermodynamic cycle could be increased by tapping a small amount of steam from the receivers between the steam engine cylinders to pre-heat the feedwater before it entered the boiler.

The Nordberg pumping engine's efficiency was tested at 185.96 British Thermal Units per horsepower minute, establishing a world record.[50]

The illustration at the right is from an article by Robert Thurston in the Transactions of ASME in which he reported on the tested record efficiency of the Nordberg quadruple expansion steam engine. An indication of the size of this engine can be seen from the circular stairwell.

Fig. 20.—Quadruple-Expansion Engine.

Bruno Nordberg filed for a patent for his innovation in 1899; it was granted in 1903.

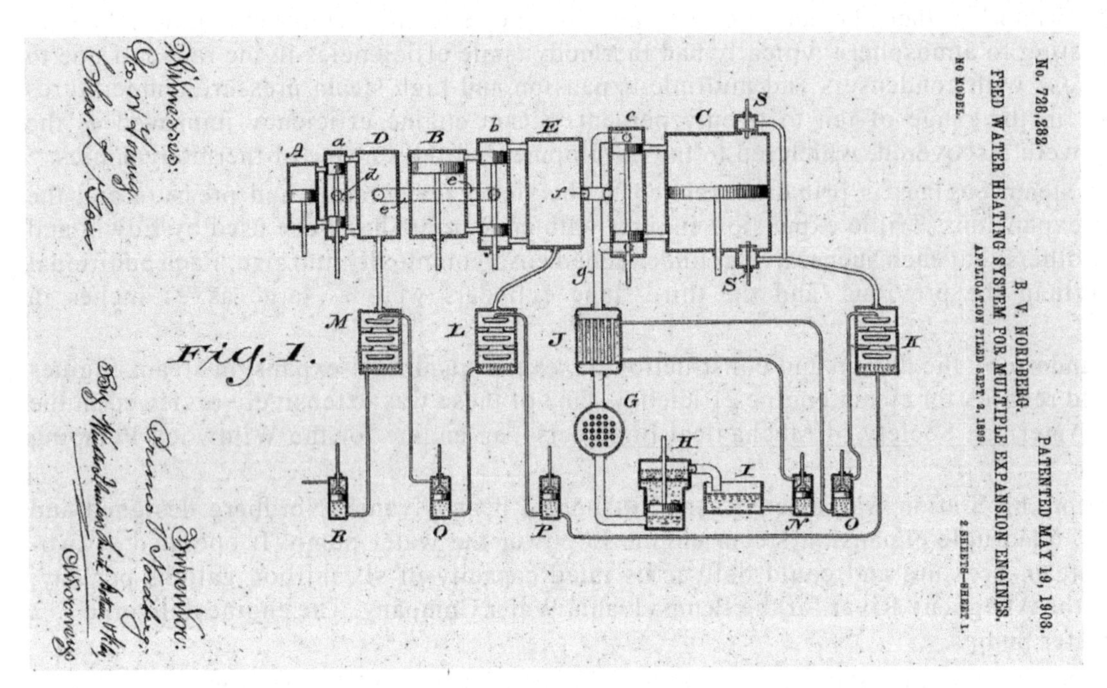

In this drawing, from Nordberg's patent application, a small amount of steam is extracted from receivers D, E and from the low-pressure cylinder C, to heat the boiler water in feedwater heaters K, L and M. The patent covered all multi-expansion steam engines—the figure is for a triple-expansion engine. Drawing rotated for clarity.

In 1902, Nordberg built a one-thousand horsepower quadruple-expansion steam engine that drove an air compressor.[51] Employing a higher throttle pressure (250 psi), it used only 169 BTUs per horsepower minute,[p] a new world record for steam engine efficiency that stood for years.[52]

As business expanded, employment increased to three hundred and Nordberg began planning to build his own factory. The Nordberg Manufacturing Company followed the Chicago, Milwaukee and St. Paul Railway tracks south and located a site for its new facilities at Chase and Oklahoma Avenues in 1901. The Nordberg story is continued in Chapter 5.

[p] The tested results were approximately 25 percent efficient—extremely high for a steam engine.

FILER & STOWELL

Filer & Stowell was first established as the Cream City Iron Works in 1865 by John Maxwell Stowell. Stowell, a native of New York, moved to Milwaukee in 1856.

Delos L. Filer became a partner of Cream City Iron Works around 1883. Prior to that, Filer was president of the Pere Marquette Lumber Company of Ludington, Michigan. Filer's lumber company used Cream City's sawmills, manufactured in Milwaukee by Stowell. Filer also did his banking in Milwaukee, finding it easier to travel across Lake Michigan by boat, than it was to travel by land to Detroit in the 1880s. It was on one of these trips that Delos Filer met John Stowell. They eventually became business partners, incorporating as Filer Stowell & Co, around 1883. Later, the company name was finalized as The Filer & Stowell Company.

By 1886, the company occupied six lots at the corner of Clinton and Florida Streets—essentially across the street from the E.P. Allis's Reliance Works.

Above illustration of Filer, Stowell & Company's Cream City Iron Works, is from Milwaukee Monthly Magazine, January 1875.

As noted earlier, the company produced sawmill machinery on a contract basis for George Madison Hinkley, before Hinkley left to join the E.P. Allis Company.

Filer & Stowell continued to specialize in sawmill machinery—manufacturing edgers that sawed flooring from cants,[q] saw guides, shingle mills and a circular saw for sawmills. They also manufactured steam locomotives and pumps. A Filer & Stowell advertisement from 1894 lists all manner of pumps being manufactured, including beer pumps for the brewing industry.

Stowell served as Mayor of Milwaukee between 1882 and 1884 and was a member of the Wisconsin legislature for a number of years.

In 1902, the company purchased land in the Town of Lake—near the relocated Nordberg facility—and built a new factory there. The Filter & Stowell story continues in Chapter 5.

[q] A 'cant' is a partially sawed wooden log.

THE MENOMONEE FOUNDRY

A number of foundries operated in the Milwaukee area in the mid-1800s. Almost all were established to provide castings to Milwaukee manufacturers.[r] One of the few foundries that broke this mold (pun intended) was William B. Walton's Menomonee Foundry. It was established in 1849 at the southwest corner of South Water and Reed Streets in Walker's Point and produced miscellaneous iron products for stoves, plows, ornamental architectural elements, and railings. Soon the company began manufacturing stationary steam engines.[53]

By 1851, W.B. Walton's company began manufacturing locomotives. In that year, it began construction of the first locomotive manufactured west of Cleveland for the Milwaukee and Mississippi Railroad, the predecessor to the Chicago, Milwaukee, St. Paul and Pacific Railroad (the Milwaukee Road). The locomotive was built under the engineering direction of James Waters and the manufacturing supervision of James Sheriffs. It was fittingly named the *Menomonee*.

Sketch of what is believed to be the first Milwaukee-built locomotive, the *Menomonee*—completed in 1852 by W.B. Walton's Menomonee Foundry. Image taken from List of Locomotives, issued October 1, 1886 by the Chicago, Milwaukee and St. Paul Railway. Image from the Wisconsin Historical Society WHS-62051.

The locomotive was completed in October 1852. The shops of the Milwaukee and Mississippi Railroad were originally located at Humboldt and North Avenue in the City of Milwaukee. In order to deliver the locomotive to the tracks of the railroad, the Menomonee Foundry contracted with a heavy moving contractor named John Miller—also known as 'Long John' because of his 6 feet 9 inch height. George Richardson, former librarian of Milwaukee's Old Settler's Club published an account of the move in a pamphlet published in June 1921 entitled *Diamond Jubilee.* Richardson worked for John Miller at the time and his account of the move is interesting.

[r] James Sheriffs presented an account of Milwaukee's early foundries in a periodical entitled, *The Foundry,* 1902, Volume 24, No. 3, p. 111. He noted that Milwaukee's first foundry was built by Downey & Mosely in the early 1850s. Sheriffs' account named thirty-seven foundries operating in the city in 1902.

> *The first locomotive (was) 'inside connected,' that is, the machinery, cylinder, etc., was all underneath the boiler, except the parallel rods connecting the two pair of driving wheels.*
>
> *On October 15, 1852, Long John with his crew of a dozen men and several yoke of oxen, began laying temporary tracks from a point at the foundry near which is now located the scales of Seeboth Brothers, and thence to Reed Street, on Reed to the bridge over the Menomonee River—then a float bridge. No trouble was experienced until the bridge was reached. At that time Reed Street was just about wide enough for ordinary wagons to meet and pass, and the locomotive and its tracks occupied the whole street. At the bridge all the power of men, block and tackle, as well as oxen, was needed to enable us to get the locomotive up the incline. The engine's weight was about twenty-six tons, and under it the bridge barely escaped sinking, but it was safely landed on the north side of the river and placed on the track, located about seventy-five feet away from the bridge, and here my connection with it ceased.*

In 1853, Walton established the Menomonee Locomotive Company to concentrate on the manufacture of steam locomotives. The company produced fourteen locomotives; seven for the Milwaukee and Mississippi Railroad with the remainder to the La Crosse & Milwaukee, the Lake Shore Road (Green Bay, Milwaukee & Chicago), the Milwaukee & Horicon Road, and to the Milwaukee & Watertown Road.[54]

The company ran into financial difficulties and ceased operation by August 1855. Walton went on to form the Bay State Foundry and Machine Shop with William Goodnow. His partner, Lewis L. Lee established the Globe Iron Works of Milwaukee to manufacture farm implements and corncob mills, but his establishment was destroyed by fire in 1860.

James Sheriffs established a foundry known as the Vulcan Iron Works, which was eventually renamed the Sheriffs Manufacturing Company. The enterprise focused a good part of its business on marine engines and boilers. In 1876, Sheriffs introduced a proprietary propeller that became known as *Sheriffs' Propeller Wheel.* It was used extensively on the Great Lakes, and reportedly came to have a worldwide reputation. The business continued until Sheriffs death in 1887.[55]

FRANK TOEPFER'S MACHINE SHOP

Frank Toepfer operated a machine shop in a red brick building on National Avenue in the former Town of Lake, now Milwaukee. Frank Toepfer's machine shop played a role in the development of what may have been the first gasoline motor-car produced in the United States.

Early Picture of Frank Toepfer's machine shop, taken about 1902. It was located on National Avenue. Source of photograph: Wikipedia Commons.

Toepfer built two 'horseless' carriages for a dentist, Dr. Christian Linger. Both were manually operated; one worked like a railroad handcar, apparently using a reciprocating walking beam that the operator pumped. The second amazingly used a rocking chair design of some sort for locomotion, in which Linger would be seen rocking along the streets. He apparently used both to generate business for an ointment that he manufactured. Linger would often scatter pennies along while rocking his strange vehicle, to further attract attention. This garnered him the nickname of the *Penny Doctor.*[56]

FIRST GASOLINE-MOTOR CAR PRODUCED IN THE US

The novelty of the vehicles Toepfer built for Linger apparently resulted in an interest from Gottfried Schloemer for an engine-driven vehicle. Gottfried Schloemer was a coppersmith, mechanic and inventor, living on the south side of Milwaukee. He had previously patented an improved device for making the tops of wooden barrelheads (1872).

Schloemer acquired a *Sintz* gasoline engine, which was produced by the Sintz Machinery Company of Springfield, Ohio, formed in about 1885. The engine was based on a Dugald Clerk design. Clerk was a Scottish engineer who patented his engine in the 1870s. Sintz used the design to manufacture a two-cycle gasoline engine for marine applications, as well as for sale to others.

It is reported that Schloemer designed the vehicle and Frank Toepfer machined the components. The single cylinder engine was mounted below the driver's seat, powering a belt-drive system. The engine had a rudimentary make-and-break sparking mechanism designed by Schloemer. He also designed the carburetor, which he patented.[57]

The vehicle was only partially successful, but was improved a year later with the assistance of Schloemer's son, Andrew, who was an apprentice at Toepfer's machine shop.[58]

Schloemer and his son tested the automobile in Milwaukee, and later also made improvements to the steering and added other features such as brakes. Schloemer attempted to market his *motor wagon* and eventually entered into an agreement with two Milwaukeeans to provide sufficient financing. Plans were begun in 1892 to manufacture the vehicle in quantity, but when a financial panic ensued in 1893 his investors withdraw from the effort.

The original Schloemer 'Motor Wagon' was preserved. It is now located in the Milwaukee Public Museum and is on display in the *Streets of Old Milwaukee*. Author's Photograph

Some reports place the date of Schloemer's vehicle as being built in late 1889 and tested the following year, although Schloemer later recollected that it first ran in in 1892. In either case, it would have predated the Duryea automobile, which was first demonstrated on September 21, 1893, in Springfield, Massachusetts. The Duryea, built by Charles and Frank Duryea of Springfield Massachusetts, is considered the first commercially successful gasoline-powered automobile in the United States.

Frank Toepfer and his sons continued in business producing other manufactured goods, as well as parts for automobiles. They supplied parts for the Milwaukee-built Eclipse automobile, built by the Krueger Manufacturing Company, and perhaps other early automobile companies.[s]

[s] A similarly named Milwaukee company of the time was W. Toepfer & Sons, which manufactured sheet metal components for tanks and similar applications. The individuals may have been related, but it does not appear that there was an ownership connection between the companies.

This 1905 Eclipse *Tonneau* was produced by the Krueger Manufacturing Company. Some components were manufactured by Frank Toepfer

GOTTFRIED SCHLOEMER was born near Cologne on the Rhine River in 1842. He immigrated to the United States with his family when he was three. The family arrived in Milwaukee in 1846 and bought farm land on Beloit Road.

Gottfried left the area to work for a time in the copper mines of the Upper Peninsula of Michigan before moving back to Milwaukee where he worked as a cooper in their shop.

While working with Frank Toepfer, the two built a 'motor wagon' which has been hailed as the first workable gasoline-engine automobile ever built in the United States. His motorcar was developed ahead of a vehicle built by Charles and Frank Duryea, who are most often identified with this achievement.

Schloemer also designed and patented a *Velocipede* which was operator propelled by bouncing the seat up and down. It was demonstrated but not considered successful.

Gottfried Schloemer is pictured driving his vehicle in an 1895 floral parade. Photograph from Wiki Commons

Gottfried was married to Mary Elisabeth Schmid and they had six children together—three boys and three girls. He died while residing in Milwaukee in 1921.

C.J. SMITH AND SONS

Charles Jeremiah (C.J.) Smith became an indentured apprentice at sixteen years of age for what was then considered one of the world's preeminent engineering firms—Maudslay, Sons and Field of London, England. After his apprenticeship, Smith moved to Milwaukee and set up his own shop. After struggling for a few years, he sold his business and accepted a job in the shops of the Milwaukee and Mississippi Railroad, a predecessor of the Milwaukee Road. However, Smith's real interest was to establish his own enterprise—employing the skills that he learned in England. In 1874 he set up operations in his home on Humboldt Street before moving his shop into a basement below a small carriage works factory on Milwaukee's North 2nd Street. As business increased he relocated to Walker's Point.

As business grew, C.J.'s sons joined the enterprise and Smith aptly named the growing concern C.J. Smith & Sons. In 1895, C.J.'s second oldest son Arthur O. Smith (known as A.O. Smith), who was trained as an architectural engineer, helped to design and build a five-story factory directly across the street from E.P. Allis's Reliance Work. From this new factory the company produced components for buggies, as well as other hardware products. A.O. Smith soon joined the family company, as had his two brothers previously

In 1889, the company developed a process that formed steel tubing, presumably by drawing thin sheets of steel through a series of rollers or a cone shaped opening to form a shaped cylinder, and then welding the edges together. While not the first to develop the process,[t] it appears that Smith was able to refine manufacturing such that the company could economically produce steel tubes in quantity. He initially developed the technique for making shoe lasts, but soon adopted the product for manufacturing bicycle forks. Partially as a result of this innovation, the company went on to become the largest producer of bicycle parts in North America.

In 1899, Smith sold the family business to the American Bicycle Company—known as the *Bicycle Trust*. The family presumably had an ownership position in the Trust. In any case, the Smith family retained management of a branch of the Trust, then known as *Smith Parts*. A.O. Smith served as its manager.

Recognizing the increasing potential of the automobile, A.O. Smith also began to develop a more suitable frame for automobiles. In 1899 he developed the world's first pressed-steel automotive frame, applying techniques he had acquired to form steel sheets into the desired shapes. The pressed-steel frame was lighter, stronger and more flexible than conventional auto frames—and less expensive. By 1902 Smith Parts was selling automotive frames to the Peerless Motor Car Company. The following year they had contracts for frames with six auto companies.

A.O. knew he had a winner on his hands. In 1903, he quit his position with American Bicycle Company and purchased Smith Parts from them. C.J. Smith passed away in 1904 and his son Arthur O. incorporated the new company as the A.O. Smith Company.

---

[t] The distinction for the process for forming and butt welding metal sheets into steel tubing probably belongs to Comelius Whitehouse of England, who in 1825 refined a process developed by James Russell the previous year. C.J. Smith likely became familiar with the basic process during his apprenticeship in England.

Worlds' First Pressed-Steel Automobile Frame

In 1899 Arthur O. Smith built a press to form automobile frame side members from sheets of steel, applying techniques the company had perfected in manufacturing bicycle parts. The resulting automobile frame was much lighter in weight, stronger and more flexible than conventional auto frames—and less expensive. It was the world's first pressed-steel automobile frame.

The Peerless Motor Car Company placed the first order for Smith's pressed-steel automobile frames in 1902. Studebaker, Cadillac, Packard and Locomobile were soon to follow.

In this picture, Arthur O. Smith, wearing the white hat, is shown looking at a finished frame for Cadillac. Photograph courtesy of A.O. Smith Corporation.

As the orders mounted, keeping up with the demand from the growing automobile industry was becoming an almost insurmountable challenge. Even though the company was manufacturing thousands of automobile and truck frames every year, production was very labor intensive.

Henry Ford soon heard word of A.O. Smith's growing expertise for shaping steel into automotive frames. Ford visited Milwaukee in 1906 and ordered ten-thousand steel frames for his Model N touring car. This was a huge order for the company, which had previously been producing only ten frames a day. To meet the demand, A.O. Smith and his engineering team retooled existing presses to produce two corresponding halves of an auto frame simultaneously. They arranged the presses to form a continuous assembly line—the first such automotive frame assembly line in the world. With the increased efficiency, they were able to deliver the full order of ten thousand frames to Ford by August of that year, which allowed Ford to introduce his popularly priced vehicle by year-end 1906.

The company's success in meeting the Ford order attracted additional automobile manufacturers. Within four years, the company became the country's largest automotive frame manufacturer. Business was so good that they were forced to turn down orders.[59] A.O. Smith decided to build a new factory—selecting a 135-acre site on Milwaukee's north side. The A.O. Smith story is continued in Chapter 7.

In this photograph, A.O. Smith (right) is seen riding with Henry Ford. They appear to be driving in a Ford Model S, which was an adaption of the Ford Model N. The Model S was the last automobile produced by Ford for the US market that was a right-hand drive vehicle. It was sold between 1907 and 1909—just prior to Ford's introduction of his iconic Model T. Photograph courtesy of A.O. Smith.

Charles Jeremiah Smith

Right: Charles Jeremiah (CJ) Smith

Charles Jeremiah (C.J.) Smith immigrated to the United States in 1840 from his native England. He was twenty-three years old. During the voyage, it is reported that he rescued a young woman who fell overboard while the ship was somewhere near the mouth of the St. Lawrence—a woman who later became his wife, Mercy Johnson.

In 1874, after working for a time at the Milwaukee and Mississippi Railroad, Charles started his own company with the two youngest of his four sons. He named the company C.J. Smith and Sons. The company developed a process in 1889 that formed steel tubing, presumably by drawing thin sheets of steel through a series of rollers or a cone shaped opening to form a cylinder, and then welding the edges together. He adopted the product for manufacturing bicycle forks. By the end of the century, C.J. Smith and Sons was the largest bicycle parts manufacturer in the world.

Arthur O. Smith

To meet the demand for bicycle parts, C.J. called on his second oldest son Arthur, who was trained as an architect, to design a new manufacturing plant in Milwaukee. After the building was completed, Arthur (who liked to be called A.O.) decided to stay with the company.

Arthur O. (AO) Smith

In 1899, Smith sold the family business to the American Bicycle Company—a trust that acquired many of the United States bicycle companies. When the Trust ran into financial difficulties, Smith began to diversity into over ventures.

Recognizing the opportunities afforded by the rapidly growing automotive industry, A.O. Smith decided to establish a new venture devoted to the manufacture of automobile frames. Applying techniques the company had developed to shape steel into various parts, Arthur developed the first pressed-steel automobile frame. The frame was both lighter and stronger than the conventional I-beam frames being used in the automobile industry at that time. The introduction of this technology led to rapid company growth. In 1904, Arthur renamed his company the A.O. Smith Company.

A.O. Smith died in 1913. He was buried in Milwaukee's Forest Home Cemetery.

KEARNEY & TRECKER

While a young lad in Iowa, Edward J. Kearney developed an interest in engineering because of his conversations with a German blacksmith. He studied mechanical engineering at Ames College, graduating in 1893 when he took a job with Kempsmith in Milwaukee as a draftsman.[60] Kempsmith was a Milwaukee manufacturer of milling machines started in 1888 by Frank Kempsmith.

Theodore Trecker moved to Milwaukee with his mother in about 1886 and eventually apprenticed with the Wilken Manufacturing Company—a predecessor of Filer & Stowell. Upon completion of his three-year apprenticeship, he also obtained a job with Kempsmith, where he met Edward Kearney.

Both men moved through the ranks at Kempsmith—Trecker becoming plant superintendent and Kearney head of the engineering department. In 1898, the two decided that they could do better by going into business for themselves, and started up their company in a building formerly used as a store at 271 Lake Street in Walker's Point. Continuing their prior working arrangement, Kearney assumed responsibilities for engineering and the office and Trecker handled shop manufacturing.

Kearney & Trecker's first factory was located at 271 Lake Street in Milwaukee's Walker's Point.

Initial business was slow. It is reported that their first customer was a lady who asked if they could repair the axle on a baby carriage.[61] In addition to serving as a repair shop, they began manufacturing lathes and bottle washing machines, but they eventually began manufacturing milling machines of their own design.

A few months after the new firm was established, Henry Crandall, superintendent of the Milwaukee Harvester works, approached the company to see if they could design a special machine tool to help produce a farm mower. Crandall told Kearney and Trecker that he could find nothing on the market that would meet his requirements. Kearney and Trecker quickly designed a suitable machine tool for the purpose, but as they were nearing completion, they ran out of funds. Kearney approached local banks, who turned him down. Fortunately, Milwaukee Harvester advanced the firm $750, which allowed them to complete the machine and pay their employees.

Crandall was pleased with the new machine tool and told an associate at the J.I Case Company in Racine about the new device, causing Case to purchase two for its use. With these orders, the firm was up and running.

Seeing that E.P. Allis was building a new factory for its expansion in West Allis, Kearney & Trecker purchased property on National Avenue in that community in 1901 and built a new factory there. The Kearney & Trecker story is continued in Chapter 8.

GEORGE MEYER MANUFACTURING COMPANY

George J. Meyer was born in Green Bay in 1871. His first job was in the shops of the Milwaukee Road in Green Bay. It is reported that he rode his bicycle to Milwaukee in 1897 to find employment. He took a job with C.J. Smith & Sons making bicycle and baby carriage hardware. He is said to have been working at a machinists' bench when he built his first bottle washing machine out of wood and metal parts that came out of the old scrap yard of Seeboth Brothers.[62] He soon left to form his own company.

Meyer's first shop was at 271 Lake Street[u] in Walker's Point. This is the same shop that Kearney & Trecker used earlier to start its machine tool business.

In 1904, Meyer built his first bottle washing machine, soaking bottles in a caustic solution to clean and sterilize them, prior to reuse. It took him another ten years to perfect bottle washing machinery.

The George Meyer Manufacturing Company did good business providing machinery for use in Milwaukee's brewing industry. Prior to Meyer's patented inventions, recycled beer bottles were washed by hand. This was true also for soft drink (soda) bottles, milk bottles and similar containers.

With the success of his bottle washing equipment, Meyer also developed equipment for filling and pasteurizing bottles.

In order to meet the company's need for expansion, Meyer moved the company to Cudahy, Wisconsin. The company sold its products internationally, becoming the world's largest bottling machine manufacturer.[63]

George J. Meyer died in 1945, at which time his son George L.N. Meyer succeeded him. George L.N. continued the tradition of designing bottle-cleaning machinery—and had several patents in his name.

The company was acquired in 1968 by Automatic Sprinkler Corporation of America.

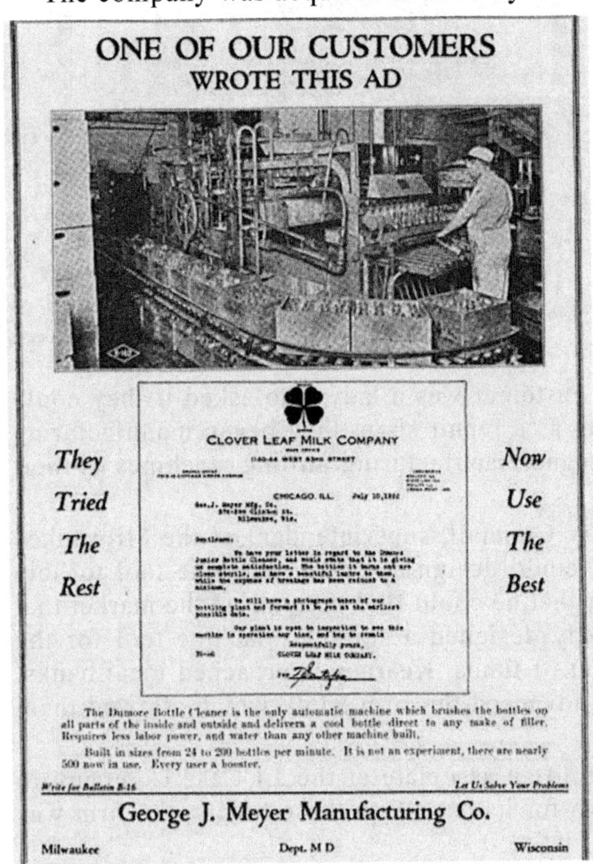

Early advertisement of the George J. Meyer Manufacturing Company

u Lake Street is now known as Pittsburgh Avenue. It was located in the Walker's Point neighborhood of the Menomonee Valley.

This photograph of a George J. Meyer bottle washer being used at a Kentucky Coca-Cola bottling plant was taken in 1932 by Lafayette Studios. Photograph courtesy of the University of Kentucky's Special Collections Research Center

WOLF AND DAVIDSON COMPANY

Milwaukee once had a thriving shipbuilding industry. During the days of the wooden-hulled ships, more ships were built in Milwaukee than in any other spot on the Great Lakes.[64] The first vessel built in the city was the schooner *Solomon Juneau*, built for Milwaukee's first mayor and launched in August 1837. James Monroe Jones established the first shipyard in Milwaukee and soon moved it to the mouth of the Milwaukee River to an island that now bears his name—Jones Island.

However, William H. Wolf established the City's most formable shipyard. Wolf was born in Germany and came to Milwaukee in 1849. He worked for a time in James Monroe Jones's shipyard as foreman. Wolf established his own firm in 1858 with partner Theodore Lawrence. The partners sold the company to Ellsworth and Davidson in 1863 but, in 1868, Wolf returned to Milwaukee and bought out Ellsworth to go into partnership with Thomas Davidson. Davidson had also worked for a time in Jones's shipyard.

Their company occupied eleven acres with one stationary dock and none floating docks serviced with a large steam derrick.[65]

Wolf and Davidson built a number of sailing vessels, including the schooner *Moonlight*—the largest constructed in Milwaukee at the time at 777 gross tons. However, the company became most known for their wooden steamboats—the largest of which was the *Ferdinand Schlesinger* which measured 2,607 gross tons with a keel length of 306 feet. The steamers of the day had two stacks abreast (side-by-side rather than fore-and-aft). The *Schlesinger* traveled the Great Lakes for 27 years, but sunk in Lake Superior during a bad storm. Among other steamers built by Wolf and Davidson were ships that carried their names, the *Thomas Davidson* and the *William H. Wolf*, along with the *Roswell P. Flower*.

An accident occurred during the launching of the *Wolf*, which resulted in two fatalities as described in the caption of the picture on the next page.

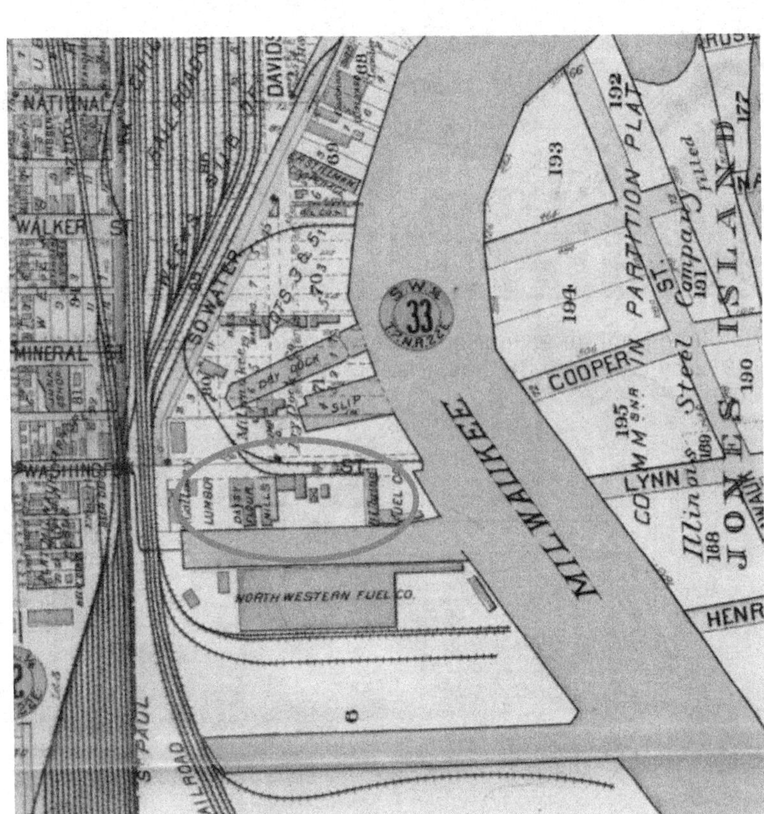

Wolf and Davidson's shipyard was located near the junction of South Water Street and Washington, providing them with access to a slip across from the North Western Fuel Company.

This photograph shows the launching of the wooden hull for the steamboat *William H. Wolf* on August 6, 1887. Had the photographer taken the photograph a few seconds later, it would have captured the wave caused by the launch, which washed over a section of the dock on which as many as seventy-five spectators stood. The dock's support beams failed, throwing the spectators into the water. Two died and several others were injured. Photograph from the Great Lakes Marine Collection of the Milwaukee Public Library.

As shipbuilding moved away from wooden hulls and toward larger vessels of iron plate and eventually steel, Milwaukee's wooden shipbuilders were no longer able to meet the market demands. Shipbuilders in the Lake Erie ports were closer to the Pittsburgh steel district. There were notable exceptions; during First World War, thirteen boats were launched in Milwaukee by Fabricated Ship, a subsidiary of Lakeside Bridge and Steel, and four wooden submarine chasers were built by the Great Lakes Boat Company. During the Second World War, Froemming Brothers established a shipyard in Bay View. Its accomplishments are covered in Chapter 5.

OBENBERGER DROP FORGE

Forging is one of the oldest known metalworking processes. The blacksmith shapes iron and steel using a hammer and anvil—a basic forging process. If done properly, forging can produce a piece that is stronger than an equivalent cast part. As the metal is shaped during forging, its internal grain deforms to follow the general shape of the part—resulting in a product that has improved strength characteristics.

By the 1800s, industrial forging had developed well beyond the blacksmith's hammer. A weight was raised and then *dropped* on the hot workpiece to form it according to the shape of a die. These *drop forges* were generally powered by steam. The lifted weight (or hammer) was typically in the thousands of pounds.

Milwaukee industry was fortunate to have the services of several competent drop forges run by the various members of the Obenberger family. George Obenberger came to Milwaukee from Bavaria in 1854 and set-up a general blacksmith business in Walker's Point on what is now the corner of Barclay Street and Pittsburgh Avenue.[66] Over time, George's sons, brothers and cousins also got into the business—either jointly with him, or setting up their own shops in Walker's Point.

This early advertisement features just one of the Obenberger companies—in this case Joseph Obenberger & Son, Machine and Forge Works.

Perhaps the largest of these companies was the Obenberger Drop Forge, originally located at 222-224 Lake Street (approximately Barclay Street and Pittsburgh Avenue). John Obenberger founded the company in 1902. Obenberger soon decided to specialize in forgings for the developing automotive industry. The crankshafts and connecting rods used in their internal combustion engines benefited from the extra strength of forgings. Other automotive components, such as drive shafts, also required the higher strength from forged steel.

John Obenberger purchased a fifteen-hundred-pound steam hammer and began the production of forged axles. Business increased considerably and Obenberger Drop Forge moved to Cudahy in 1913 to accommodate the expansion in business. The history of Obenberger, which later was renamed Ladish Drop Forge, is continued in Chapter 5.

CHRISTENSEN ENGINEERING

Niels Anton Christensen immigrated to the United States in 1891 from Denmark, initially residing in Chicago. He became a leading draftsman at Fraser and Chalmers in Chicago. While inspecting the new Cicero & Proviso electric railway system at Oak Park near Chicago, Christensen observed an accident that resulted in two deaths and a great many injuries. The accident was attributed to the inability of the handbrake to stop the car adequately. Christensen noted that the handbrake was a simple mechanical adaption of a handbrake used on horse-pulled cars, relying on the operator's muscle and sometimes amplified by the electricity that ran the streetcar. Since the loss of electricity often occurred during a streetcar accident, Christensen reasoned that a better system was required to ensure reliable braking. He set out to design a power brake for electrically powered cars, developing the Christensen scheme of air brakes.

His streetcar air braking system used an electric motor to drive an air compressor, both of which were totally enclosed in a metal case containing oil. They system generated and stored compressed air, which would control the streetcar's brakes even upon loss of electricity.[67] Christensen obtained US and European patents for his system and its various components. The system was tested on two streetcars in Detroit in 1893 and greatly increased the ability to stop the cars quickly, when necessary—reducing stopping distances. While the tests were successful, an economic downturn that year prevented Christensen from going into business to manufacture the system.

As noted earlier, Christensen likely met the principals of the E.P. Allis Manufacturing Company while working on the electrical systems for Chicago's Columbian Exposition. Christensen decided to accept a job offer from Allis and relocate to Milwaukee, working for E.P. Allis as a mechanical engineer.

At E.P. Allis, Christensen was placed in charge of the design of blowing engines for steel mills. During the evenings and weekends, he continued to refine his air brake system for electric streetcars and trains. His system was tested on two cars on the Milwaukee Street Railway System. He succeeded in simplifying his system, developing a triple valve brake. In 1896, he obtained financial backing that allowed him to make an experimental test apparatus. He also secured patents on the new valve mechanism. In early 1897, he founded Christensen Engineering Company, which initially operated in the Menomonee Valley at 718 Hanover. His operations were co-located with the Seamless Structural Company at the corner of Hanover and Burnham Streets. Later that year, the company was awarded the contract for the South Side Elevated Railroad of Chicago, having formerly been tested on some cars on the Metropolitan Elevated Railroad Company. The success of this contract resulted in many new orders and resulted in the Christensen air brake being adopted as necessary equipment on most electric railways of the era.[68]

In addition to its adoption in the United States, the Christensen air brake system was utilized in Australia, on the surface and underground lines of Paris, France and many other French cities, on nearly all the electric railways of Italy, and on some systems in Germany, Norway, Sweden, Russia, Canada, Mexico, South American republics, China, Japan, South Africa, and in England.[69]

As demand for Christensen's air brake system grew, the company quickly outgrew its factory space in the Menomonee Valley. His financial backers were also interested in broadening the scope of the business to include general electrical manufacturing, which would require additional space. Christensen began making plans to relocate his factory to Milwaukee's north side—acquiring property and starting construction. It appears that the large capital requirements of the new factory resulted in the formation of a joint-stock company. However, by 1903, Frank Bigelow, Samuel Watkins and Henry Goll, owned majority interests in the company and renamed it the National Electric Company. Christensen continued to retain the patent rights to the air braking system and related equipment, for which the company agreed to pay royalties, but no longer retained an ownership position in the company and retired as general superintendent.

Christensen Air Brake System

The Christensen Air Brake System for electric railways and streetcars was a significant advance in the safety of electric rail transportation. It reduced braking distances significantly, especially under adverse situations, and decreased the number of accidents. It was especially helpful for streetcar use, where the streets were typically shared with automobiles and other traffic, as well as crossing pedestrians. The operator had to bring the streetcar to a stop frequently to avoid accidents.

The control apparatus was installed on both ends of streetcars, to permit operation going in both directions without necessitating the rotation of the streetcar. The *motorman's valve* had several positions that the operator could move through by moving the handle left to right: a quick release position, a slow release and running position, a normal stop position and an emergency stop.

First tested on the Detroit streetcar system, Christensen air brakes were soon used by streetcar companies through the United States, and eventually in many parts of the world.

The Christensen story continues in Chapter 6, along with that of the National Brake & Electric Company.

Brake controls on the St. Charles streetcar in New Orleans. The Christensen-style air brake is shown with red body and adjacent pressure gauge. The motor-man controlled the speed of the streetcar with the left hand, and the brake with the right hand. Author's photograph, 2016.

COMPRESSION RHEOSTAT/ALLEN-BRADLEY

The need for a better mechanism to control its electric motors provided the inspiration for young Lynde Bradley to develop a better motor controller. Milwaukee's Pawling & Harnischfeger Company's innovative overhead crane was covered earlier in this chapter. It was the first crane design to use three electric motors to control movement and lift operations, setting up a need for a controller to control movement precisely.

Bradley was only 14 years old when he developed his first compression rheostat.[70] Bradley's carbon disc compression-type motor controller enabled precision control of the electric motors used for lifting loads, as well as for the movement of industrial cranes. In 1903, Lynde Bradley and Dr. Stanton Allen founded the Compression Rheostat Company to market the device. They were joined by Lynde's brother, Harry, the following year. The team initially operated out of space in the Pfeiffer & Smith machine shop, located on Clinton Street (now South First Street) just north of Greenfield Avenue. Soon they also rented office space on the second floor of a grocery store also on Clinton Street.

There was a large and growing market for an effective means of controlling electric motors. The Compression Rheostat Company developed motor controls for many other applications, including medical and dental equipment.

The first offices and drafting room of the Compression Rheostat Company was on the second floor of this building, located on Clinton Street (now 1st Street) and one block north of Greenfield Avenue in Milwaukee's Walker's Point neighborhood.

Initially, manufacture of their rheostat controls was outsourced under contract to the American Electric Fuse Company, which had a factory in Muskegon Michigan. Lynde Bradley worked on prototypes for various controls. The contract with American Electric Fuse was soon terminated for lack of payment of royalties and other discrepancies. In 1909, Allen and Bradley moved all manufacturing operations into rented space in the Charles Pfeiffer and George H. Smith Machine Shop located on Milwaukee's First Street near Greenfield.

The name of the company was soon changed to the Allen-Bradley Company. In late 1916, following the death of George H. Smith, Allen-Bradley purchased the machine shop from Charles Pfeiffer. Its product line grew to include automatic starters and switches, circuit breakers, relays and other electric equipment. Sales increased dramatically during the First World War due to government orders.

In the 1920s, the company developed a line of miniature rheostats to support the rapidly growing radio industry. It also developed foot-controlled switches for sewing machines made by Singer and other manufacturers.

This photograph shows a number of female employees at Allen-Bradley, including Julia Polczynski (presumably the woman facing the camera). Julia was the first female to work for the company. Source: Wiki Commons

Despite the pressures of the Great Depression, Lynde Bradley supported an aggressive research and development approach intended to "develop the company out of the Depression." Lynde Bradley's R&D strategy was successful. By 1937, Allen-Bradley employment had rebounded to pre-Depression levels

During the Second World War, eighty percent of the company's orders were war-related. The company experienced heavy demand for its industrial controls and for its electrical components or "radio parts," which were used in a wide range of military equipment. Allen Bradley expanded its facilities several times to meet wartime production requirements.

Compression Rheostat Company's Crane Controls

One of the Compression Rheostat Company's first commercial products was a controller for operating industrial cranes. Bradley's carbon disc compression-type motor controller enabled precision control of the electric motors used for movement of the cranes, as well as for lifting its loads.

Lynde Bradley first experimented with a compression rheostat device when he was 14, building one in 1893 by sawing disks from an electric arc-light electrode. He placed a stack of eleven disks into a wooden tube and applied compression with a borrowed clamp. He found that the device "gave almost infinite control of the motor speed."

Bradley obtained a patent for his "electric current controller" in 1903. A crane controller using the compressed rheostat principal was demonstrated at the St. Louis world's Fair in 1904. It was one of the first commercially manufactured crane controllers ever produced—a 3.5 horsepower Type A-10 controller.

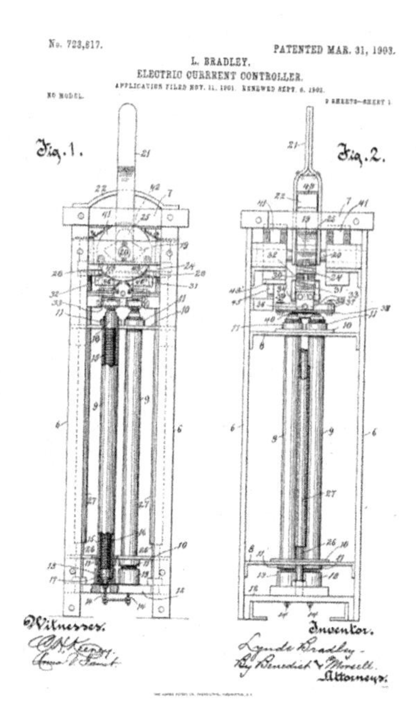

A patent drawing from Lynde's 1893 Electric Current Controller

Lynde Bradley spent several years perfecting the device, designing an apparatus to test the resistance of the carbon discs, and another to test their strength—all to improve the quality and consistency of the controllers.

The individual rheostat stacks could be easily removed for cleaning and maintenance, as shown in the image below.

Lynde Bradley's assistant, Roderick Amdt, is pictured removing a carbon pile rheostat from a crane controller that had been returned for factory maintenance.

Allen-Bradley Clock Tower

In the early 1960s, Allen-Bradley capped off an expansion of its facilities with a large, four-faced clock tower. The resulting clock was determined by the Guinness Book of World Records to be "the largest four-faced clock in the world."[v] Each face has a diameter of 40 feet, 3-1/2 inches. It was unveiled to the public on October 31, 1962, and quickly developed into a Milwaukee icon.

The mechanism was designed by Ray Ellsworth, an engineer in Allen-Bradley's special machine design department. Originally from Michigan, Ellsworth came to Milwaukee and worked for Allis-Chalmers, Harnischfeger and Nordberg before joining Allen-Bradley. He holds the patents on the mechanism, along with Gerald Lohf and Philip F. Walker.

The tower and faces of the four-sided clock were designed by architect Fitzhugh Scott.[71]

The Allen-Bradley Clock Tower is the largest four-faced clock in the world. Photograph courtesy of Allen-Bradley.

[v] The Allen-Bradley Clock Tower lost this distinction in August 2010 when the Mecca Clock in Mecca, Saudi Arabia was completed. The A-B clock is currently the World's second largest.

LYNDE BRADLEY was born in Milwaukee in 1878, the oldest son of Henry C. and Clara Bradley. As a result of the family's difficult financial situation at the time, Lynde dropped out of high school to take a job with the Julias Andrea and Sons Bicycle Shop, which in addition to repairing bicycles also fixed locks and installed electric doorbells. While working at the company, Lynde also tested his own innovations, which included an electrically operated wind indicator—essentially a weather vane that used electromagnets to indicate wind direction. Bradley obtained a patent for the device in 1894, while he was only 16 years old. However, it doesn't appear that this was Lynde's first invention. He is reported to have built a compression rheostat device the preceding year. It consisted of a stack of eleven 1/8th thick carbon disks that he sawed from an arc-light electrode. The disks were placed in a wooden tube and compression was applied with a borrowed pattern maker's clamp.

In 1898, Lynde left Andrea to go into business for himself, opening an X-ray laboratory in Milwaukee. His business failed by 1900, but through it he met Dr. Stanton Allen, an orthopedic surgeon who was impressed with Bradley's innovations.

Lynde Bradley took a job with the Milwaukee Electric Company, which provided him with valuable insights into the problems with crane motors controls. He tried unsuccessfully to convince them to use his compression rheostat device for the application. Lynde resigned in early 1901 and, with the backing of Dr. Allen, he constructed a rheostat control for crane motors. Taking it to his old employer for testing, it was used successfully to control the hoist motor of their shop crane. Lynde then proceeded to build several crane controllers for testing at other Milwaukee manufacturing companies, including Allis-Chalmers, Falk, Chain-Belt, and Cutler-Hammer.

Dr. Allen and Lynde Bradley formed the Compression Rheostat Company in December of 1903, and signed an agreement with the American Electric Fuse Co. to manufacture their rheostat controls. Harry Bradley, Lynde's brother, joined the enterprise in 1904. The contract with American Electric Fuse was soon terminated, however, leading to a lengthy legal battle. In 1909, Allen and Bradley moved their manufacturing operations into rented space in the Pfeiffer and Smith Machine Shop, located on Milwaukee's 1st Street near Greenfield Avenue.

Compression Rheostat soon changed its name to the Allen-Bradley Company. In late 1916, Allen-Bradley purchased the machine shop from Charles Pfeiffer. Allen-Bradley experienced significant growth during the First World War, leading to additional employees and the need to enlarge the manufacturing space. It purchased land for its expansion on West Greenfield Avenue, adjacent to its existing plant.

In 1920, the company introduced a 'miniature' graphite compression resistor. The new rheostat turned out to be an ideal noiseless, step-less control adjustments for early radio sets. Sales of the device soared. During the Depression Allen-Bradley continued to invest on research and development—introducing electric motor control devices and a much smaller carbon composition radio resistor.

When the Second World War broke out, Allen-Bradley's electronic resistors were in extremely high demand for walkie-talkies, radar equipment, military radios and other military products. A-B responded by expanding its manufacturing capabilities significantly.

Lynde Bradley died in 1942. His brother Harry succeeded him and led the company until his retirement in 1960.

EVINRUDE DETACHABLE OUTBOARD MOTOR COMPANY

Ole Evinrude was born in Norway in 1877. His family settled in Wisconsin when he was five. He never went beyond third grade, yet he developed a natural mechanical ability while growing up on the farm. At 16, he apprenticed with Fuller and Johnson in Madison, Wisconsin, a farm machinery manufacturer.

Nights after work, Ole studied mathematics and engineering on his own using borrowed texts from the library. After a number of jobs, he finally settled in Milwaukee, accepting an offer by the E.P. Allis Company in 1900 to act as head of its pattern shops and as an engineering consultant.[72]

Ole was a tinkerer, working nights in the basement of his boarding house at the corner of Florida and Grove—just four blocks west of E.P. Allis. His principal after-work interest was building gasoline engines. One of his first engines was a 4-cylinder, air-cooled engine, which he installed in a friction-drive automobile that he also designed.

After this engine proved out well, Ole went into business to build and sell it. His business, the Motor Power Equipment Company, failed and Ole went back into the pattern making business on the second floor of John Obenberger's forge shop on Lake Street.

It is reported that an unsuccessful, heroic attempt to row across Waukesha County's Okauchee Lake, buy an ice cream cone for his fiancé, Bess, and return the treat to her before melting, inspired Ole Evinrude to find an easier, more logical way to propel a boat across the water. While the story might be folklore rather than fact, Ole eventually decided to use some of his leftover engine parts to build a motor for a row boat—and in the process developed a successful outboard motor.

Encouraged by Bess, whom he had married by then, he improved it and then built parts for 25 engines. In 1909, Ole leant the first of his engines to a friend who used it at Pewaukee Lake and came back with orders for ten outboard motors and stories about the excitement it caused.

Billed initially as a detachable rowboat motor, the engine weighed 62 pounds and sold for $62. They sold so quickly that Evinrude immediately obtained financial backing and formed the Evinrude Detachable Outboard Motor Company, locating his factory at 228 Lake Street—just a few blocks north of the E.P. Allis Company. He moved to larger quarters on Reed Street in 1911 and to a new building on Walker Street in 1913.[w]

Evinrude's outboard engine was not the first to be designed and sold for powering small recreational boats. In 1896, the American Motors Company (not to be confused with the Wisconsin-based automobile company that had the same name) introduced an outboard motor. However, it was unreliable and not a commercial success. In 1906, Cameron Waterman of the Detroit Michigan area began manufacturing an outboard engine that he dubbed the "Porto." It appears that they produced as many as a thousand engines per year between then and 1914, but most sales appear to have been local and the engine had little impact.

As a result, Evinrude's engine is considered the first practical and reliable outboard sold commercially in the world. The simplest form of his outboard motor was a 2-stroke internal combustion engine that was powered by gasoline with oil added as a lubricant. Built of steel and brass, it had a crank on the flywheel to start the two-cycle engine.

Ole Evinrude began to look for a business partner in order to grow the business. In 1911, he sold a half interest in the company to Chris Meyer, president of the Meyer Tug Boat Lines and formed the Evinrude Detachable Row Boat Motor Company.[73]

In 1914, Evinrude sold his remaining half interest to Meyer to help care for his wife, who was experiencing health issues. The two traveled extensively. However, Ole continued to work on ideas for improving the outboard engine. His sale of the company to Meyer stipulated that he had to stay out of the business for five years. As that five-year period ran out, Evinrude was ready to introduce a new engine made largely of aluminum.

---

[w] Lake Street is now known as Pittsburgh Avenue. It, along with Reed Street and Walker Street were located in the Walker's Point neighborhood of the Menomonee Valley.

Meyer was not interested in the new engine, causing Evinrude to start another company which he called the ELTO Outboard Motor Company—ELTO standing for Evinrude Light Twin-Outboard. His new company quickly surpassed the original Evinrude Detachable Motor Company in sales.

In 1926, ELTO introduced a lightweight two-cylinder engine and two years later a four-cylinder engine (the Quad).

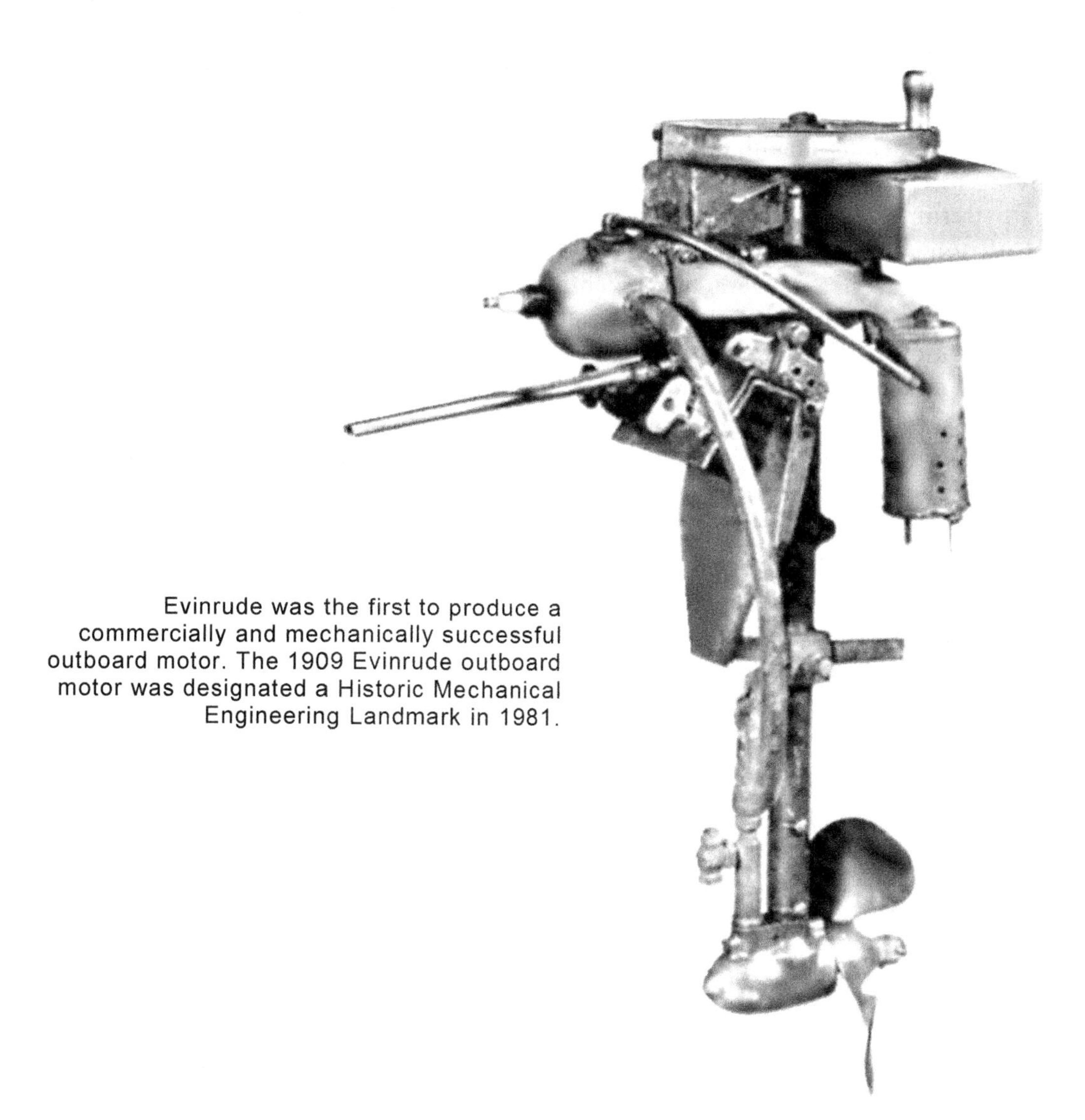

Evinrude was the first to produce a commercially and mechanically successful outboard motor. The 1909 Evinrude outboard motor was designated a Historic Mechanical Engineering Landmark in 1981.

OLE EVINRUDE was born Ole Andreassen Aaslundeie in 1877. He and two siblings emigrated from Norway to the United States in 1881 with their parents. Ole eventually adopted the "Evinrude" surname, which came from the 'Evenrud' farm in Vestre Toten where his mother was born.[74]

The family settled on a farm in Ripley Lake near Cambridge, Wisconsin. At age sixteen, Evinrude went to Madison, where he worked in machinery stores and studied engineering on his own. He became a machinist while working at various machine tool firms in Pittsburgh and Chicago, before moving to Milwaukee and accepting a position at the E.P. Allis Company as a patternmaker.

In 1900, Evinrude co-founded the custom engine firm Clemick & Evinrude, as well as several joint ventures to manufacture automobiles, which eventually failed. In 1907, he invented the first practical and reliable outboard motor, which was built of steel and brass, and had a crank on the flywheel to start the two-cycle engine.

Two years later, the Evinrude Detachable Outboard Motor Company was founded in Milwaukee. The simplest type of engine the company produced was a 2-stroke internal combustion engine that was powered by a mixture of gasoline and oil.

By 1912, the firm employed 300 workers. It is reported that Evinrude allowed Arthur Davidson tinker in his Milwaukee machine shop. It is known that Evinrude assisted Harley-Davidson with the design of the carburetor for their first motorcycle engine.

Ole Evinrude had at least fourteen patents to his name. Many were for various improvements in outboard motor design, including water-cooling mechanisms. He also had patents for various power shop tools and a motorized bicycle.

Evinrude sold his company in 1913, in order to care for his wife. The two traveled extensively.

In 1919, Evinrude invented a more lightweight two-cylinder motor and founded the ELTO Outboard Motor Company for its manufacture. (ELTO was an acronym for "Evinrude Light Twin Outboard.") Although the company faced stiff competition from other companies, such as Johnson Motor Company of South Bend, Indiana, Evinrude's company survived. Through acquisitions, he eventually forming the Outboard Marine Corporation.

His wife Bess died in 1933 at only 48 years old, and Ole Evinrude died the following year when he was 57 years old. After Ole Evinrude died, his son, Ralph Evinrude, took over day-to-day management of the company, eventually rising to Chairman of the Board.

The company is now called Evinrude Outboard Motors and is owned by Bombardier Recreational Products.

WILLIAM S. SEAMAN & COMPANY

It could be argued that the William S. Seaman Company doesn't belong in a book about Milwaukee's machines. Indeed, a company that specialized in the manufacture of furniture might not seem to fit alongside concerns that made items like steam engines, hoists and locomotives. However, furniture companies sometimes morph into other industries. That certainly was the case for Seaman. Seaman transformed from furniture maker into an automobile body manufacturer and, as automobiles became more sophisticated, significant engineering features were incorporated into the bodies of their vehicles.

The Milwaukee roots of the Seaman enterprise go back to the earliest days of the community. In 1846, William Seaman's father, Alonzo, brought his family to Milwaukee and established a cabinetmaking practice. The business grew rapidly to meet the needs of the young community. Following Alonzo's death, William, who was the oldest son of the family, established his own furniture manufacturing company in 1891. He named it William S. Seaman & Company and the factory was initially located on what is now Bruce and First Streets in Walker's Point.

Looking to diversify his furniture business, he developed a wooden telephone booth for the Western Electric Company of Chicago, and eventually produced telephone switchboards. In 1906, a fire destroyed the Seaman plant, but the company leased a much larger factory at 8th and Virginia Streets. The new factory was larger than current activities required and William Seaman looked for additional business to help fill the plant.[75] In the summer of 1909, the Petrel Motor Car Company moved into the Seaman factory.

The Petrel Company was established in Kenosha the preceding year—founded by Samuel Watkins of the Beaver Manufacturing Company, which built internal combustion engines, and John and Harry Waite, who had built a few early automobiles. Their new vehicle was named "Petrel" after a seabird, because of its "ability to make good speed over very rough roads."[76] In addition to leasing space for assembly of the vehicle, Seaman obtained an order to build the bodies. The auto bodies were constructed of wood on a steel chassis with steel fenders.

The Petrel did not do well financially. Ford's Model T had come out in 1908 and it was priced so aggressively that other manufacturers were not able to compete. The Petrel Company went into bankruptcy proceedings in early 1910. An effort was made to resurrect the company by Milwaukee-area investors, and for a time vehicles were made under the F.S. Motor Company brand. However, this effort soon failed as well.

While less than a thousand vehicles were built, it launched Seaman into the automobile body business. Seaman, the furniture manufacturer, needed to learn to also deal with metals quickly. Seaman hired engineer Edward Hardtke to oversee automotive operations. Hardtke appears to have been a good choice. In addition to providing overall engineering support, he created a number of innovations associated with automotive windows and doors and had several patents to his name. At about the same time, William Seaman became ill and his sons took over operation of the company. Irving Seaman, a graduate of the University of Wisconsin-Madison, became chief executive of the concern.

Over the next decade, Seaman produced passenger car bodies for numerous automobile manufacturers including Cadillac, Case, Chalmers, Chicago Electric, Columbia Taxicab, Franklin, F.W.D., Hudson, Jonas, King, Kissel, Lafayette, Locomobile, Marmon, Mitchell, Moline-Knight, Oakland, Packard, Pierce-Racine, Rambler, Regal, and Stevens-Duryea. The company also made winter tops for Ford and Cadillac, using a patented metal-reinforced wooden framing system that Seaman started using in 1910.[77] In addition to production bodies, Seaman also made custom coachwork for various other companies in the Midwest.[x]

[x] Between 1896 and 1930, there were over 1,800 Automobile manufacturers in the United States. Many of the manufacturers purchased parts from other companies and assembled them. It was inevitable that there would be a consolidation of companies and indeed, by 1930 there were only 60 or so active car companies.

The body for this 1915 Chalmers was produced by W.S. Seaman. It is pictured outside of Seaman's factory at Virginia and 8th Street in Walker's Point. The building still stands, now occupied by Aurora Healthcare.

Of all of the companies Seaman worked with, none would be more important than its relationship with the Thomas B. Jeffery Company of Kenosha, Wisconsin—producer of the Rambler automobile. Jeffery had previously manufactured the Chicago-based Rambler bicycle. In 1900, presumably following the formation of the 'Bicycle Trust,' Jeffery decided to go into the automobile business. In that year, he bought the Sterling Bicycle Company factory in Kenosha to serve as his automobile production facility. By the end of 1902 he had produced fifteen hundred automobiles, one-sixth of all existing motorcars in the United States at the time, making the Jeffery Company the second largest auto manufacturer at that time (Oldsmobile was first).[78] Soon, the Jeffery Company was Seaman's largest customer.

In 1916, Charles W. Nash, former president of Buick, acquired the Thomas B. Jeffery Company and renamed both the automobile and the company after himself. In 1919, the Nash Motor Company purchased a half-interest in the Seaman Body Corporation and the coachbuilder devoted its operations almost exclusively to Nash. As Nash sales grew, the size of Seaman's operations increased accordingly. In 1936, Nash bought out the remaining Seaman family interests, although Seaman family members were still active in the management of the firm for a while. Seaman Body became to Nash what Fisher Body was for the General Motors Corporation—a division responsible for all aspects of automobile body production.

By 1938, Seaman Body Division had exceeded the capacity of its Walkers' Point factories and a decision was made to build a new modern plant on Milwaukee's north side at Capital Drive and North Richards Street. The Seaman Body story is continued in Chapter 6.

A 1935 Nash automobile. Engine and chassis were produced in Kenosha, while the body was manufactured by the Seaman Body Company in Milwaukee

MECHANICAL APPLIANCE/LOUIS ALLIS COMPANY

In 1901, Thomas Watson acquired a small former shoe factory located on Hannover Street (today's 3rd Street) in Milwaukee's Walker's Point, and established the Mechanical Appliance Company. In spite of its name, Watson's intention was to make electric motors.

His cousin, Louis Allis, became interested as well. Louis had left his father's company, the Edward P. Allis Company, because of ill health. But, as his health improved, he began to explore investment and employment opportunities. He invested in his cousin's firm and, in 1903, was elected its president. Watson, an electrical engineer, concentrated on engineering the company's line of direct-current electric motors.

The timing was good for the company. The availability of distributed electric power was transforming manufacturing processes. Electricity allowed manufacturing factories to locate electric motors where needed to drive machinery. It eliminated the requirement to power all machinery from cumbersome line-shafts and belts.

This photograph shows a milling room prior to the availability of electric motors and associated controls. The machines were all laid out to provide access to the overhead line-shafts. Each machine was driven by one or more belts. The line-shafts were typically driven by a steam engine.

The company initially manufactured small electric motors and dynamos ranging in size from 1/8th to two horsepower. In 1908, the young firm brought out its first alternating current electric motor. All motors displayed the Watson name.

Importantly, Watson also led the team of engineers that went into factories and assisted in their conversion from line-shaft power to electric motors.

The company quickly outgrew its original manufacturing facility and relocated to Bay View in 1906—on the site of the Allis family's homestead farm.

An early advertisement for Watson Electric motors manufactured by The Mechanical Appliance Co., which eventually was renamed Louis Allis Co.

In 1922, the company officially changed its name to "The Louis Allis Company." The company also came out with a series of multi-speed squirrel cage motors in that same year.

For a time, Louis Allis also manufactured exhaust and ventilating fans—all presumably driven by its own electric motors. However, the company eventually settled on producing specialty electric motors exclusively.

In the 1940s the company designed and produced the world's first integral-horsepower, explosion-proof electric motors. Because of this design, they manufactured almost all of motors used to drive gun turrets for the United States Navy during the Second World War. The Navy preferring Louis Allis's steel motor casings since they wouldn't send out shrapnel if hit by enemy fire.

Eventually, Louis Allis moved operations to Birmingham Alabama, where the company continues to manufacture a full line of electric motors.

THE LAWSON AIRPLANE COMPANY

Milwaukee's Menomonee Valley would seem to be an unlikely place to build airplanes, yet at least two aircraft companies had ties to the Menomonee Valley.

In fact in 1919, Alfred W. Lawson, founder of the airplane manufacturing company proclaimed, "The future of the airliner is assured. We have already demonstrated the practicality of it. Milwaukee will soon be the center of the greatest aerial passenger factory in the world if the people here will continue to support the Lawson airline."

Alfred W. Lawson came to Milwaukee in the spring of 1919 to obtain financial backing to build a large passenger-carrying aircraft, which he labeled an "air liner." Lawson had formerly headed a company located in Green Bay, Wisconsin, which was organized during the First World War to manufacture training planes for the U.S. Army. The war had ended before production of military aircraft could be started, so Lawson turned to the commercial side of aviation.

LAWSON 'AIRLINER'

Obtaining limited financial support, Lawson began manufacturing parts for an airliner at the Cream City Sash and Door Company on West Pierce Street in the Menomonee Valley. The airliner was assembled in the auto pavilion at State Fair Park. It was towed, minus wings, to the old Zimmerman farm at Lisbon Road and Lovers Lane Road, now Currie Park, which was the site of Milwaukee's first municipal airport. Lawson had previously convinced Mayor Hoan of the need for setting aside the area for an airfield.

On August 28, a trial flight was made to Ashburn Field, which was located on Chicago's south side. The airliner was the largest commercial aircraft constructed in the United States at the time. It had a wingspan of 95 feet and its loaded weight was almost seven tons. It could accommodate 16 passengers and 2 pilots.

Elated by the performance of the aircraft, Lawson continued on to Toledo, Cleveland, Buffalo, Syracuse, New York and Washington D.C. While landing in Syracuse, with a female reporter as one of the eight passengers, the first accident occurred when one of the landing wheels dropped into a small ditch, and caused the airliner to up-end on its nose. The slow speed at which the accident occurred saved the passengers from injury, and the crushed nose section of the craft was repaired on-site.

On September 13, flying from Syracuse to Long Island, the plane covered 312 miles in two hours 47 minutes, averaging 111 miles per hour—setting a record for an airplane of its size. A forced landing at Collinsville, Pennsylvania because of severe weather over the mountains again damaged the airliner. The craft was dismantled and shipped by rail to Dayton for repairs. On October 24, the trip was resumed with a flight to Speedway race track in Indianapolis, and a return to Milwaukee on November 14.

After the 2,000-mile journey, Lawson purchased the vacated Fisk Rubber Company building at 9th and Menomonee Avenues in South Milwaukee to begin the manufacture of larger, more powerful airliners. The new design had a wingspan of 120 feet, and used three Liberty engines. The craft, dubbed the *Midnight Liner,* was designed for twenty-four passengers and a load of mail. The night version was to include sleeping berths and a bathroom with shower!

On July 10, 1920, Lawson was awarded the first airmail contract ever granted by the US government to a private individual; $685,000 for the annual movement of airmail over three different eastern routes.

Delays, and the financial recession that began in the fall of 1920, taxed Lawson's enterprise to the utmost to provide funds to meet payroll and other expenses. The first *Midnight Liner* was finally completed on December 9, 1920. Because of inclement weather, however, its maiden flight had to wait. Lawson's efforts were turned to trying to hold off his creditors.

As the financial situation heightened, Lawson decided that he would fly his airliner from a space near the factory, rather than make a costly move to Hamilton Field (now Gen. Mitchell Field). The prepared strip was only about 300 feet long. After days of waiting for favorable wind conditions, Lawson finally gave the order to attempt take-off on May 8, 1921. The airborne craft did not clear an elm tree, which caused the airliner to crash land. The pilots were unhurt but the stockholders refused to put up additional funds and the company went bankrupt. The company folded in 1922, and the assets were auctioned off.

The Lawson Airliner was the world's largest commercial aircraft in 1919.

ALFRED W. LAWSON was born in London, England in 1869. While an infant, his family immigrated to Ontario, Canada, before settling in the Detroit area in 1872.

After an early career in minor-league baseball, he launched a popular aviation magazine entitled *Fly* in 1908, which was eventually renamed *Aircraft*. He was the first advocate for commercial air travel and coined the term "airline." In early 1913, he learned to fly monoplanes, becoming an accomplished pilot.

In 1917, Lawson began to seek support for establishing his own aircraft company. Businessmen from Green Bay, Wisconsin offered to invest and the Lawson Aircraft Corporation was established there to build training planes for the Army. Later that year, utilizing the knowledge gained from ten years advocating aviation, he built his first airplane, the Lawson Military Tractor 1 (MT-1) trainer. He secured a contract from the Army and built a refined version, the Lawson MT-2.

After the First World War, Lawson undertook an effort to build America's first airline. He secured financial backing and relocated his company to Milwaukee. In 1919, he built and demonstrated a biplane airliner, the 18-passenger Lawson L-2. It was demonstrated in a two-thousand mile, multi-city tour that traveled from Milwaukee to New York and back, stopping at numerous cities along the way. It created a good deal of publicity, which enabled Lawson to obtain funds to build a 26-passenger airliner that he called *Midnight Liner*. In 1920, the company secured government contracts for three airmail routes. Because of financial difficulties, as well as a crash of his maiden aircraft, the company went into bankruptcy.

In 1926, Lawson started a project to build a much larger airliner, the 56-seat, two-tier *Lawson* super airliner. Unfortunately the aircraft crashed on takeoff on its maiden flight. He was unable to secure additional financial backing.

Lawsonomy

In the 1930s, Lawson began promoting health and dietary practices, and claimed to have found the secret of living to 200. He also developed his own highly unusual theories of physics—publishing numerous books on his concepts all set in a distinctive typography.

He later called his own philosophy, Lawsonomy, and founded the Lawsonian religion. During the Great Depression, he published a populist economic theory of "Direct Credits," believing that the government should replace banks as the provider of loans to business and workers. His rallies and lectures attracted thousands of listeners, but by the late 1930s the crowds had dwindled.

In 1943, Lawson founded the University of Lawsonomy in Des Moines to spread his teachings. After IRS investigations into its financial practices, the school was closed and finally sold in 1954, the year of Lawson's death. A farm near Racine, Wisconsin, is the only remaining university facility, although a tiny handful of churches may yet survive. A large sign, reading "University of Lawsonomy," was a familiar landmark for motorists passing by on I-94. A storm in 2009 apparently damaged the sign and it no longer exists.

Alfred W. Lawson is pictured second from the right, along with his employees. They are standing in front of his South Milwaukee factory with his new airplane, the Lawson *Midnight Airliner* L-4. When introduced, it was the largest passenger carrying heavier than air aircraft in the world. The cabin was 65 feet long and the wing spread 120 feet. It was powered by three Liberty motors aggregating 1,200 horsepower. It carried enough fuel for a fifteen-hour flight, making it capable of flying from New York to Chicago without refueling.

ELECTRONIC SECRETARY INDUSTRIES

In this day of smartphones with amazing capabilities, and cordless telephones with stored directories and digital answering machines, it seems quaint to recognize an early achievement in telephone devices—the automatic telephone answering machine. Until 1982, the American Telephone and Telegraph Company (AT&T) and its subsidiaries (collectively called the "Bell" system) was regulated as a "natural" monopoly. AT&T prohibited its customers from connecting phones not made or sold by its subsidiary, Western Electric, without paying fees. If a customer wanted a type of phone or another device not leased by the local Bell subsidiary, the customer had to purchase the device at cost, give it to the phone company and then pay a "rewiring" charge and a monthly lease fee in order to use it. Third-party equipment could not be connected to the AT&T telephone system until 1968.[79]

While Western Electric maintained an extensive research laboratory, the lack of competition stymied innovation by others. AT&T argued that answering machines could harm the phone system as well as violate people's sense of privacy, and persuaded the Federal Communications Commission to prohibit the devices. Joseph Zimmermann found a way around this limitation and developed the first practical and commercially available automatic telephone answering machine. His innovation helped to win some of the crucial legal and regulatory battles to allow the use of answering machines in the United States.[80]

Joseph Zimmermann said that he got the idea to develop an automatic answering machine because he could not afford to hire a secretary to answer calls to his air-conditioning and heating company. He developed the device in his basement in 1948; it was patented the following year.

In 1949, Zimmermann formed Electronic Secretary Industries, along with partner George W. Danner, to manufacture and market the automatic telephone answering machine. The two-man Electronic Secretary operation grew to sixty employees. Their facility was located at 805 South 5th Street in Milwaukee's Walker's Point. The company was sold to General Telephone and Electronics, Inc. in 1957, but Zimmermann continued to work for the company following the sale. By 1963, the company had one hundred and sixty employees and annual sales exceeded $4 million. It appears to have been the most advertised brand of answering machine and tens of thousands were sold.

Joseph Zimmermann, Jr. is pictured with his wife Helen adjacent to the Electronic Secretary company car in 1950. Photograph courtesy of the Zimmermann family.

Development of the Automatic Telephone Answering Machine

Zimmermann got around AT&T's restrictions that prohibited attaching 'foreign' (not made and sold by the Bell System) devices to the telephone system by developing a mechanical arm that picked up the telephone receiver when a phone call was received. While he recognized it as a crude solution to the problem, it worked and survived efforts to prohibit its use. He named his device the Electronic Secretary Model R1. It used a 78-rpm record player that played a greeting once a phone call was initiated, and used a wire recorder to record messages from callers. At the end of thirty seconds, the recording ended and the phone was mechanically returned to its 'hook.'

While earlier attempts to develop a telephone answering machine, at least one as early as the 1890s, they were impractical devices and were not developed commercially. AT&T is reported to have developed a device, but kept it concealed over fears that it would result in fewer telephone calls.[81]

This original model of the Electronic Secretary is in the collection of the Milwaukee Public Museum. It weighed 80 pounds. Image courtesy of the Milwaukee Public Museum, catalog number H52547.

Joseph Zimmermann, Jr. in 1958. Photograph courtesy of the Zimmermann family.

JOSEPH J. ZIMMERMANN, JR. was born in 1911 in Milwaukee, Wisconsin. He graduated from Marquette University in 1935 with a degree in electrical engineering. During the Second World War, Zimmermann served in the US Army Signal Corps. He was among the first of the soldiers that landed on Omaha Beach in Normandy, France, during the D-Day invasion.

Returning after the war, he became the owner of an air-conditioning and heating company and got the idea for an automatic answering machine because he could not afford to hire a secretary to answer calls while he was out of the office.

Tinkering in his basement, he built a device that answered the telephone by picking up the receiver. A 78-rpm record was used to play a greeting. If the caller wished to leave a message, it was recorded on a wire recorder located on top of the machine. The device was designed to get around the prohibition that the American Telephone and Telegraph Company exerted over attaching "foreign" devices on their system by automatically lifting the handset from a stand AT&T telephone. The answering machine was not connected to the telephone line. Zimmermann's solution was functional and many thousands were sold.

Zimmermann established Electronic Secretary Industries to manufacture, sell and market the product, along with businessman George W. Danner. AT&T attempted to block their attempt to introduce the product to the market. A lengthy legal battle ensued and Electronic Secretary Industries ultimately prevailed. The rights to the machine were eventually acquired by General Telephone and Electronics (GTE), which continued to market the device. The terms of the sale required GTE to continue to manufacture the device in Waukesha.

Joseph Zimmermann had at least a dozen patents, including a magnetic recorder to monitor heart patients, a security device that would automatically dial a telephone number and convey information in the event of an emergency, and a mobile radio-telephone automatic recording device. He also developed a cordless telephone prototype and worked on heart monitors with GE-Healthcare. Throughout his adult life he maintained a lab in his basement to continue to explore new ideas.

Zimmermann was awarded Marquette University's Professional Achievement Award in 1977.

Joseph married Helen Marie Puccinelli of Milwaukee in 1950, whom he originally met while teaching Sunday school. They had one child, Joseph James Zimmermann, III, who was born in 1958.

Joseph J. Zimmermann, Jr. died on March 31, 2004. He is buried in St. Mary's Cemetery in Elm Grove, Wisconsin.

CHAPTER 3: MILWAUKEE'S MENOMONEE VALLEY

In 1869, the Wisconsin State Legislature authorized a "system of water channels, canals or slips in the Menomonee Valley," which before then was largely marshland.[82] The channeling of the Menomonee River allowed industrial Milwaukee to expand into the valley area, which had formerly been unsuitable for development.

The Chicago, Milwaukee, St. Paul and Pacific Railroad, better known as the *Milwaukee Road,* was the first major company to move into the valley once the land was channeled and drained. The railway's contribution to the area's manufacturing was enormous because it provided ready transportation from firms located in Walker's Point to the rest of the country.

The Milwaukee Road was soon followed in the Menomonee Valley by the Milwaukee Harvester Company and by Falk. They were eventually joined by other companies, to include Cutler-Hammer, TL Smith, Chain Belt, Koehring Machine Company and Hamilton Metalplane.

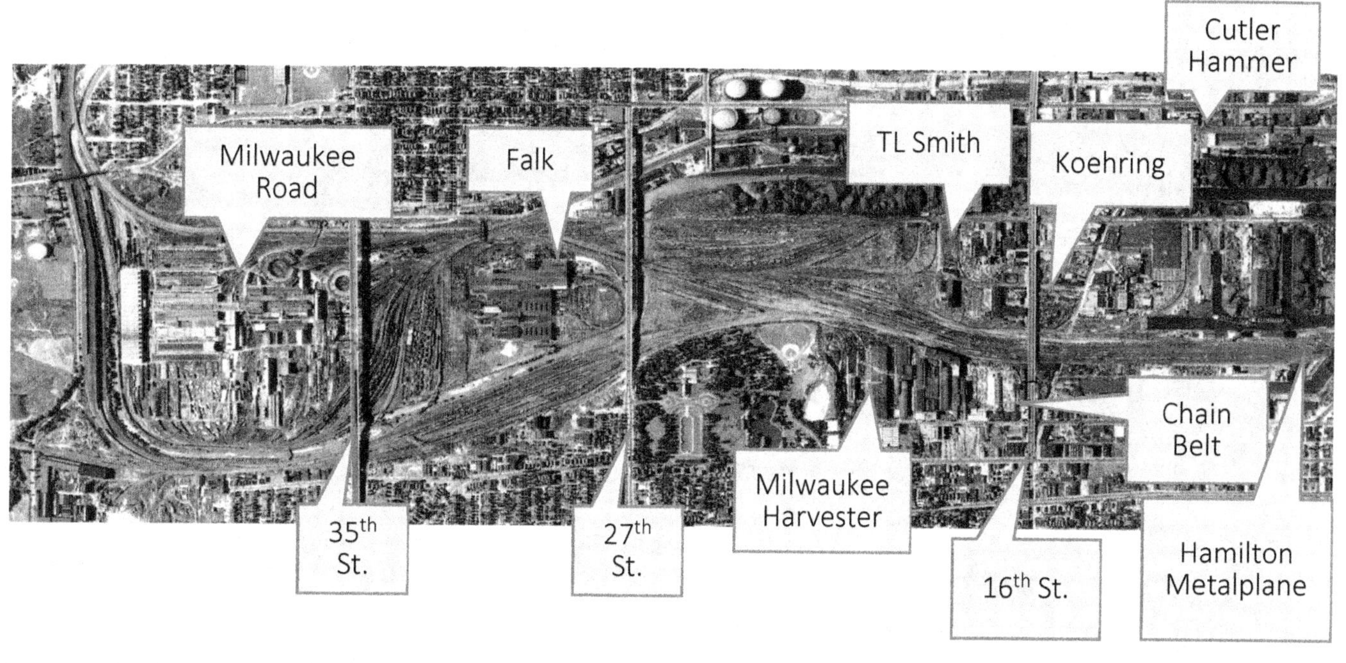

The approximate locations of area manufacturers that moved into the Menomonee River Valley are noted, superimposed over a 1937 aerial picture of the valley area. The Milwaukee viaducts that traverse the valley are also noted.

This 1872 Illustrated Map, looking from Lake Michigan to the west, shows Walker's Point in the foreground to the left of the river, and the Menomonee River valley above it with its recently dug water channels. Much of the land in the Menomonee Valley was still marshland in this view. The map is a portion of a map created by Milwaukee Lithographing & Engraving Company. Map courtesy of the Library of Congress.

THE MILWAUKEE ROAD

In 1847, a charter was granted to the Milwaukee and Mississippi Railroad Company to build a railroad from Milwaukee to the Village of Waukesha, twenty miles away. As noted earlier, the first president of the new railroad was the mayor of Milwaukee, Bryon Kilbourn.[83] The railroad expanded quickly, installing 90 miles of track by the end of 1853 and 620 miles by the end of 1857.[84]

Following a series of mergers, the company's name eventually morphed to become the Chicago, Milwaukee, St. Paul and Pacific Railroad. For simplicity, the company will be referred to by its common name—*the Milwaukee Road.* Despite its common name, the company was headquartered in Chicago. However, its operational headquarters and manufacturing shops were located in Milwaukee.

The railroad's shops were originally located at Humboldt and North Avenue in the City of Milwaukee. In the early 1880s, however, the company moved its shops to the west end of the Menomonee River Valley, where they were more accessible and provided the needed room for the growing railroad. Eventually the shops occupied 160 acres between the main line of the railroad and the Menomonee River—an area that was originally marshland and had to be filled in.

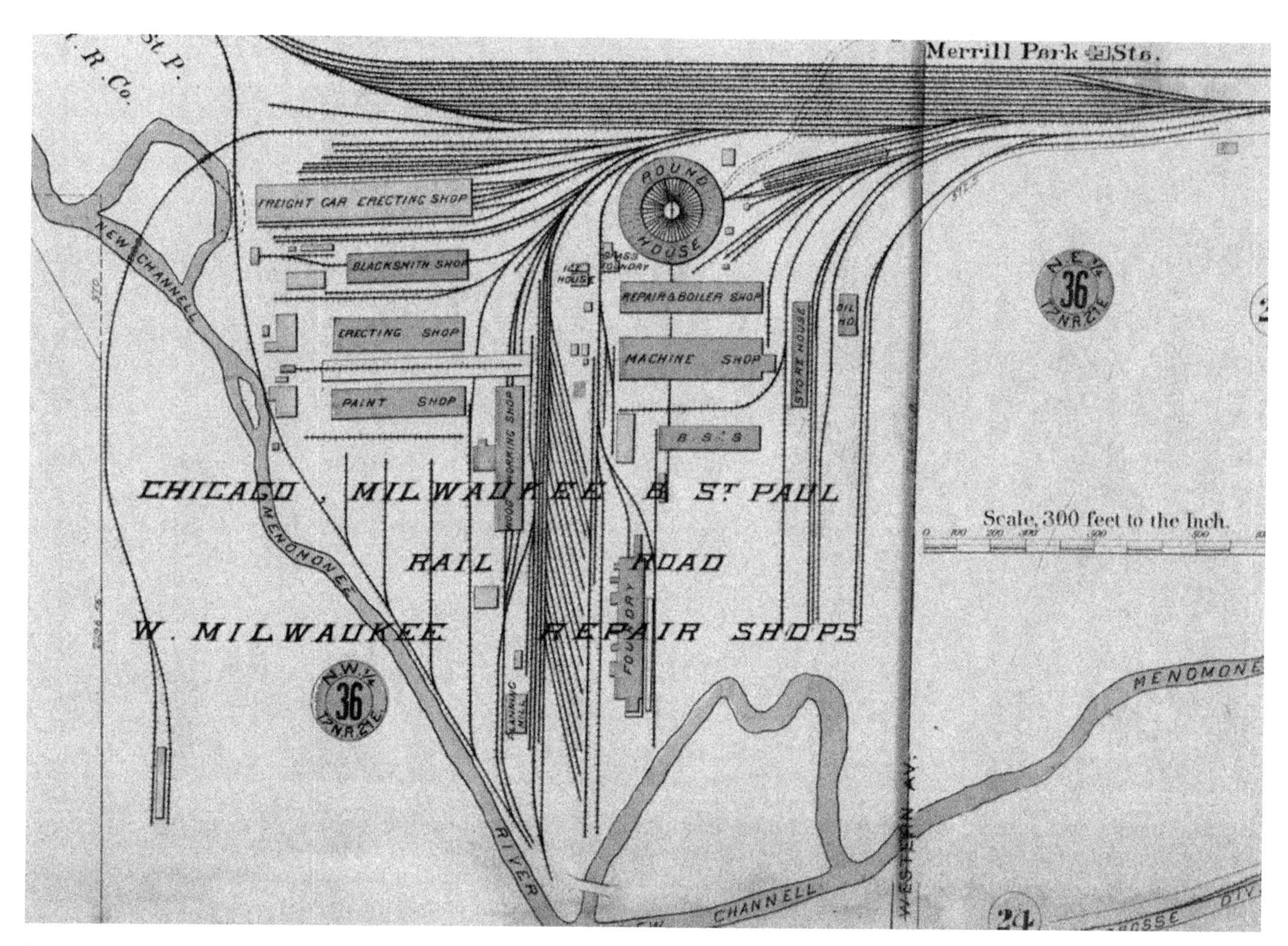

This 1892 fire insurance map shows the shops of the Milwaukee Road, along with the functions of the various buildings.

By 1882, the Milwaukee Road began construction of boxcars and passenger cars at its Milwaukee shops.

At peak capacity in the 1890s, the shops completed a new locomotive engine every three days. Over 500 steam engines were built at the complex in the west end of the Menomonee River Valley.

During the 1920s, the shops ceased locomotive construction to concentrate on the manufacture and service of rail cars. During this era, an average of twenty-eight freight cars was manufactured there daily.[85]

This interesting photograph of the interior of the Milwaukee Road's locomotive factory shows a number of early locomotives lined up (on the right) with various levels of completion. It is believed the photograph was taken in 1884. Photograph from the Historic Photo Collection of the Milwaukee Public Library.

MILWAUKEE ROAD'S INNOVATIONS

The Milwaukee Road is credited with many railroad firsts, a good number of which were designed to increase passenger comfort. For example, it was the first railroad to equip passenger cars with coil springs, the first to illuminate trains with electricity and to heat passenger cars with steam. It also pioneered passenger coach windows that opened with the turn of a crank and was the first to introduce air-conditioning to passenger coaches and add electric refrigeration to dining cars.[86] It also introduced the all-welded steel passenger car.

Engineer Karl F. Nystrom is credited with many improvements in railcar design, including features that increased safety, provided better riding qualities, and reduced weight, noise levels, and rolling friction. Nystrom joined the Milwaukee Road in 1922 as engineer in charge of car construction. He pioneered the development of the all-welded steel passenger and freight cars. By employing high-tensile steels in an all-welded design, the Milwaukee Road was able to achieve significant reductions in weight—some of its railcars were as much as 4.5 tons lighter than similar cars of riveted construction.[87]

Many of the Milwaukee Road's innovations were designed to increase speed.[88] For example, it was the first to equip passenger cars and locomotives with roller bearings to reduce rolling resistance. It also was the first to install automatic lubrication on its locomotives to allow them to make longer runs, which reduced the need to stop for "oiling."

During this period of innovation, the Milwaukee shops attracted delegations of railroad managers and mechanical engineers from virtually every part of the world to view what the inventive ingenuity its engineers were achieving.[89]

THE HIAWATHA PASSENGER TRAINS

In the 1930s, the Milwaukee Road was locked in a tough competition with the Chicago and North Western Railroad (C&NW) for travel between Chicago and the Twin Cities, especially for business travelers. The competition for customers led to a battle to advertise the fastest times for traversing the route. In Milwaukee, the C&NW's railway station was just a few blocks away from the Milwaukee Road's, which led to the local excitement for this all-out speed battle between the competing railroads. The C&NW developed its "400" passenger train, later named the *Twin Cities 400* because of its claim that it could complete the 400 miles between Chicago and Minneapolis in 400 minutes.

The Milwaukee Road's entry in this battle was its line of *Hiawatha* Passenger Trains. Four generations of *Hiawatha* equipment were introduced: in 1933-34, 1935, 1937-38 and in 1947-48. The 1933-34 trains, known as the *Twin Cities Hiawathas*, were powered by locomotives manufactured by the American Locomotive Company (known as Alco) in Schenectady, New York. These engines with their 4-4-2 wheel arrangement[y] were known in the industry as *Atlantics*—a design popular earlier in the century. The Milwaukee locomotives were built with enormous 84-inch diameter driving wheels. While not particularly powerful, they were exceedingly fast. On July 15, 1934, a steam locomotive pulling five coaches with roller bearings on a scheduled run between Chicago and Milwaukee traveled the 85 miles in 67 minutes. The train hit a maximum speed of 103.5 miles an hour and averaged 92.1 miles an hour for 53.6 miles—a world record for sustained speed by a steam-powered train at the time.[90]

The streamlined locomotive concealed its steam boiler and attendant accessories behind its metal casing.[91] Its sleek appearance suggested speed even when motionless. When placed in service, it was the first locomotive scheduled for daily operation at speeds in excess of 100 miles per hour.[92] The Hiawatha made the Milwaukee to

---

[y] The notation for designating locomotive types in the United States is to number the wheel sets, starting at the front of the engine, known as the Whyte notation. A 4-4-2 designation means that there was an initial set of four wheels, followed by another set of four, followed by a final set of two wheels. You can verify this by referring to the pictures of the

Chicago run in 59 minutes, averaging 86.96 miles per hour start-to-stop, which included several intermediate stops and speed restrictions.[93] In a speed test in 1935, the Hiawatha attained 112.5 mph.

Picture of the Hiawatha Locomotive and its tender followed by passenger cars, leaving the Milwaukee Road depot in 1935. Picture from the Historic Photo Collection of the Milwaukee Public Library.

Stylized 1939 advertisement featuring a streamlined 4-6-4 class F7 steam locomotive. Source: Wikipedia Commons

Area residents lined the road to watch the *Hiawatha* pass by during its maiden run in 1935. The train routinely covered the 85 miles between Milwaukee and Chicago in 75 minutes. In a speed test that year, the *Hiawatha* attained 112.5 mph. Photograph from the Historic Photo Collection of the Milwaukee Public Museum.

While the locomotives were built elsewhere, the passenger cars were manufactured by the Milwaukee Road at its Milwaukee shops. The overall streamlined design was also the responsibility of the Milwaukee Road. Nystrom oversaw the overall design, assisted by Charles Bilty, a mechanical engineer in the locomotive department.

Successive generations of *Hiawathas* were even more striking in appearance. Most dramatic, perhaps, was the "Beaver Tail" observation cars of the late 1930s and the "Skytop Lounge" observation cars by industrial designer Brooks Stevens in the 1940s.

The *Hiawathas* of 1937-38 were powered by high-speed, streamlined 4-6-4 "Baltic" or "Hudson" type steam locomotives, also built by Alco referred to as class-F7 locomotives. They are major contenders for the fastest steam locomotives ever built, as they ran over 100 miles per hour (160 km/h) daily. One run in January 1941 recorded by a reporter for *Trains* magazine saw 110 mph achieved twice—in the midst of a heavy snowstorm. These engines also had 84-inch drivers and, with six driving wheels, they had much greater steaming capacity than the previous *Hiawathas* and thus the ability to haul longer, heavier trains. These were the last high-speed trains in the United States to use steam power.

KARL FRITJOF NYSTROM was in born Asba Bruk, Sweden in 1881. He graduated in 1904 as a mechanical engineer from the Mining School at Filipstad, Sweden. To pay for his education he worked summer jobs in machine shops in Stockholm and in steel mills in other parts of the country. Following graduation, he went to Germany to study high tensile steel in Germany, but before finishing his studies he decided to seek employment in Pittsburgh, the center of the US steel industry. He worked as a "blueprint boy" and then as an engineer at the Midland Steel Company, before obtaining employment for Pressed Steel Car Company as a draftsman.

His work at Pressed Steel Car apparently caused Nystrom to decide to focus his career on railcar design and manufacture. It appears that he may have decided to leave Pressed Steel during the Pressed Steel Car strike of 1909[z] because he joined Pullman in the later part of that year, where he was co-designer of the first steel sleeping car. He also established the specifications for the first all-steel railway post office car.

He moved through a rapid succession of railcar jobs, working for the Southern Pacific during the electrification of its Oakland-Alameda line, where he designed and built their first electric interurban cars for the American Car & Foundry Company, for the Acme Supply Company, for the Grand Trunk (the Canadian National) and for the Canadian Pacific.

In 1922, Nystrom was appointed engineer of car design for the Milwaukee Road and worked there for the rest of his career. He quickly rose through the ranks, eventually becoming chief mechanical officer of the company.

Nystrom is named on almost 100 patents.

He was an innovator in many aspects of railcar design, but his principal distinction was the development of welded lightweight freight and passenger railcars, his design of car wheel assemblies (trucks) for passenger railcars which were considered the smoothest riding in the industry, and for the cars for the Hiawatha trains where he played a leading role in designing the well-known bay-window cabooses. He also perfected steam jet air conditioning for passenger cars.

Nystrom served for several years on the board of supervisors of Marquette University, where in recognition of distinctive work he received an honorary Ph.D. in mechanical engineering from that university in 1941. He was a consultant for the War Department Transportation Corps and a member of the War Production Board. In 1945, he was elected a Fellow of the American Society of Mechanical Engineers. He retired on January 31, 1949, and died on June 5, 1961.[94]

Karl Nystrom at his desk, referring to his a slide-rule for making calculations

---

[z] The Pressed Steel Strike was also known as the "1909 McKees Rocks strike." The walkout drew national attention when it climaxed on August 22nd of that year in a battle between strikers, private security agents, and the Pennsylvania State Police. At least 12 people died, and perhaps as many as 26.

MILWAUKEE HARVESTER COMPANY

Milwaukee Harvester Company was located in the Menomonee Valley at the west end of Park Street. It was organized sometime around 1883, but its history goes back much earlier. It can trace its roots to an 1850 Beloit company, where Israel S. Love built approximately fifty "Fountain" reapers. In 1852, the partnership of Love & Otis was formed to build self-rake reapers. Eventually the partnership was renamed Love & Stone and, in 1857, a new firm named Parker & Stone was organized. It appears that Parker had the rights to manufacture the Appleby binder, which was also known as the "Beloit" reaper.

An artist's rendering of the Milwaukee Harvester Company, located in the Menomonee Valley.

By 1881, the partnership changed its name to Parker-Dennett Harvesting Machine Co., but it was short-lived because in 1883 the stockholders of the Milwaukee Harvester Company acquired Parker and Dennett's firm and moved operations to Milwaukee.

In addition to harvesters and twine binders, the company manufactured corn harvesters and husker/shredders. The company employed five to six hundred workers and about two hundred and twenty traveling salesmen and agents by 1886.

The company used illustrated catalogs, printed in various languages, to advertise its products to farmers. For example, the adjacent whimsical covers are from an 1895 catalog that was printed almost entirely in Norwegian. It was published and distributed in areas of Wisconsin and Dakota where there were large populations of immigrant, Norwegian-American Farmers. It is titled simply "Aarligt Katalog af Milwaukee Harvester Kompagnie" (which translates "Yearly Catalog of Milwaukee Harvester Company") on the title page and "The Circus on the Farm" (in English) on the front cover.

The catalog is written and illustrated almost like a children's book, showing various horse-drawn farm machinery, including the Milwaukee steel binder-harvesting machine and the Milwaukee chain drive mowers. Interspersed with the illustrations of farm machines are comic and/or charming scenes depicting interactions between the circus animals and the circus performers with the farmer, his family and his farm machinery. The catalog, which was likely published in multiple languages, provides some interesting insights into the efforts the company took in order to introduce its products to a rural population.

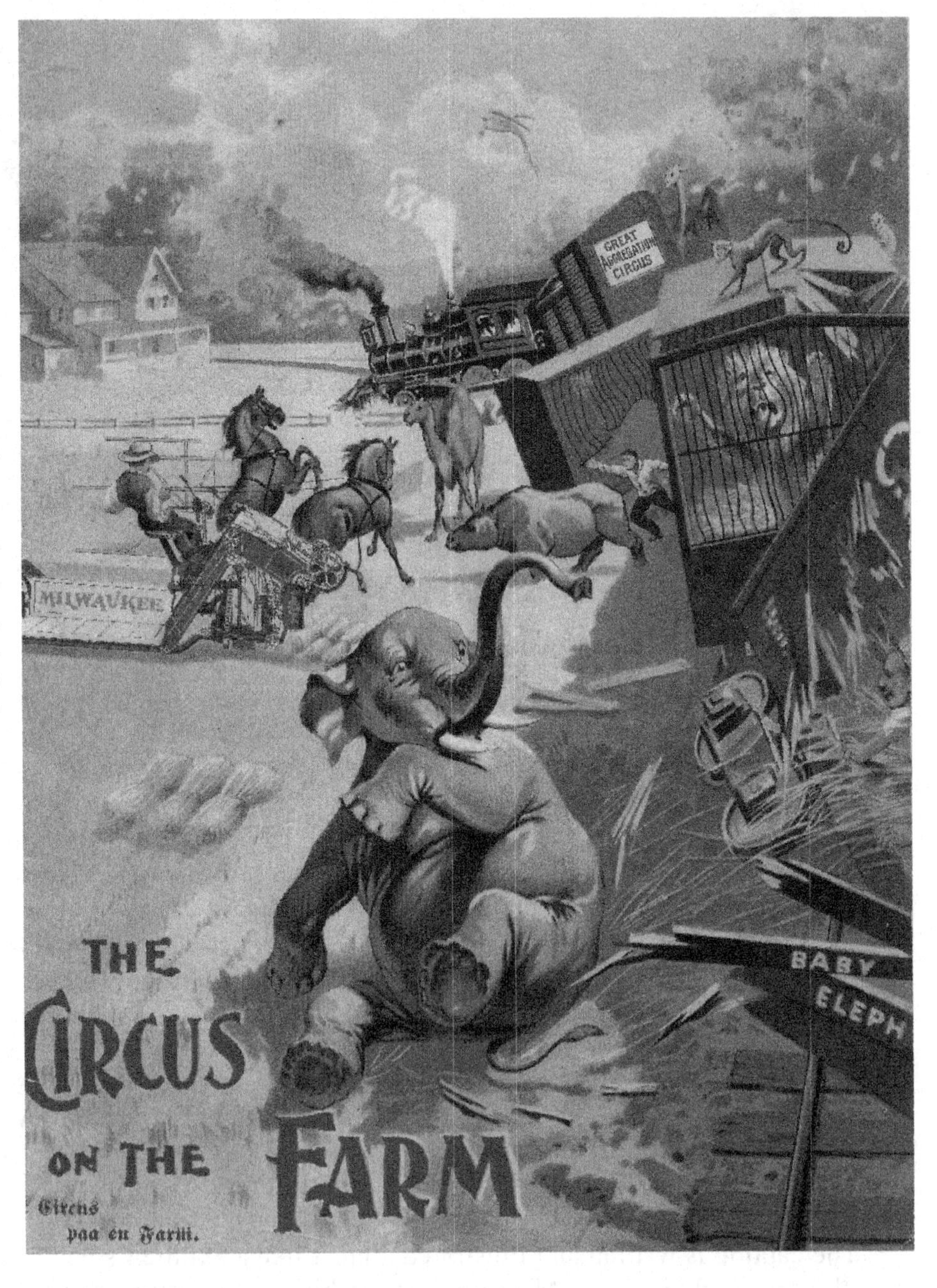

These chromolithographs were produced in 1895 for the annual catalog of the Milwaukee Harvester Company. The cover includes an image of a farmer using his Milwaukee horse-drawn grain binder and mowing machine (at left) while a train derailment occurs, causing circus animals to be released. Note that the binder's "wind-board" proudly displayed the name Milwaukee. From the author's collection.

In 1902, the Milwaukee Harvester Company's stockholders participated in a large merger with four other agricultural implement companies, including McCormick Harvesting Machine Company and Deering Harvester Company, to form International Harvester Company. The newly merged concern dominated the market for harvesters and agricultural mowers.

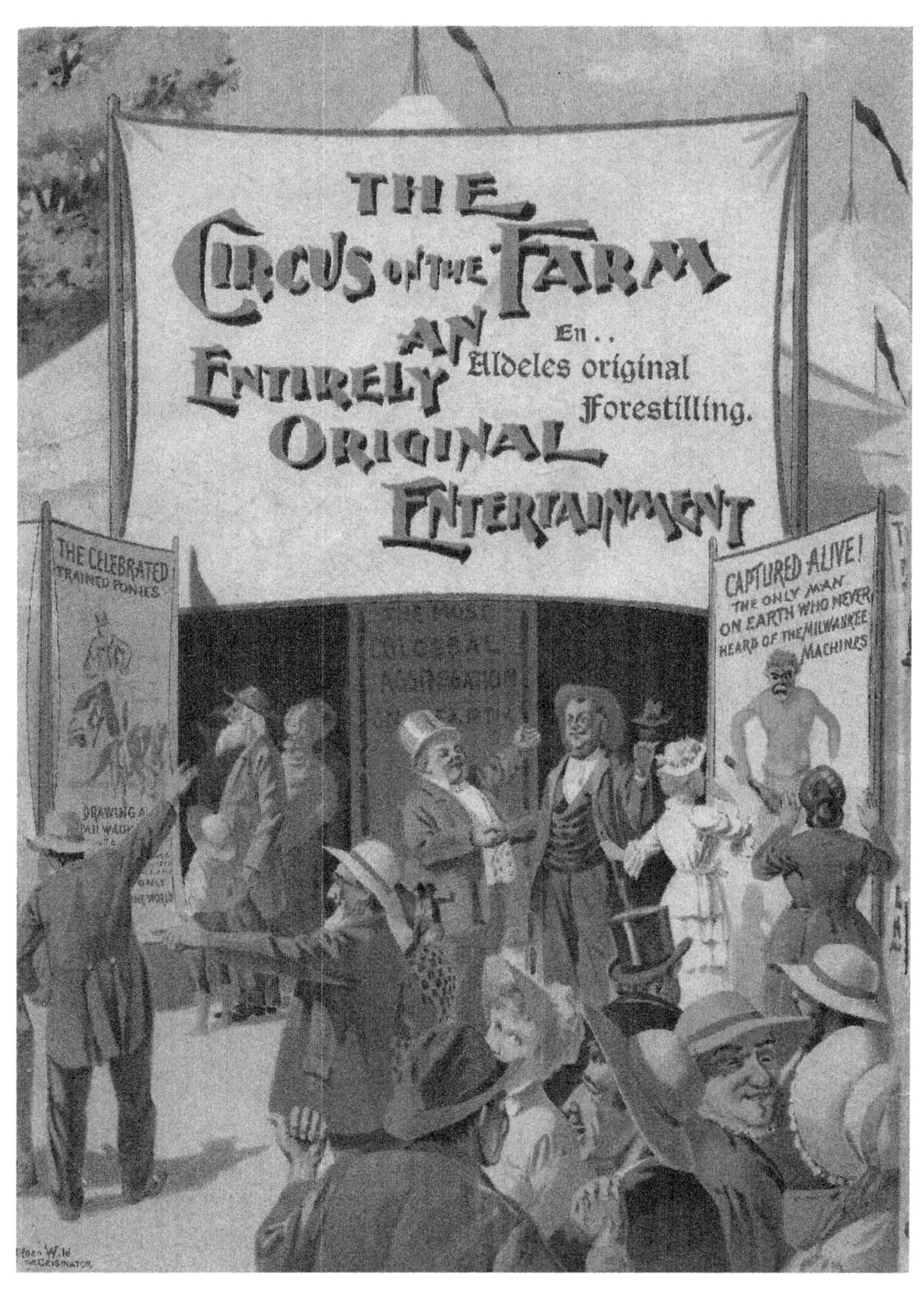

This illustration is from the back of the 1895 annual catalog of the Milwaukee Harvester Company, continuing the *Circus on the Farm* theme. From the author's collection.

Milwaukee Harvester Company No. 5 Mower

In 1895, the Milwaukee Harvester Company introduced its No. 5 Mower. Presumably there were other previous models. The No. 5 included a patented design "to facilitate the lifting of the cutting apparatus of a mowing-machine and for floating it over the surface of the ground while mowing." It was claimed that the arrangement and construction of the hand and foot levers and of the elastic connection between the lifting mechanism and the frame, assisted in the lifting of the finger-bar by levers and improved the floating of the cutting apparatus over the surface of the ground.

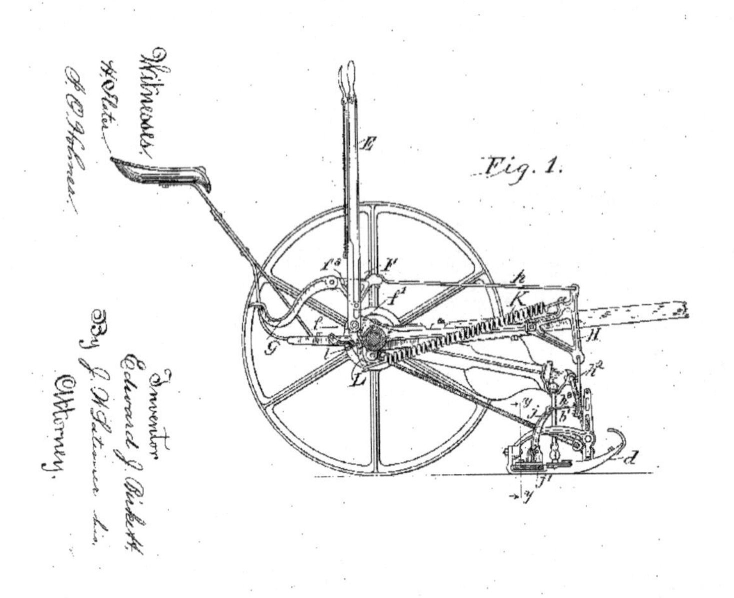

The patent drawing (tipped on its side to better show the mower). This is Patent US547411, issued to Edward J. Birkett of Milwaukee.

A vintage No. 5 Mower of the Milwaukee Harvester Company, on display at the Stuhr Museum of the Prairie Pioneer. Photograph included with their permission.

In using this mower to cut hay, the farmer would hook it up to a team of two horses or to a tractor. The farmer would lower the sickle bar so that it was nearly parallel to the ground at the desired height. As the mower was pulled forward, guards (or teeth or fingers) guided the hay to triangular cutting blades that ran along the length of the oscillating sickle bar. Blades connected to the oscillating sickle bar, moved from side to side as they cut. The cut hay fell to the ground, where it was then raked up into piles with a horse-drawn hay rake. The hay would then be gathered using pitchforks (or a mechanical hay loader) onto a wagon for conveyance to storage until use as animal feed.

It appears that a large number of these mowers were constructed and sold between 1895 and 1902.

FALK MANUFACTURING COMPANY

Franz Falk of northern Bavaria immigrated to the United States in 1848 at the age of 24, with the intention of opening a brewery. After spending time in Cincinnati, he settled in Milwaukee. In 1856 Franz Falk, along with partner Frederick Goes, bought land in the Menomonee Valley to locate a brewery, which when opened was appropriately named the Bavaria Brewery. When Franz died in 1882, his sons Louis, Frank, Otto, and Herman took over the business. The brewery suffered several misfortunes. It was destroyed by a fire in 1889 but was rebuilt in 1892. Subsequently, it was destroyed again by fire. The sons must have had enough because they sold the brewery to Frederick Pabst.

With his share of the brewery money, Herman Falk started his own business producing wagon axle couplings. When that business failed, Herman rented the former blacksmith shop of the family brewery from Pabst. He started his new business as a job shop—performing miscellaneous repair work for a number of clients. Eventually, he began focusing on electric street railways. It is likely that his repair business resulted from his recognition of a problem with streetcar rails. He observed that the tracks tended to wear and develop a depression adjacent to where the rails joined one-another. The repair was costly, involved tearing up the rails, sawing off the worn ends, and relaying the track "after preparing the shortened rails, so that they may again be joined by means of bolts and fishplates."[95]

Working with Albert Hoffmann, a Milwaukee electrician, Falk developed a portable cast welding machine that could be transported to the site on wheels to repair the rails at a lower cost. The invention, which was patented, along with other patents for methods to effect the rail repairs, helped Falk incorporate Falk Manufacturing Company on May 23, 1895. During the next seven years, the Falk process was used to repair over one million track joints on streetcar rail systems.[96]

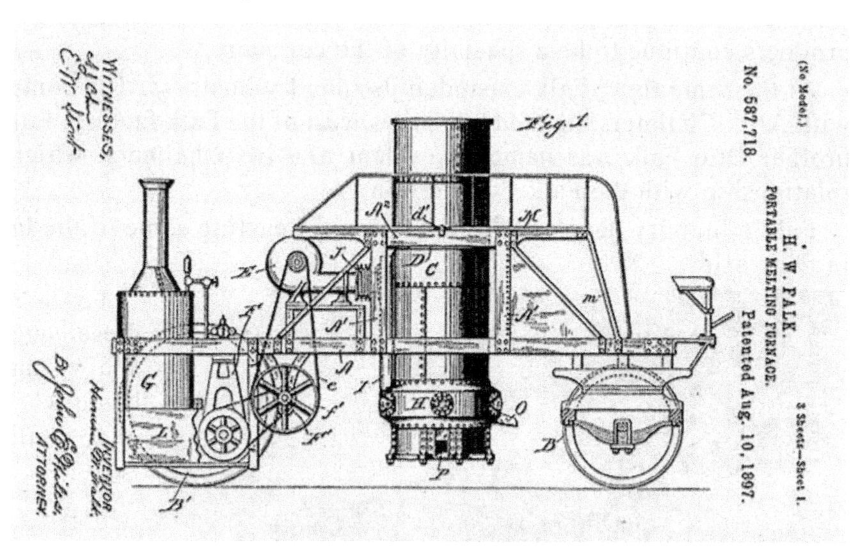

Patent drawing from Falk's portable welding apparatus. Note the steam engine used for driving a blower for the furnace used to melt steel for welding to streetcar rails. Later units used electricity to melt the steel alloy used for the rail repairs, powered directly by the railway's overhead wires.

A young Milwaukeean named Julies Peter Heil joined Falk as an apprentice during that time, working on electric streetcar rail projects in St. Louis and New York. He eventually left to form his own company, the Heil Company—more on that later in this chapter.

As business increased, Falk began construction of a seventy-thousand square foot facility in Milwaukee's Menomonee Valley, which was completed the following year. The company concentrated on its cast-welding equipment for the repair of streetcar rails and operated a foundry for casting various metal parts. However, it also engaged in other street railway construction and special track work and developed an early oil switch.

In 1898, Falk invested in Western Gear, a small Milwaukee company that manufactured motor gears and pinions principally for electric streetcars. The company was a good match for Falk's streetcar repair business. He soon acquired the firm, but continued to operate it from its small factory at Milwaukee and Michigan Streets.

By 1910, Falk's Menomonee Valley factory had expanded to 250,000 square feet and employed over a thousand people during its busiest months. But its streetcar rail business was declining. Falk lost a patent battle and, with the loss of patent protection, competitors had entered the field. Newer welding technologies were also becoming available.

The market for precision gearboxes, however, was increasing dramatically. Electric motors and steam turbines were replacing steam engines. They both operated at high rotative speeds and often required gear drives to match the requirements of the application.

In a visit to Europe, Herman Falk became aware of new machines for manufacturing herringbone gears. A herringbone gear is a specific type of double-helical gear—essentially two gears that are mounted side-to-side. Their advantage is that the side thrust of one-half of the gear is balanced by the thrust on the other half. Helical gears also are ideal for high torque applications because more than two gear teeth are engaged at any moment in time. However, they are more difficult and expensive to manufacture—requiring specialized tools.

Caspar Wüst-Kunz of Seebach Switzerland, principal of a company named C. Wüst and Company (eventually renamed Wüst AG), developed a machine to 'hob' herringbone gear teeth.[aa] Falk acquired the United States rights to use his patent and hired English engineer Percy C. Day to oversee the production of herringbone gears at his Milwaukee factory. As a result, Falk became the first in the United States to manufacture herringbone gears.[97]

The need for heavy-duty, precision gear reducers provided the company with growing product lines. For example, herringbone gears made it possible for steam turbines to be used in marine propulsion applications—mating high-speed steam turbines shafts to lower speed propeller shafts. Falk has produced some of the largest gearboxes in the world, and for some of the most demanding applications. Custom products continue to be a specialty of the company.

At the same time, Falk expanded its foundry business. The company developed a supplier relationship with Allis-Chalmers, located two miles west of the Falk factory. Following its 1912 bankruptcy, Herman's brother Otto Falk was named president of Allis-Chalmers, which undoubtedly helped to foster A-C's relationship with the Falk Corporation.

Falk's foundry developed a reputation for casting some of the largest and most complex steel castings in the world.

For a time, Falk also manufactured marine diesel engines, such as the one shown at left. While marine diesels were a complementary business to the company's marine reduction gears, Falk soon exited the business.

Falk manufactured this 16-cylinder, double-bank diesel engine coupled to a gear reducer for large marine applications in the 1920s. Photograph courtesy of Rexnord Corporation.

[aa] Hobbing is a machining process for cutting gear teeth, as well as for similar applications. It is a specialized type of milling machine. The gear teeth are progressively cut into the workpiece by a series of cuts made by a cutting tool called a 'hob'. Compared to other gear forming processes it is relatively inexpensive but still quite accurate.

Falk Herringbone Gears

When English engineer Percy C. Day arrived in Milwaukee in mid-1910, he immediately got to work on developing the tooling to manufacture herringbone gears. He was able to make full use of the technology developed by Swiss engineer Caspar Wüst-Kunz in designing a specialized milling machine to hob the gears.

Similar to the United States patent that Caspar Wüst-Kunz took out on his hobbing machine, Day's hobbing tool simultaneously machined the teeth on two sides of the gear blank cylinder. A pair of helical milling hobbers moved on traversable tool slides that were equally spaced from the spindle of the blank holder. Each hobber machined gear teeth in opposite directions. Caspar Wüst-Kunz's 1904 patent drawing is shown below.

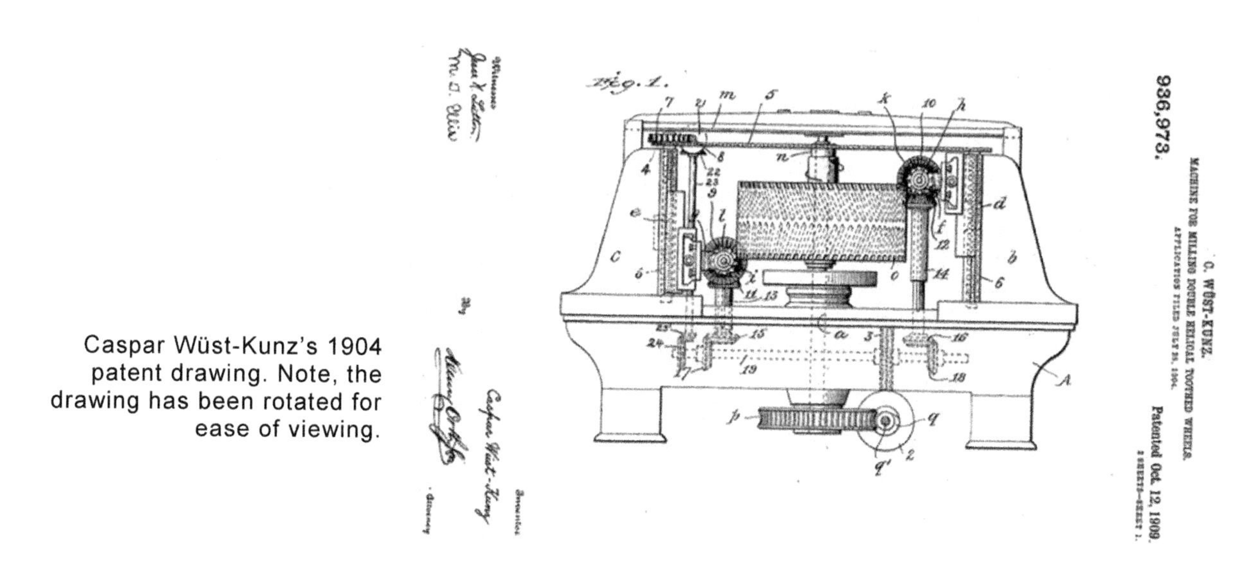

Caspar Wüst-Kunz's 1904 patent drawing. Note, the drawing has been rotated for ease of viewing.

Using hobbers refined by Percy Day, Falk was able to introduce its first herringbone gears to the United States market in 1911. Initial sales were slow since the US market was unfamiliar with the advantages of herringbone gears. However, the ability of Falk's herringbone gears to meet the duty requirements of some of the most severe applications soon resulted in sales to such customers as Carnegie Steel, Anaconda Copper, Allis-Chalmers, and General Electric.[98]

Day also began designing herringbone gearboxes for use by the United States Navy, including three battleships and a destroyer in 1913.

Soon Percy Day developed hobbing machines for larger herringbone gears. In 1915, he designed and built a machine for hobbing herringbone gears that was the largest in the world. It provided Falk with unique manufacturing capabilities, allowing it to exploit new markets.

The herringbone gear hobbers designed by Percy Day produced precision gearing—gears that were uniform and accurate and mated precisely with their counterparts in the gearbox or drive train. The market soon discovered that Falk herringbone gears were durable, efficient and relatively quiet operating. While more expensive to produce than spur gears, they proved their value to industry and established Falk as the pre-eminent manufacturer of precision, heavy-duty gears.

Falk had the largest machine in the world for hobbing herringbone gears, shown in this 1915 photograph. Photograph courtesy of Rexnord Corporation.

Falk Gear Reducers for US Navy Applications

While Percy Day was employed by The Power Plant Company in England, he designed and built a marine gear drive for Charles Parsons' *Vespasian*. Parsons was an Anglo-Irish engineer, best known for his invention of the compound steam turbine. He worked as an engineer on dynamo and turbine design, and power generation, with great influence on the naval engineering fields. The *Vespasian,* originally an old cargo ship, was acquired by Parsons' to demonstrate the use of steam turbines for maritime use.

Parsons replaced the original 750 horsepower, triple-expansion engine on the *Vespasian* with steam turbines. In order to permit the steam turbines to match the much lower speed required by the propellers, Parsons installed gear drives designed by Percy Day. They performed flawlessly and demonstrated that steam turbines could greatly increase the speed of large ships. As a result of Parson's demonstration, steam turbines began to be installed on most large ships—especially those ordered by the military. His turbines also began replacing steam engines in electric power plants.

Percy Day's experience with the Vespasian project was fortunate for Falk. The United States had been upgrading its navy fleet under President Theodore Roosevelt. During Roosevelt's tenure, the US Navy went from the sixth largest in the world to second only to the British Navy. It included sixteen new battleships that made up Roosevelt's "Great White Fleet."

The push by the superpowers to increase naval capacity continued, as Germany began expanding its Navy to rebuff what it saw as a threat from British Naval superiority. As a result of this activity, Falk received its first orders for gearboxes from the United States Navy in 1913 for three battleships and a destroyer.[99]

First World War

While the United States didn't enter the First World War until 1917, Congress authorized the US military to build up its military in 1915 under the banner of "preparedness." One hundred and fifty-six warships were authorized. In January 1917, Falk received an order of sixteen gear drives for eight destroyers from Bethlehem Shipbuilding. A few months later, sixteen more gear drives were ordered.

Falk was at full capacity and it quickly proceeded to increase its shop floor space, building what became known as Shop 2. By the end of 1917, Falk received a single order for 180 gear drives for ninety destroyers on an accelerated schedule. Falk immediately began construction of a third shop—a duplicate of Shop 2. At wartime peak, Falk was completing one destroyer drive a day. By the completion of the war, Falk built 336 massive gear drives for the Navy.[100]

The wartime buildup significantly increased Falk's gear making capacity, and its reputation. By the end of the war, it had the largest herringbone gear factory in the United States.

Second World War

Similar to the build-up that occurred prior to the Second World War, overseen by President Teddy Roosevelt, President Franklin D. Roosevelt worked to increase the Navy's arsenal in the years leading up to World War Two. Falk received orders in 1934 for gear drives for the aircraft carrier *Ranger*, and over the next few years for the carriers, *Yorktown, Enterprise, Wasp* and *Hornet*. During this time, Falk also supplied gear drives for fifteen cruisers and twenty-eight destroyers.[101] But this military business was nothing compared to the demand that occurred following the Japanese attack on Pearl Harbor.

Falk had already begun to expand capacity—adding a shop in the space between Shops 1 and 2, and expanding its weld shop and foundry. Following the United States' direct entry into the war, Falk added Shop 4, known as "the Navy building," to help meet the demand.

During the Second World War, Falk drives turned the propellers on twelve aircraft carriers, twenty-seven heavy cruisers, 184 destroyers, hundreds or cargo ships and tugboats, and 1,024 Landing Tank Ships (known as LSTs).

Design and manufacturing of the gear drives for the LSTs was a herculean effort. As the Allies began planning to retake France, it became evident that they would need a huge fleet of landing craft to convey troops and supplies into areas without harbors. A landing craft was designed to meet the specific requirements for the amphibious landings. The design requirements for the gear reducers for the LSTs was particularly important. The ships were designed specifically to land on beaches, off-road equipment and personnel, and then quickly withdraw off the beach, requiring that the gear reducers be able to quickly move from forward to reverse mode.

Falk had gained prior experience with reverse gear drives when it designed gear reducers with pneumatic clutches for the diesel powered *Bull Calf*—a towboat used on the Mississippi River.[102] Because of this experience, Falk was selected to provide the gear drives and clutches for the LSTs. Falk engineers, overseen by its chief engineer Walter Schmitter, designed a propulsion reversing reduction drive to meet the requirements. A pilot model was rushed into production and performed well—no design changes were required for the production gear drives.

The LST gear reducers needed to reduce the shaft speed of the twin General Electric V-12 diesel engines that powered the propellers, provide for both a forward and a reverse drive direction, and also provide a means to disconnect the drive from the engines. The Falk drives were designed to handle nine-hundred horsepower at 744-RPM full engine speed.

A key element in the LST gear reducers was Falk's *Airflex* clutch, patented by engineer Thomas L. Fawick. Fawick was one of the incorporators of Twin Disc Clutch Company in Racine Wisconsin. He sold his interest in Twin Disc in 1936 and organized the Fawick Clutch Company, which he moved to Cleveland in 1942. Joining forces with the General Tire and Rubber Co. of Akron, he designed a clutch with an inflatable rubber tube molded inside a steel drum that, when pressurized, locked to the main pinion shaft and produced forward motion. When unpressurized, the engine's power was diverted through a set of reversing gears. The clutch enabled the LSTs to change propeller direction in seconds.

Picture of the Falk gear reducer equipped with an Airflex rubber clutch (shown at left), as tested in 1939 for use on a 130-ton towboat for use on the Mississippi River. The clutches were tested successfully through 1,400 consecutive reversals simulating full ahead to full reverse in less than three seconds. Author's collection.

A picture of one of the production versions of the Falk reversing gear reducers developed for use on US Navy LSTs. Falk produced over a thousand of these drives for the Navy during the Second World War.

HERMAN WAHL FALK was born in Milwaukee in 1867, the fifth son of Franz Falk. He graduated from the Allen Military Academy of Chicago. After graduation, he worked for a time in his father's brewing business. In 1892, he rented space from the Pabst Brewing Company for a small job-shop—a shop that he eventually expanded into the Falk Corporation, one of the largest manufacturers in the world of precision industrial gears. Under his leadership, the company introduced herringbone gearboxes, along with the necessary tooling to manufacturing these complicated gears.

In 1948, the University of Wisconsin conferred its honorary doctor of law degree on Falk. In 1950, the National Metal Trades Association bestowed him with its Industrial Relations Achievement Award and, in 1956, the National Conference of Christians and Jews presented him with its Distinguished Service Award.

Herman Falk died in 1957 at the age of 73.

PERCY CRUNDALL DAY was born in England in 1874—the son of John Day. He attended Central Technical College in South Kensington, London.

Day worked as an engineer and works manager at Acetylene Illuminating Company in Foyers Scotland and in London between 1896 and 1902. He was employed as chief engineer for the Phoenix Process Trust between 1902 and 1905, before accepting a position as chief engineer for the Power Plant Company of London, where he worked until 1910. While employed at the Power Plant Company, Day designed the world's first commercial-scale hobbers for machining herringbone gears.[103]

In 1910, Percy C. Day was hired by the Falk Company to assist its development of herringbone gears and their applications in industry. He became a recognized expert in the technology for creating gears and helped to make Falk a prominent manufacturer of precision, heavy-duty gears for industry.

Percy Day married Beatrice Cockrill in 1910. They had two children, William and Marjorie. Day was a member of ASME, as well as the University Club and the Milwaukee Athletic Club.

T.L. SMITH COMPANY

Thomas L. Smith was born in England in 1855. He came to Wisconsin with his parents when he was four. As a youth, he learned the machinist's trade in his dad's Watertown machine shop and foundry. In 1873, he enrolled in a collegiate course at Iowa State College at Ames and graduated in 1877. Upon graduation, he taught mathematics and bookkeeping at Iowa State, before enrolling at Massachusetts Institute of Technology in Boston, where he completed his engineering education.

After completing college, Smith returned to Wisconsin and started employment with the Whitehill Sewing Machine Company. When Whitehill went out of business, Smith worked successive jobs with D.J. Murray Manufacturing Company in Wausau, Wisconsin, Filer & Stowell Manufacturing Company, Kempsmith Manufacturing Company, C.J. Smith & Sons, and Pawling & Harnischfeger.

Smith briefly established his own machine shop in Reedsburg, Wisconsin, where he invented a flexible-arm wood carving machine. He established the Milwaukee Carving Company to use that machine to produce woodcarvings. However, he soon sold the rights to the device. In 1898, Smith helped to organize a school of engineering and mechanical drawing in Milwaukee.[104] There he met D.W. Cutter of the Northwestern Tile Company, who interested him in concrete construction. The following year, Smith invented a tilting concrete mixer and made a working model. Cutter agreed to purchase the machine when it was completed later that year. Smith's first machine was a chain-driven mixer, mounted on a wagon and rotated propelled by a steam engine.

Smith's first concrete mixer, from a company catalog.

Recognizing that there was a significant market for concrete roads and other construction uses, Smith began designing various mixers for the construction industry—building models and obtaining patents. He was eventually issued several dozen patents, many of which dealt with refinements of mixing devices that would assure rapid and thorough mixing of products. His machines developed a wide reputation for mixing concrete, as well as mixing other materials such as fertilizers, chemicals and baking powder.

Smith entered into an arrangement with Doelger and Kristen, Milwaukee machinists, to build his machines. The machines were manufactured on credit until Smith's customers made payment. In this way, Smith was able to build up capital until he could organize his own manufacturing facility. He eventually established the Smith Machine Company, along with two cousins as partners. They initially operated in a small shed, but soon rented space at the Grant Marble Company, 27th Street and Canal Street in the Menomonee Valley. Sales were made under the name, "The T.L. Smith Company," with offices in the Majestic Building, 231 West Wisconsin Avenue.

Growth was rapid and, in 1905, Smith organized the T.L. Smith Company to handle sales and finances, although manufacturing was carried out by the Smith Machine Company. Eventually, however, Smith Machine Company and the Wisconsin Foundry Company, which produced the casting for the mixers, were folded into T.L. Smith Company. In 1905, Smith also acquired controlling interest in the Sterling Wheelbarrow Company.

Smith used a double conical drum with a patented "end-to-center" mixing mechanism that was claimed to be faster and provide better mixing results than other mixers. The company sold mixers in a range of sizes from the very small concrete construction mixers used for home construction, to much larger mixers used for large construction projects such as hydroelectric dams.

Smith's innovations also included machines for mixing and heating asphalt, for road paving and various rock and ore crushing machinery.

In 1906, Thomas Smith and Paul W. Post incorporated the Smith & Post Company to manufacture and sell rock and ore crushing equipment. The newly formed company acquired the rights to manufacture machines under the Symons patents and built a factory in Milwaukee to manufacture gyratory crushers. Post sold his interest in the concern to Smith in 1910, and the company was renamed the Smith Engineering Works of Milwaukee. The company was located at 532 East Capitol Drive.

In about 1916, the T.L. Smith Company purchased land for a new factory for its concrete mixers, in order to increase capacity to meet the demand. The new plant was built at what is now 2835 North 32nd Street on Milwaukee's northwest side. The story of the T.L. Smith Company continues in Chapter 7.

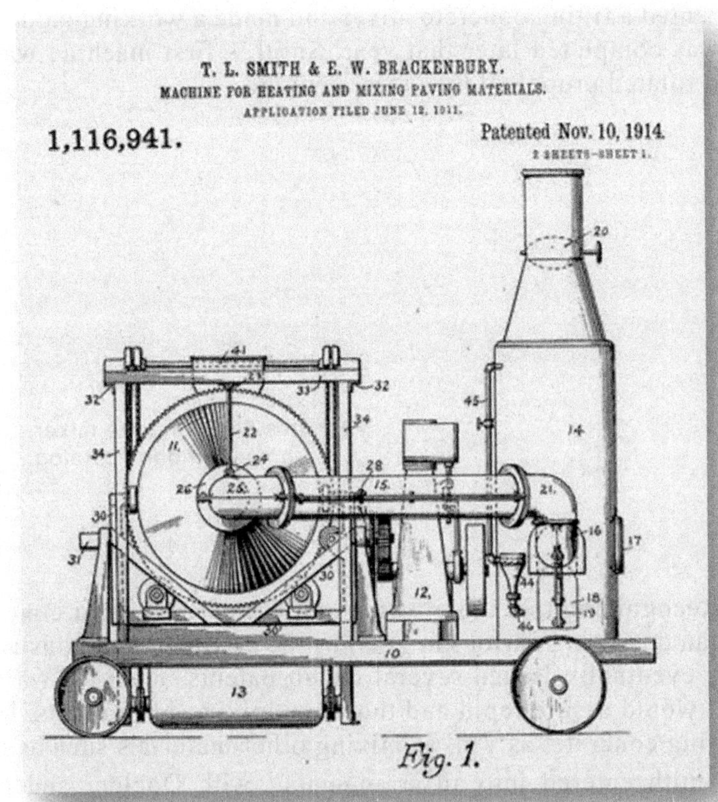

Patent drawing from a machine developed by Thomas L. Smith and E.W. Brackenbury for heating and mixing asphalt road materials

THOMAS L. SMITH was born in Bolton, England on June 6, 1855. He was only four years old when brought by his parents to Milwaukee, where his father secured a position in the car repair shop of the Milwaukee Road. His early education was acquired in the parochial school of St. James Episcopal Church.[105]

The family moved to Watertown, Wisconsin, when Thomas was young, where his father operated a machine shop and foundry. Thomas learned the machinist's trade working in his father's shop, before entering Iowa State College at Ames in 1873 to take a collegiate course in mechanical engineering. Smith graduated in 1877 with high scholastic records. He was immediately appointed as an instructor in mathematics and bookkeeping—a position he retained for several years. He then went to Boston and completed his engineering education at the Massachusetts Institute of Technology.

After leaving college, Smith moved to Milwaukee, where he was employed for several years in the engineering department of the Whitehill Sewing Machine Company of Milwaukee. He was also identified at various times with D.J. Murray Manufacturing Company of Wausau, Wisconsin, Filer & Stowell Manufacturing Company, Kempsmith Manufacturing Company, C.J. Smith & Sons, and Pawling & Harnischfeger.[106]

In the 1880s. Smith established his own machine shop at Reedsburg, Wisconsin. From a financial standpoint this enterprise was not successful but during this period he invented and built the first flexible arm wood carving machine. The sale of this invention provided Smith with resources to establish several important industries.

In 1898, Smith invented the tilting cement mixer, which he patented and sold under the name T.L. Smith. He established the Smith Machine Company to manufacture his line of mixers. He also developed and patented rock and ore crushing machinery. In 1906, along with Paul W. Post, he organized the Smith & Post Company, which eventually became known as the Smith Engineering Works of Milwaukee. The company built a factory and took over the manufacture and sale of gyratory crushers under the Symons patents.

Smith married Anna A Shillcox of Watertown, Wisconsin, soon after completing his education. They had three sons and one daughter. He was a member of St. James Episcopal Church, and had membership in the Masonic Lodge and the City Club. Smith was director of the Milwaukee School of Trades. In 1920, Iowa State College conferred its honorary degree of Doctor of Engineering upon him.

Thomas L. Smith died on April 29, 1921, leaving his three sons to carry forward the enterprises that he founded.

THE CHAIN BELT COMPANY

Christopher Warren LeValley moved to Milwaukee in 1884, to establish his own establishment. He had previously worked as general manager of the St. Paul Harvester Company, in St. Paul Minnesota, where he had patented a harvester, as well as an improvement for linked chain drives. It appears that his interest in developing chain drives for various applications caused LeValley to relocate to Milwaukee, where he initially outsourced production of his linked chains to Pawling and Harnischfeger and rented space in their Walker's Point factory.

In 1892, LeValley incorporated his enterprise as the Chain Belt Company. In addition to producing chain belt, the company also manufactured various products made up of malleable iron, including iron buckets, bolt clippers, sprocket wheels, and elevating and conveying machinery.

LeValley continued to find new uses for his chain belts and developed new designs for these applications, many of which were patented by him.

His new company quickly applied chain belts to drive conveying equipment, in addition to machinery. In the later part of the decade, the company manufactured chain-driven material handling conveyors and bucket elevators and installed such equipment for the Milwaukee breweries.

By 1896, the Chain Belt Company employed over two hundred people at its 145,000 square foot factory. It was reported to have, "the most modernly improved machinery for making the specialties" that were manufactured there.[107]

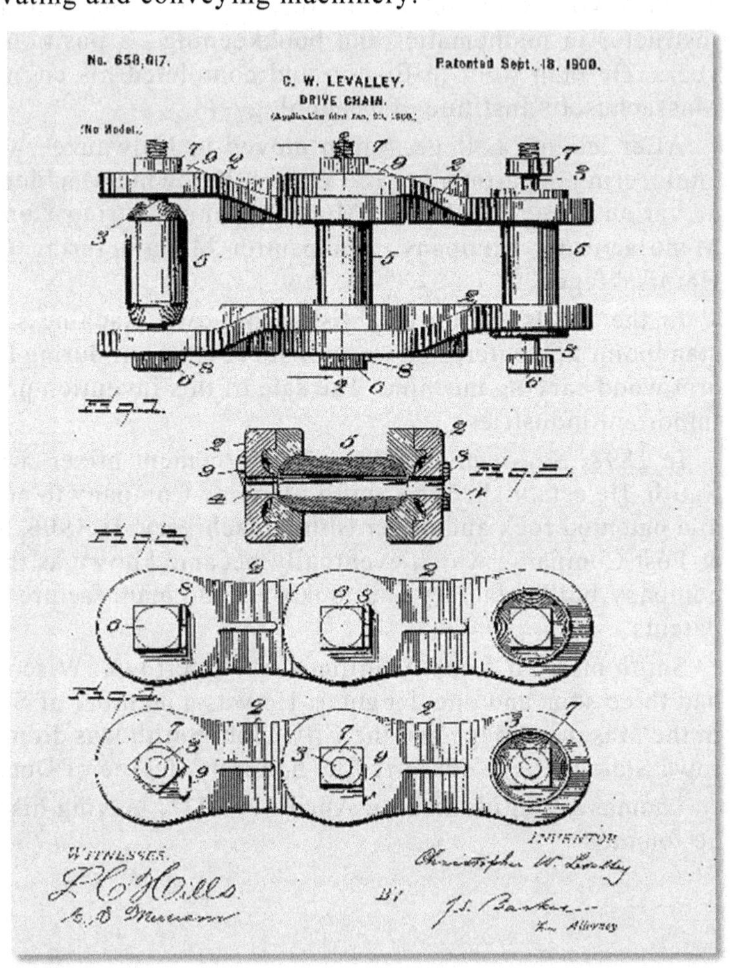

Patent drawing by Christopher W. LeValley for an early chain-belt drive. LeValley had over 100 patents to his name for various mechanical devices.

In 1907, LeValley conceived of the idea of driving concrete mixers with one of his steel chains around a cast steel drum. The idea of powering a concrete mixer wasn't new, having been developed in Milwaukee by Thomas L. Smith in 1899. However, LeValley's use of a steel chain drive for the mixing drum was reliable. His chain-belt mixer was named the *Rex Mixer* in 1914. It was so popular that the company's name was later changed to *Rex Chain-Belt.*

Rex Chain-Belt later put a concrete mixer on a truck bed, allowing the aggregate and concrete to be mixed in transit to the job-site. This had the added advantage of eliminating the need to separately deliver aggregate and cement to the work location. As a result, this process was quickly and widely adopted.

In 1910, the company began building traveling water screens for power plants and other applications. This led Rex Chain-Belt to enter into the manufacture of equipment for water and sewerage treatment plants. Two years later the company produced its first road paver.

This interesting photograph shows a worker assembling heavy-duty linked chain. It was taken in Chain Belt's "Plant 7" in 1948. Photograph courtesy of Rexnord Corporation.

LeValley built his own Chain-Belt factory in the Menomonee Valley. As shown here, it was adjacent to the 16th Street Viaduct. Picture from the Historic Photo Collection of the Milwaukee Public Library.

In this photograph, women are shown working the sand casting department of the Chain-Belt factory. Apparently their work was overseen by male supervisors. Picture from the Historic Photo Collection of the Milwaukee Public Library.

Rex Chain-Belt Concrete Mixer

In the first decade of the 1900s, Christopher LeValley designed a concrete mixer using a cast iron drum that was rotated using a chain drive. The mechanism was powered by a boiler and steam engine. LeValley dubbed his mixer the REX, to point out that his mixer was the "King" of concrete mixers. The new mixer was popular and the name stuck—since then REX has always been part of the corporate name.

In 1912, the company also began offering road pavers and other construction machinery. A REX paver was used to pave the first concrete road in Milwaukee County—a ten-foot strip of concrete to replace the Janesville Plank Road. In large part because of the new product offerings, sales increased over $1 million in 1913.

This photograph is of an early 'Rex' concrete mixer. The boiler and its stack, which powered a small steam engine to turn the mixer, are shown on the left. The mechanism was all mounted on a horse-drawn wagon. Picture from the Historic Photo Collection of the Milwaukee Public Library.

Chain Belt Traveling Water Screens

As noted earlier, Chain Belt began manufacturing water screens for industrial applications in 1910. The application was apparently a logical one for the company since the continuously looped screens were well adapted to the company's chain drives where ruggedness and durability were required. The water screens were initially designed to remove anything greater than 3/8^{ths} of an inch in size from the water stream. They were used to protect pumps, condensers and other equipment located downstream. They became known as "traveling water screens."

The first known Chain Belt traveling water screen was installed at the downtown Commerce Street Power Plant of the Milwaukee Electric Railway and Light Company. Installed in 1911, it was designed to protect the water pumps used to cool the condensers for the steam engines. A second traveling screen was added in 1923, presumably when the plant's steam engines were replaced with steam turbines. These traveling water screens remained in operation well into the 1970s, reflective of their robust design.

Over the years, Chain Belt engineers filed for numerous patents for improvements to the basic traveling water screen design. The individuals involved in these innovations include Benjamin S. Reynolds, Gustav R. Roddy, Reginald J. Hickman, George B. Welser, Jr., Francis P. Gary and Lloyd G. Bleyer.

This product line and its related innovations eventually caused the company to enter into the manufacture of equipment for water and sewerage treatment plants and resulted in the beginning of its process equipment division.

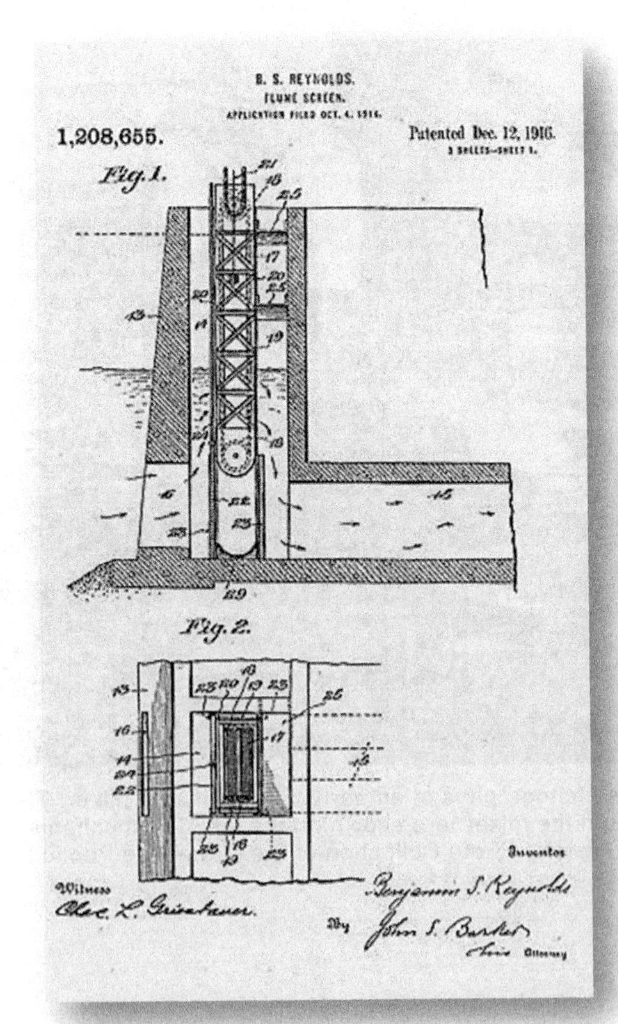

This is a drawing from the company's earliest known patent for innovations in traveling water screen design, showing the basic configuration of such screens.

Non-Metallic Conveyor Chains

In the 1950s, Rex Chainbelt developed the concept of non-metallic (thermoplastic) conveyor chains. The company was granted its first patent on a thermoplastic chain on November 3, 1959, for a "Plastic Flat-Top Conveyor Chain," designed explicitly for use in beverage and food processing plants. The non-metallic chain was used on conveyor lines for transporting processed materials to their various stations for sterilization, filling, closing, sealing, and final packaging.

A major feature of this invention was "to provide a flat-top conveyor chain that can run satisfactorily without lubrication," to prevent contamination of such articles by grease or other lubricants and by foreign materials absorbed by the lubricants. With the need for absolute cleanliness in the processing of foods, beverages, pharmaceuticals, and other such products, the development of a non-metallic conveyor chain offered obvious advantages.

Rex Chainbelt was the first to develop a line of non-metallic conveyor chains. The product was designed for use in beverage and food processing industries.

Other significant benefits of these non-metallic conveyor chains were the reduction of the noise levels in plants with high-speed conveyors and the reduction of damage to transported containers of almost any material.

Due to the differences in the mass/strength ratios of plastic chains versus those of metallic chains, the development was not just a simple matter of substituting plastic materials in existing metallic chains. Each new plastic chain application has its own peculiar design parameters, and development and testing programs were conducted prior to their introduction.

The use of non-metallic conveyor chains has continued steadily without public fanfare. Today, such conveyor chains are produced by the company in many varieties and sizes for use in environmentally sensitive industries and pollution abatement installations.

Christopher LeValley. Photograph courtesy of Rexnord Corporation

CHRISTOPHER WARREN LEVALLEY was born in Manchester, Connecticut in 1833. When age fourteen he moved to Hartford and served an apprenticeship in a machine shop. He served in the US Army during the Civil War, following which he took a position in St. Paul, Minnesota as superintendent of the St. Paul Harvester Company. He later became its general manager. While there, he recognized the desirability of a positive drive for machinery used for harvesting crops.

In 1884, LeValley came to Milwaukee and established his own enterprise in late 1891, incorporating it in February 1892 as the Chain Belt Company. In addition to supplying chain belt drives, the company also manufactured malleable iron buckets, elevator arms, bolt clippers and other goods made of malleable iron. In 1907, LeValley conceived the idea of driving a concrete mixer with a steel chain and using a cast semi-steel drum. These ideas were incorporated in what was known then as the Chin Belt mixer—which was later renamed the *Rex* mixer.

During his career, LeValley was granted one-hundred and eighteen patents for agricultural implements, as well as different types of chain-driven mechanisms. At least five of his patent models are maintained by the Smithsonian Museum.

LeValley was president and general manager of Chain Belt Company from 1891 until 1916, at which time he was named chairman of the board. He served in that capacity until his death in 1918 at the age 82.

LeValley was an accomplished musician and composer. He obtained copyrights for several songs, including *Dream Little One* in 1915 and a fifth verse to the national hymn *America* in 1917.

He made many gifts to charitable institutions, including $100,000 to the Milwaukee Foundation in 1916 (equivalent to $2.3 million in 2016). LeValley was inducted into Wisconsin's Industrial Hall of Fame in 1966.

KOEHRING MACHINE CO.

Koehring Machine Company was founded in Milwaukee in 1907 by Philip Adolph Koehring along with his brothers William and Richard. The company brought its first product to market in that year—a steam-driven portable concrete mixer.

It is reported that Philip Koehring, who lived in Kiel, Wisconsin at the time, saw one of T.L. Smith's concrete mixers on a job site in Milwaukee. On his return home, he built a tin model of his own design. He and his brothers organized the Koehring Machine Company in Milwaukee to manufacture their version of the concrete mixer.

Koehring was born on a farm near Kiel. He worked for a time in the Kiel furniture factory, before coming to Milwaukee, where he worked in a small shop under the 16th Street viaduct in Milwaukee's Menomonee Valley. He stated, "Every cent he could scrape together went into" his paving mixer.[108] A prototype was taken for a test on Milwaukee streets, but made so much noise that the police ordered the trial stopped. Koehring continued to refine his machine, producing a suitable machine for the market. Eventually, Koehring machines were sold worldwide.

In 1921 Koehring developed a line of cranes that later evolved into crawler cranes for rough terrain.

Koehring added two more advancements to improve the mechanization of concrete paving: self-propelling crawler tracks and internal combustion engines in place of steam engines. With over fifteen mixing and paving patents to his name, combined with his manufacturing expertise, Koehring helped to transform the world of concrete mixing and paving.

KOEHRING PORTABLE CONCRETE MIXER

Philip Koehring and G. H. Miller are credited with developing the first dry batch paver. His innovation streamlined the flow of materials for roadwork. As the paving mixer was pulled forward, it mixed concrete and then distributed the final mix at the rear of the machine.

The innovation led to the development of the ready mixed industry for the paving of concrete roads. Koehring steam-powered concrete "pavers," mixed concrete on-site and moved with the other paving machines as the work progressed. The machine gained wide acceptance as the preferred method of producing concrete for road pavement.

Milwaukee's T.L. Smith, Koehring Machine Company, Rex Chain Belt Company, Kwik-Mix of Port Washington, Gilson Manufacturing Company of Fredonia and the Leach Company of Oshkosh dominated the US production of concrete mixers and related equipment. In the 1940s, the combined companies satisfied seventy percent of the US concrete mixing market.[109]

Wisconsin-made mixers were employed for the construction of the New York subway system and for most of the dams built in the United States in the 1930s. They were also used extensively in other countries, to repave roads in Europe following the First World War and to rebuild the infrastructure of Tokyo following the earthquake of 1923.[110]

By 1928 Koehring had outgrown its small shops in the Menomonee Valley and moved to a new facility at North 30th Street and Concordia—on what is now known as Milwaukee's 30th Street Industrial Corridor. Koehring's shops were just south of A.O. Smith's factory and close to the T.L. Smith Company factory.

This Koehring mixer is shown being used at a job site. Photograph from the Historic Photo Collection of the Milwaukee Public Library.

AMERICAN RHEOSTAT/CUTLER-HAMMER

It has been noted previously in this book that the availability of reasonably priced steam engines transformed industry, which previously was dependent upon hydropower to run machinery using line-shafts and belts. A similar transformation of industry occurred when electric power became readily available. Equipment could then be located to take advantage of the material flow of the parts being manufactured. Importantly, it also allowed machinery to be remotely controlled from a central location.

The adoption of electric motors for industry and other applications created an immediate need for devices to control those motors. Frank Rogers Bacon of Milwaukee decided to meet that need.

Bacon was a student at Princeton in 1892 when his mother's death caused him to return to Milwaukee to work in his father's grain business, the E.P. Bacon Co. Frank observed the need for electrical equipment and, in 1896, he and an inventor named Lucius Gibbs formed a company named the American Rheostat Company. Their initial product was an electric motor starter that had been recently patented by Gibbs.

The switch that controls an electric motor is generally known as a motor controller or motor starter. While a simple switch is adequate for very small motors, larger more powerful electric motors require devices that are able to handle the required voltage and currents. They are often provided for remote operation, overload protection, speed control, and other functions. Importantly they typically are designed to open upon loss of power, so that the machinery they are connected to will not automatically restart when power is again available, which would be a safety concern.

While there was an immediate demand for American Rheostat's new motor starter, their initial design had a number of glitches that needed to be resolved. Gibbs quickly lost interest and sold his interest in the concern to Bacon. Frank Bacon was able to find another partner, Frederick L. Pierce, who provided needed financial backing to get the new product established.

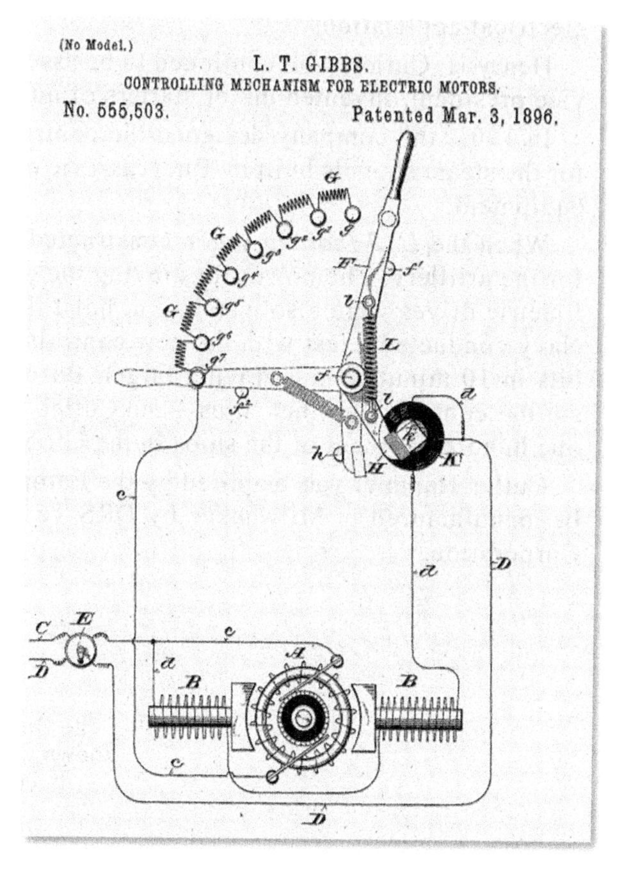

Right: Patent drawing, filed by Lucius T. Gibbs, for a "controlling mechanism for electric motors"—known today as an electric motor starter.

When the improved motor starter entered the market, Bacon discovered another problem. American Rheostat's starting box was similar to one manufactured by Cutler-Hammer Manufacturing Company in Chicago. Cutler-Hammer had been formed in 1893, specializing in electric starters, speed regulators and field rheostats, and had introduced an electric motor starter box designed and patented by Harry H. Blades of Detroit. Cutler-Hammer's new product achieved almost instant success.

Soon, Frank Bacon filed a lawsuit against Cutler-Hammer, claiming patent infringement.[bb] Cutler-Hammer countersued. As messy and lengthy legal battle faced both companies, in mid-1897, Bacon approached Harry H. Cutler with a deal to end the lawsuit. They reached an arrangement under which American Rheostat bought out Cutler-Hammer, its plant and all rights to its patents—but retained the Cutler-Hammer name. Bacon became president of the newly merged company, and Cutler was named plant manager and chief engineer. Operations were initially established in Chicago, but moved to Milwaukee in 1899 where facilities for expansion were considered better. The original Milwaukee plant was located in the first two floors of the Badger State Shoe Company at the corner of 12th Street and St. Paul Avenue, just north of the Menomonee River. This location was used by the company until 1975.[111,cc]

The newly merged company quickly went to work at developing other electrical control devices, including crane controls, motor controllers for metal plants and for continuous rolling mills, and motor controls for use on ships. They also developed switches for more mundane applications such as for lamp sockets (for light bulbs),[dd] as well as circuit breakers for some of the most demanding medium-voltage electrical applications.

Henry H. Cutler, who continued to be associated with Cutler-Hammer until 1915 as chief engineer and vice president, invented motor starters of numerous configurations, resulting in dozens of patents.

In 1901, the company designed the control equipment for the Panama Canal—providing the controls for the steam shovels built by Bucyrus-Erie of South Milwaukee, and designing the original Canal control equipment.

When the *USS Indiana* was reconstructed in 1904, Cutler-Hammer was invited to design the controls for the artillery. The power for moving the gun turrets was changed from steam to electric motor drives. Electric drives were also installed to hoist the ammunition and fire the guns. Following renovations, the Navy conducted a test with the new controls. During the first day of trials, the USS Indiana recorded 10 hits in 10 minutes on a moving target. Based on the results of the test, the Navy decided to implement similar controls for other ships—and Cutler-Hammer marine controls were favored. Since the 1904 trials, one hundred percent of the ships in the US Navy fleet have used Cutler-Hammer equipment.

Cutler-Hammer was acquired by the Eaton Corporation in 1978. Controls for the US Navy continue to be manufactured in Milwaukee by DRS Technologies, which acquired this line of business from Eaton Corporation.

The Cutler-Hammer label is known worldwide because of its line of motor controls.

---

bb It would be interesting to know more about Frank Bacon's claim, since Harry Blades had acquired a patent for his electric motor switch prior to Lucius Gibbs. In fact, Blades is generally credited with the adaptation of the "no voltage release electro-magnet" with starting rheostats for electric motors in 1895, which was adopted universally. Ref: *The Electro-Magnet is a Veritable Aladdin's Lamp,* by George J. Kirchgasser, Popular Electricity and the World's Advance, Volume 5, May, 1912, p. 455.

cc Note: a substantial portion of the plant had to be relocated in the 1950s because of freeway construction

dd Charles J. Klein of Cutler-Hammer patented more than a dozen electric switch configurations for lamp sockets in the early 1900s.

The Automatic Motor Starter

After restructuring in 1898, one of Cutler-Hammer's first products was an automatic motor starter patented by Harry Cutler in 1900. It allowed electric motors to be operated remotely, using a feedback system that indirectly sensed the motor speed, which was used to regulate a series of solenoid switches, progressively removing the starter resistances from the circuit.

The automatic motor starter opened the door to controlling electric motors remotely, laying the foundation for the modern motor control industry. Hearing of Cutler-Hammer's innovation, other companies adopted their own versions and automatic motor starters throughout the industry—largely replacing manual motor starters.[112]

Shown here are controls for the first electrically driven steel mill constructed in the United States in 1903. It incorporated Cutler-Hammer controls.

THE HAMILTON METALPLANE COMPANY

At the end of the First World War, Thomas F. Hamilton purchased the aviation department of the Matthews Brothers Woodworking Company of Milwaukee. Matthews Brothers manufactured wooden propellers and pontoons, in addition to an extensive line of architectural wood products.

Hamilton had previously been in charge of the propeller department. He established his own company, Hamilton Aero Manufacturing Company of Milwaukee and continued the manufacture of wooden propellers. In 1926, he hired two aeronautical engineers, James S. McConnell and James Cowling Jr., to design an all-metal airplane.

Their first aircraft, labeled the Model H-18, was a monoplane with an aluminum skin that consisted of evenly spaced V-sections, which were crimped into the flat stock to provide rigidity. The first flight of this new aircraft occurred on April 2, 1927.

The plane was appropriately christened *Maiden Milwaukee* by Hamilton's seven-year-old daughter. It was one of the first all-metal airplanes in the United States.[113]

In spite of early successes of this plane in various air races and reliability flights, the company had trouble attracting orders. This was attributed to the lack of side windows and a limited cabin size. Hamilton pressed on with the design of a new aircraft, to be designated Model H-21. In order to help launch this new craft, the airplane business was separated from the propeller operations and the Hamilton Metalplane Company was founded.

A building was acquired on Park Street (now West Bruce Street) in the Menomonee Valley, and Dr. John Akerman was hired as chief engineer. Akerman had formerly worked on the Ford Trimotor, as had McDonnell. The H-21 was powered by a 425 horsepower Pratt & Whitney Wasp engine. It had a metal Hamilton adjustable propeller. The first plane was completed in April of 1928.

Eventual production versions were dubbed the H-45 and H-47. The H-47 had a more powerful engine—a P&W Hornet which developed 525 horsepower. The planes were designed for airline passenger service, carrying a pilot and seven passengers.

The first production versions were delivered to Universal Airlines, Northwest Airways, and Wren Alaska Airways. The pilots who flew these aircraft regarded them as rugged, reliable airplanes of great endurance.

In 1929, Hamilton Aero and Hamilton Metalplane were both acquired by the United Aircraft & Transport Corporation, which was formed by the merger of Pratt & Whitney, Sikorsky Manufacturing Company, Chance Vought Aviation Corporation, and the Boeing Airplane Company of Seattle.

The company became known as the Hamilton Metalplane Division of the Boeing Airplane Company, a division of United Aircraft & Transport Corporation. In 1930, the Hamilton Aero Mfg. was consolidated with the Standard Steel Propeller Corporation of Pittsburgh, becoming Hamilton Standard, and its Milwaukee plant was closed.

A similar fate awaited Hamilton Metalplane; on October 11, 1930, the Park Street plant was closed with operations dispersed to Boeing plants in Seattle and Wichita.[114]

A Milwaukee-built Hamilton Metalplane is seen adjacent to a freight train near St. Paul, Minnesota. The aircraft, owned by Northwest Orient Airlines, was likely an H-47—a later model of the aircraft with a more powerful engine and a high fixed wing that increased passenger space. It appears that freight was being transferred from the Rock Island train, or the Railway Express Agency truck, to the aircraft. Photograph used with permission of the Minnesota Historical Society.

THOMAS FOSTER HAMILTON was born on July 28, 1894, in Seattle, Washington. At the age of 14, Hamilton took a part time job repairing hot-air balloons during the Alaskan-Yukon Exposition, held in Seattle. This job helped fuel his interest in aviation. Hamilton, along with friend Paul J. Palmer, began building experimental biplane gliders and flying their gliders around their neighborhood. Soon they began building propeller-driven aircraft. Hamilton built 10 to 25 early airplanes, using a combination of existing designs, modified with his own concepts.

In this 1929 photograph, Thomas Hamilton is shown outside the aircraft. Charles Lindbergh is in the front and Fred E. Weick sits in the rear of the cockpit. Photograph: Wiki Commons.

The US military was interested in his designs and during the First World War requested that he relocate to Milwaukee to take responsibility for the aviation division of the Matthews Brothers Furniture Company. The company had just obtained a contract to produce wood propellers for the Army and Navy and needed an experienced individual to oversee manufacturing.

Once the war ended, Hamilton brought Matthews Brothers' entire inventory of aviation parts and established his own company, which he named the Hamilton Aero Manufacturing Company. About this time, Hamilton met and married Ethel Inez Huges of Milwaukee.

Largely because of Hamilton's company, Milwaukee became one of the nation's major aviation hubs during the 1920s. In 1927, the company designed an all-metal airplane that he dubbed the Hamilton Metalplane H-18 and christened it *Maiden Milwaukee.* The principal designer of the aircraft was James McDonnel, who had previously worked for Stout and Ford—and incorporated similar features into the H-18, along with some new ideas. The aircraft used a tubular frame with corrugated skin and a thick single wing projecting out of the fuselage underneath the open cockpit. It was powered by a 200-horsepower J-4 Wright Radial engine and used a Hamilton metal propeller.[115]

Maiden Milwaukee received the first United States air certificate for an all-metal airplane. It was designed to carry mail, with a few passengers as an extra revenue bonus. Subsequent models of the Hamilton Metalplane were introduced called models H-45 and H-47, which were better able to accommodate passengers. Northwest Airlines purchasing a number of Hamilton airplanes to be used in their first passenger service, and Ralph Sexton bought several planes to be used for his Isthmian Airways. A few aircraft went to Alaska and Canada for use in the Arctic.

Hamilton and his wife lived in Milwaukee for approximately ten years. He was a "dollar-a-year-man" during the Second World War, running Hardman aircraft in Southern California. He was a technical assistant to the 1966 movie, "Those Magnificent Men in their Flying Machines."

Thomas Hamilton died on August 12, 1969.

CHAPTER 4: CITY OF MILWAUKEE

The City of Milwaukee grew from a combination of three separate communities. In 1818, Solomon Juneau founded the town called Juneau's Side, or *Juneautown*. Byron Kilbourn established *Kilbourntown* west of the Milwaukee River. George H. Walker claimed land to the south of the Milwaukee River, where he built a log house in 1834—this area became known as *Walker's Point*.

By the 1840s, the three towns had each grown significantly. Some intense rivalries occurred between Juneautown and Kilbourntown—much of which was caused by Kilbourn's efforts to remain independent and dominate over Juneautown. When laying out his town, Kilbourn made sure the streets running toward the river did not align with those on the east side. He distributed maps of the area that only showed Kilbourntown—implying that Juneautown did not exist. He also resisted efforts to build bridges across the Milwaukee River. The only way to travel between the two towns was to take a ferry from one to the other.[116]

In 1840, the Wisconsin Territorial Legislature determined the ferry system on the Milwaukee River was inadequate, and ordered the construction of a drawbridge. Kilbourn viewed the bridge as a threat. Furthermore, the two towns couldn't agree how to fund the cost of the bridge. Ultimately, the tension came to a head in May 1845 when Kilbourn and his supporters destroyed the west end of the bridge. While a number of east side residents gathered at the river to protest the action, violence was averted for a time. However, two weeks later a group of east siders destroyed two smaller bridges in an attempt to cut the west side off from the south and the east. A skirmish broke out in what became known as the Bridge War of 1845. Several people were seriously injured.[117]

In the aftermath of the skirmish, nearly everyone, including Byron Kilbourn, agreed that the two communities needed to work together. A committee was appointed to draft a charter. On January 31, 1846, the charter was approved and Juneautown, Kilbourntown and Walker's Point merged into a single city. Milwaukee's population at the time was about ten thousand.

At the time, the City of Milwaukee was much smaller than it is currently. It was surrounded by the Town of Milwaukee to the north, the Town of Lake to the south, Wauwatosa to the west, and the towns of Granville and Greenfield to the northwest and southwest, respectively—as shown in the map on the next page. These towns limited Milwaukee's expansion for a time; there were lengthy legal battles whenever the City of Milwaukee attempted to expand its borders.

This chapter addresses the manufacturing companies that were located in the areas of what was the City of Milwaukee in the late 1800s and early 1900s, largely as defined by the original City map. The City was limited on the west to aptly named Western Street (now known as 35th Street), North Avenue on the north, and Greenfield Avenue on the south.

This 1876 map of the Milwaukee area shows the original boundary of the City of Milwaukee, surrounded by the Towns of Milwaukee, Granville, Wauwatosa, Greenfield and Lake.

THE MILWAUKEE ELECTRIC RAILWAY AND LIGHT COMPANY

Electrification had a huge impact on industrial Milwaukee—as well as on its populace. Electricity's introduction locally mirrored its introduction nationally.

The first practical use of electricity to light US cities was developed by Charles Francis Brush, who introduced an arc lighting system in the late 1870s. Suitable mainly for exterior lighting and large public buildings, it gained limited acceptance, as well as some resistance from gas companies that wanted to preserve their franchises. Brush electric systems were installed in Fond du Lac and Janesville, and competing companies began operations in several other Wisconsin cities. But the arc lighting boom ended soon, because by 1880 Edison began introducing his electric lighting system.

The story of Thomas Edison's development of the incandescent light is well known, involving hundreds of experiments to find the right filament material to create illumination in a vacuum bulb. What isn't as well-known is his contributions to electric distribution systems.

Edison recognized that the arc lighting market was largely limited to large open-space lighting applications. He concentrated his efforts at the much larger market for lighting small spaces—homes, shops and offices. At the time, more than ninety percent of gas company revenues came from such lighting services.[118] In addition to developing an incandescent bulb that would displace natural gas lighting, Edison also addressed another major shortcoming of Brush's arc lighting system. Brush's arc lights were installed in series and a ten light circuit ran at about 500 volts—the more lights in a circuit the higher the voltage requirement. Edison's system used multiple parallel circuits, all running at a constant 100 volts. In order to distribute electricity to more than one customer using a central station, Edison invented the feeder main in which multiple local circuits were connected in series to a higher voltage transmission circuit.[119, ee]

Edison established several companies to manufacture and market his products. Until his central station and electric distribution products were tested and available, he limited sales to his Edison Company for Isolated Lighting. This company provided Edison electric systems for factories, commercial and residential buildings. All sales between 1880 and 1882, and most until 1884, were 110-volt systems and were sold by his "Isolated Lighting" company.

One other significant engineering innovation was needed for the design and construction of distribution systems—the ability to economically transmit electricity at a distance. At the relatively low voltages used by Edison, the resistance to current flow is high and it is impractical to transmit electricity over distances. George Westinghouse and William Stanley of the United States found a solution based on the work of Frenchman Lucien Gaulard and his British business partner, John D. Gibbs.

In 1881, Gaulard and Gibbs received English patents for the use of transformers on an alternating current system. Gaulard and Gibbs were not looking for an economic means of transmitting electricity, but rather a workaround for a British law that regulated the incandescent lamps that could be used.[120] George Westinghouse read about their patent, and the controversy it caused, and recognized the innovation could be applied to transmit electricity economically. He secured the United States patent rights for the technology. William Stanley perfected the transformer design needed for the application and by March 1886, they had a system operating in Massachusetts. Electric system competitor Thomson-Houston soon developed its own alternating-current system and introduced it to the market.

Thomas Edison was vehemently opposed to the use of alternating current and launched a public campaign to convince the public that such systems were dangerous. However, Westinghouse obtained the rights to provide the lighting for the Chicago World's Fair in 1893. The Westinghouse display featured a great quantity of incandescent lighting, a full line of power equipment and a street railway system. The

---

[ee] British law at the time essentially necessitated that the voltage be either 48 or 91 volts—the voltage approved for incandescent bulbs designed in England at the time. Gaulard and Gibbs proposed a way to get around this by transmitting at a higher voltage, but decreasing the voltage locally to the required lamp voltage.

displays went a long way toward convincing the public that alternating current was the future of electric power.

The market for electric lighting was a difficult one, however. Most of the demand for lighting was limited to a few hours nightly, while the capacity required to meet the demand was available full-time. Fortunately an innovation occurred in the late 1880s that significantly increased daytime loads—the development of electric-powered transportation. Soon streetcars, trolleys, electric rail systems, subways, and elevated rails were being built in most large cities in the United States. Two Americans pioneered successful electric rail systems—Charles J. Van Depoele and Frank J. Sprague. By the end of 1889, there were 154 electric street rail systems operating across the United States.

Milwaukee Electrifies

Milwaukee entered the Edison era in 1882, when Harry Sutter, a photographer, installed a small system using Edison-based electrical equipment at his shop. During the following two years, customer-owned Edison-system plants were installed at several area factories and facilities including the shops of the Milwaukee & St. Paul Railroad, the Plankinton House, Best Brewing Company and the Kerns Flour Mill.[121]

But these were all isolated electric systems. Central station power plants didn't develop immediately, due in part to the reluctance of Milwaukee's Common Council to grant franchises, even though there were a number of interested companies vying for the rights to serve the City.

This all changed when Sheridan S. Badger was able to secure a franchise to supply street lighting to the City for three years. Badger had a small electric plant serving the Industrial Exposition Building. His Badger Illuminating Company expanded its system aggressively during that time. Within months, the company relocated its central station into a former flourmill on the Milwaukee River. The company installed over 400 streetlights by the end of 1890.[122] These lights were arc lights installed on towers, so they provided widespread illumination to the downtown area. Badger Illuminating sought to expand into incandescent lighting, but the Milwaukee Common Council did not grant him a franchise.

Other developers were also clamoring for the opportunity to provide electric service to the City. The debate at City Hall continued, swaying from a fraction that wanted municipal ownership, to another that favored free competition. Eventually, Milwaukee decided to grant franchises to anyone who applied. As many as six companies were serving downtown Milwaukee in the early 1890s—with overlapping service areas. Perhaps predictably, the situation rapidly deteriorated. Multiple electric companies often served customers in the same city block. The electric lines were a tangled mess, resulting in poor service and occasional sabotage.[123]

Henry Villard waded into this chaotic environment, with the intention of acquiring all of the competing companies and attaining a monopoly. Villard was a friend and backer of Thomas Edison and was the president of the Edison General Electric Company, the predecessor of the General Electric Company. He organized the Edison Illuminating Company of Milwaukee, obtaining funding from his North American Company. Not only did he acquire the largest competing electric companies, he also purchased two of the most important horse-drawn streetcar systems in the city. However, as quickly as he acquired companies, new central stations and street railways were organized. The price of acquiring all competitors became excessive. While Villard eventually achieved almost full control of the electric and street rail companies in Milwaukee, his enterprise fell into receivership. However, North American, as the sole bondholder, reorganized the company into The Milwaukee Electric Railway and Light Company, which was generally known by its abbreviation—TMER&LCo.

Villard selected John I. Beggs to run the Milwaukee company. Beggs had been previously tapped by J.P. Morgan as manager of the Edison Electric Illuminating Co. of New York. During his five-year tenure in New York, Beggs built the Pearl Street Station and the 26th Street Station. He worked closely with Thomas A. Edison and consequently became one of the small group known as Edison Pioneers.[124]

In addition to being president and general manager of TMER&LCo., for a time Beggs was also responsible for the North American Company's interests in Cincinnati, Ohio, and later St. Louis. He divided his time between these cities, although his principal residence was in Milwaukee between 1897 and 1911. In 1911, he built a summer residence on what became known as Beggs Isle on Lac La Belle, near Oconomowoc, and arranged to extend a trolley line to serve the area.

While Beggs was president of TMER&LCo., he built the Public Service Building in Milwaukee, which continues to serve as the headquarters of We Energies—the successor company to TMER&LCo, now part of the WEC Energy Group. Beggs also constructed the systems of interurban railways radiating from Milwaukee and oversaw the construction of the company's Oneida Street, Commerce Street, and Lakeside power plants.

Beggs was president of TMER&LCo until 1911, and then was reappointed to the position in 1920 and served until his death in 1925. At the time of his death, Beggs was an active director or officer of fifty-three companies. His funeral services were conducted in the Public Service Building auditorium by the Employees' Mutual Benefit Association.

Oneida Street Power Plant

The Oneida Street Power Plant was constructed in 1900 by TMER&LCo. The plant was built adjacent to and immediately south of the River Street plant[ff] of Edison Electric Illuminating Company, constructed in 1890. The two facilities were eventually merged into a single power plant.

The new plant had five boilers feeding steam engines that powered direct current generators for incandescent lighting, as well as for powering the bridges across the Milwaukee River. Steam from the plant also provided district heating to many Milwaukee buildings.

In 1912, TMER&LCo hired John Anderson from Union Electric Light and Power Company of St. Louis to serve as chief engineer. Anderson had an engineering degree as well as a marine background. As a result, he was quite familiar with the laborious task of stoking boilers with coal and removing the ash residue. He envisioned grinding coal into a fine powder and blowing it into boilers—burning the coal powder like fuel oil.

In 1914, Anderson received authorization to conduct experiments on the use of pulverized coal at the Oneida Street Power Plant. TMER&LCo was beginning to plan for a large new generating station and Anderson wanted to explore the feasibility of designing the new plant to use pulverized fuel. He, along with Fred Dornbrook, W.E. Schubert and Ray Mistele, took part in the research project. One of the five boilers at the Oneida Street plant was dedicated to the project. Numerous boiler and furnace variations were explored and carefully tested to find a furnace configuration that would handle the high firing temperatures and allow for near-complete combustion of the coal. Eventually, all five boilers were converted and the plant became the first central station in the United States to be equipped and successfully operated with pulverized fuel.

Tests were conducted in 1918 to determine whether burning coal in a pulverized form could conserve fuel and potentially reduce the cost of electric power. The tests were meticulously conducted, observed by independent members of the power community, and reported widely in numerous publications. The results documented the potential for increasing combustion efficiency through the efficient firing of pulverized coal. Largely as a result of these experiments, pulverized coal firing has almost completely replaced stoker firing in large central station boilers.

In June of 1948, thirty years after the development, Combustion magazine devoted an entire section to this momentous occasion. In it an article entitled "*Pioneer Work at Milwaukee*" written by Fred L. Dornbrook, chief engineer of power plants for Wisconsin Electric, discussed the significance of the tests:

[ff] River Street was eventually renamed Edison Street, taking the name from the Edison Electric Illuminating Company's power plant.

At the time of the First World War, in 1917, coal was burned in Milwaukee power plants on underfeed stokers. Efficiencies were those usual for that time, outages were frequent and accepted, and the quality of coal was becoming poorer, while its cost was increasing.

At the Oneida Street plant it was thought that better results could be obtained if the coal were burned in pulverized form. Thus came the birth of a trial installation at that plant in early 1918 on one boiler, and shortly thereafter on four more boilers. These boilers, in boiler room one, were equipped with furnaces and burners for firing pulverized coal. Pulverizing equipment was installed near the oil battery room on the third floor, and piping connected from there to burner boxes at the boilers.

By November 1919 it was possible to perform tests on the five boilers, and a test of 99 hours duration, or a total of 495 boiler hours, showed a gross efficiency of 80.67%."

Right. Pulverized coal burners and feeders in the East Wells Plant during the early 1920's. This was the first central station in the United States to be equipped and successfully operated with pulverized fuel. Designated a Mechanical Engineering Landmark in 1980.

Top. View of the boiler room, with the front face of the furnaces shown at floor level. (Note: the photograph has been intentionally flipped to provide the same orientation as the drawing at right.)

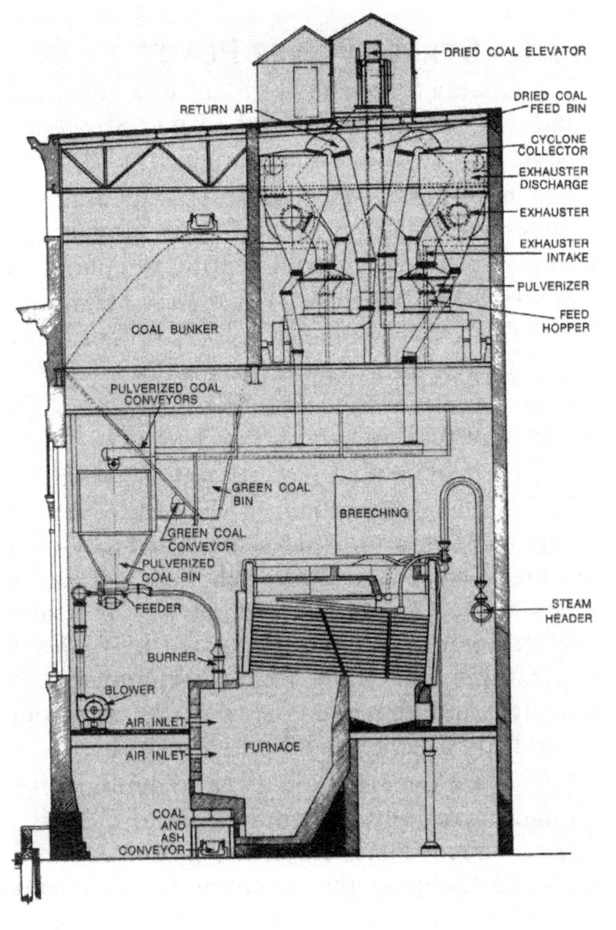

One of the problems of those early experiments was the disposal of coal ash. Deposited on the furnace bottom, ash became "sticky" and fused into a sheet covering the entire hearth. Removal was extremely difficult. The accumulation of slag, together with the erosion of sidewalls, caused many shutdowns. It became apparent that if the temperature of the ash deposit could be lowered below the plastic temperature it could be removed in a dry or powdered form. Many experiments were conducted, using different furnace configurations, to determine the best method of freeing the furnace of slag. This was finally accomplished in 1920 by the development of the so-called water-wall screen, consisting of a series of tubes through which boiler water flowed. The tubes both absorbed energy from the fire, and reduced the temperature of the furnace walls. This change in furnace design was adopted universally, to the point that modern boiler furnaces are completely surfaced with water tubes.

The results of these experiments at the Oneida Street Power Plant went largely unnoticed to society at large. However, within the power industry they resulted in dramatic changes in boiler design and power plant design. Prior to use of pulverized coal, boiler size was limited to the size of the underfed stokers—tiny compared to today's boilers. Upon the introduction of pulverized coal, boiler sizes increased dramatically. As boilers became enormous, by comparison, power plant construction costs were significantly reduced. At the same time, the increases in combustion efficiency resulted in the reduction in operating costs.

The reduction in the real cost of electricity during the period was a major driver of economic growth. A substantial portion of the reduction in the price of electricity can be directly attributed to the successful deployment of pulverized coal firing in electric utility boilers during the period.

In his book entitled *Let There be Light*, author Forest McDonald states, "As an engineering feat, the development of pulverized fuel and its attendant developments constituted a monumental achievement, ranking with Edison's lamp and multiple distribution system, Stanley's transformer, and Parsons' steam turbine as one of the four fundamental technological developments that made low-cost central station service possible. Furthermore, it was ultimately to have a vital effect on central station economics, for in reducing by two-thirds or three-fourths the amount of coal necessary to produce a kilowatt-hour, it significantly lowered the principal variable cost of producing electricity."[125]

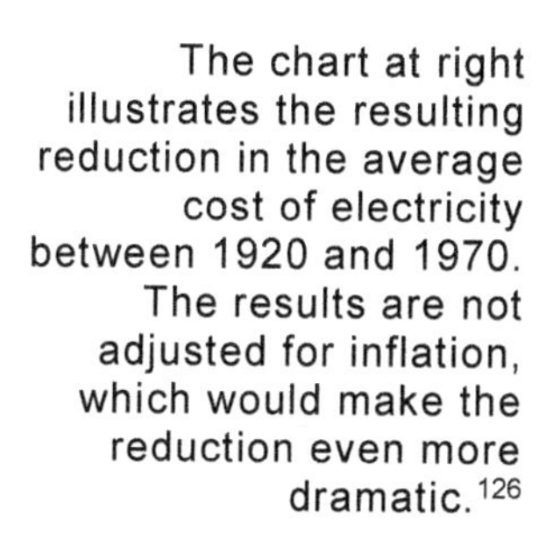

The chart at right illustrates the resulting reduction in the average cost of electricity between 1920 and 1970. The results are not adjusted for inflation, which would make the reduction even more dramatic.[126]

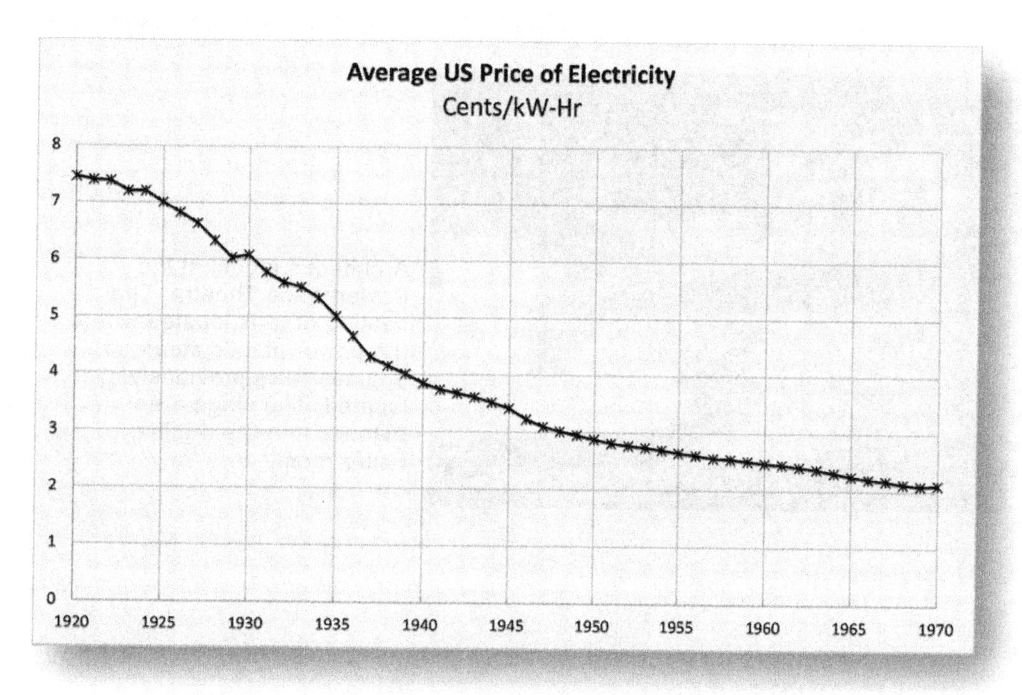

A few years after its retirement in the 1980s, the power plant was converted into a facility for the Milwaukee Repertory Theater, which is one of the cornerstones of Milwaukee's Theater District. The original boiler room is now the stage area for the Quadracci Powerhouse Theatre. The complex also houses two smaller theaters, the Stiemke Studio and the Stackner Cabaret, along with all of The Rep's rehearsal facilities, production shops and administrative offices.

A recent view of Milwaukee's Theater District showing the original Oneida Street Power Plant in the foreground. Author's photograph.

Architect's model of the Powerhouse Theatre. The seating area is located where the power plant's steam engines were previously located. The stage area extends into the original boiler room.

LAKESIDE POWER PLANT

As noted earlier, the motivation of the Milwaukee Electric Railway and Light Company to conduct experiments in the firing of pulverized coal at the Oneida Street power plant was its desire to build a major new central station power plant. The successful test results convinced the company to use the new technology at its new facility. Named the Lakeside Power Plant, this new facility was placed in service in 1921 in the City of St. Francis.[gg]

The necessity for the plant was realized as early as 1915. A site on Lake Michigan south of Milwaukee was selected, the land was purchased, plans were developed and contracts placed. The country's entry into the World War in 1917, however, made it impossible to proceed with construction.

When the war began, the utility's rated generating capacity was 77,650 kilowatts, which was a comfortable 28,000 kilowatts above the maximum demand for the railway, light and power loads. Electric demand increased rapidly, however, and by 1919 the peak load was dangerously close to the rated capacity of the entire system. The generating capacity became inadequate by the end of the summer of 1920. Area manufacturers were asked to curtail consumption that year, and an appeal went out through newspaper advertisements that asked customers to curtail consumption by twenty percent. These efforts were only partially successful. Toward the end of the year, the system was overloaded two or three times a week.

Construction on the new plant proceeded at a rapid pace and by the end of 1920, turbo-generator No. 1 was turning—just ten months after groundbreaking. Subsequent additions to the plant over the next decade brought the plant to its planned capacity of 310,800 kilowatts. The plant ultimately had twelve turbine-generators fed by twenty boilers.

The plant was an engineering marvel in its day—a model for power plants throughout the world that paved the way to modern, efficient, electric power production. The plant established and held world records for efficiency and economy in its initial decade of operation. It was known as "the world's power plant laboratory" because of its many firsts in the production of electricity. It was:

- The first power plant designed to burn pulverized coal exclusively, using the experimental results from the Oneida Street Power Plant,
- The first to introduce radiant superheaters into its furnaces,
- Among the first power plants to introduce a reheat cycle, which significantly increased the thermodynamic efficiency of the steam cycle,
- Among the first to use 1,200-pounds per square inch steam pressures,
- An innovator in the successful introduction of various metals to allow it to withstand the increased temperatures and pressures, and
- The first to use several innovative applications in valves, gauges and other equipment.

The design and reliable operation of the Lakeside Power Plant was a joint achievement of numerous employees of the company. Thousands of individuals devoted significant portions of their working lives to the economic production of electricity from the plant. The Lakeside project committee consisted of G.G. Post, John Anderson, L.C. Huff, P.H. Pinkley, J.A. Cheverton, G.W. Kalweit and F.V. Benz. The mechanical and pulverized fuel design and construction was the responsibility of John Anderson and L.C. Huff.

[gg] While the Lakeside Power Plant was not constructed in downtown Milwaukee, it was designed at the company's headquarters located at 231 West Michigan Street in the downtown area. For this reason, this plant is included in this chapter, as is the Port Washington Power Plant.

Lakeside Power Plant was located in the city of St. Francis. It was known as "the world's power plant laboratory" in the 1920s.This picture shows a portion of the turbine room, as it looked in 1984

Port Washington Power Plant

The Port Washington Power Plant of Wisconsin Electric Power Company was the most efficient power plant in the world starting from its initial operation in 1935 and continuing for approximately thirteen years thereafter. It operated at 31.6 percent efficiency during a period of time at which the national average electric power plant efficiency was well under nineteen percent. The design, characteristics, and operating experience of this power plant were reported in many technical publications. The plant served as a model for many other power stations and contributed to an increase in the efficient utilization of energy nationwide.

The Port Washington Power Plant held world efficiency records for many years. The steam-turbine generators shown here were produced by the Allis-Chalmers Corp. Designated a mechanical engineering landmark by the American Society of Mechanical Engineers on June 18, 1980.

Much of the credit for the exemplary efficiency from this station is attributed to Fred L. Dornbrook, former chief engineer of power plants for Wisconsin Electric, for his part in the plant's original design and the selection of equipment. Dornbrook's contributions were recognized by The American Society of Mechanical Engineers in 1949 when he was presented with the ASME Medal for Distinguished Service in Engineering and Science. Other personnel responsible for the plant's design and construction include G. G. Post, vice president in charge of power plants, C. F. John, assistant vice president, and M. K. Drewry, assistant chief engineer.

Each of the five coal-fired steam units produced 80,000 kilowatts. The turbine-generators, condensers, and many of the plant's pumps were produced by the Allis-Chalmers Corporation in West Allis, Wisconsin. The turbine-generators were of the tandem compound design, operating at 1800 RPM.

John Anderson was born in Aberdeen, Scotland, the son of a sea captain. He was educated in the British Government School of Science and Technology in Liverpool. Anderson worked as a marine engineer in the British Navy from 1880 to 1906, before coming to the United States. He was hired by the Union Electric Light and Power Company of St. Louis to serve as their heating superintendent. He came to Milwaukee in 1912 as chief engineer at The Milwaukee Electric Railway and Light Company.

In 1914, Anderson received authorization to conduct experiments on the use of pulverized coal at the Oneida Street power plant. TMER&LCo was beginning to plan for a large new generating station and Anderson wanted to explore the feasibility of designing the new plant to use pulverized fuel. He, along with Fred Dornbrook, W.E. Schubert and Ray Mistele, took part in the research project.

Anderson later became the company's vice-president in charge of power production—a position he held until his death in 1929.

Fred Dornbrook, left, and John Anderson are pictured with Anderson's daughter Miriam at start-up ceremonies for a new unit at Lakeside Power Plant in 1926. Miriam at the time was the only woman mechanical engineering student at the University of Wisconsin.

Fred Dornbrook was born in Brandon Wisconsin and worked for a time as a marine fireman and assistant marine engineer.

One of his Dornbrook's first jobs was a fireman on the steamboat *Angeline*. The company he worked for offered an annual bonus for the boat in their fleet with the best fuel economy. The *Angeline* had the worst fuel economy in the fleet. At the time, the fine particles of coal that resulted from breakdowns in the coal lumps were considered a nuisance and were typically collected and tossed overboard.

Dornbrook experimented with throwing some of the coal fines into the boiler, noting that the steam gauge responded to the increase in momentary energy. As he began using all of the coal fines and dust, the *Angeline's* fuel economy improved to the second best in the fleet. However, one day the boiler fired so hot that it burned out the furnace wall. Rather than reprimand him, the chief engineer encouraged his continued experiments. Dornbrook was ultimately named chief engineer for the company, but eventually left marine service to join TMER&LCo. At the electric company he directed the construction of the Commerce Street Power Plant in Milwaukee in 1903. Dornbrook was named engineer of operations and maintenance of the Power Plant Department in 1914 and became named chief engineer in charge of power plants—a responsibility he held until his retirement in 1950.

JOHNSON CONTROLS

Warren S. Johnson was a college professor at the State Normal School in Whitewater, Wisconsin in the early 1880s. He didn't like the frequent interruptions by the janitor during the heating season, who needed to check room temperatures to adjust the dampers to regulate heat, and decided to do something about it. As a teacher, Johnson had developed an interest in electricity and batteries. He had a laboratory at the school where he experimented. In that laboratory he produced a thermostat, which he dubbed a 'tele-thermoscope.' It used a bi-metal coil and a mercury switch. He installed one in his classroom, to great success, and obtained a patent for the device.

While it wasn't the first bi-metal thermostat, it caught the eye of William Plankinton, hotelier and heir to the Plankinton Packing Company, who was interested enough in the prospects for the device and other ideas Johnson was developing, that he provided financial backing.

W. S. JOHNSON.
ELECTRIC TELE-THERMOSCOPE.
No. 281,884. Patented July 24, 1883.

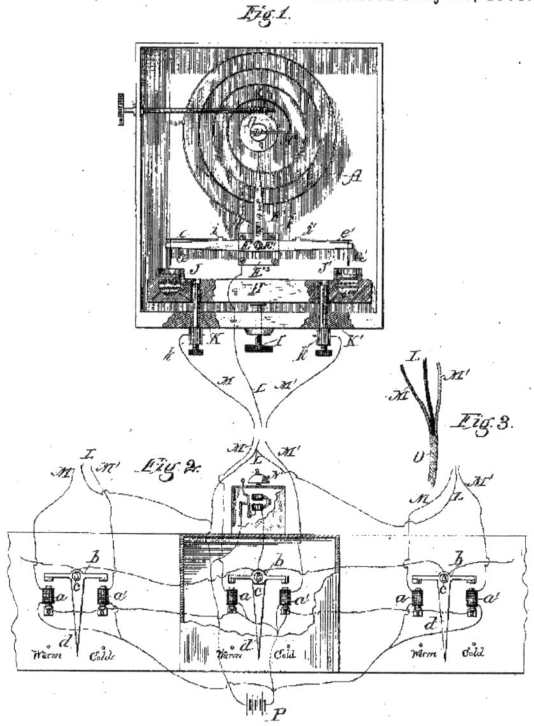

Johnson retired from teaching to pursue his scientific inventive interests and, along with Plankinton, established the Milwaukee Electric Manufacturing Company. In 1885 the company was renamed the Johnson Electric Service Company. The company initially operated in the basement of a building located off an alley south of Grand Avenue (now Wisconsin Avenue), between Third and Fourth Streets. Soon the company moved to larger quarters at the southwest corner of Grand Avenue and the Milwaukee River. Finally, in 1902, it began construction of a building on the southeast corner of Michigan and Jefferson Streets—a building that still serves as the company's building efficiency headquarters.

Johnson was given a free hand to continue to experiment and develop other devices. And invent he did. He had a prolific mind and a multitude of interests. He designed products ranging from chandeliers, door locks that worked without springs, puncture-proof tires, a bicycle seat with a pneumatic cushion, a hydraulic air compressor, a pipe coupling, a hose coupling to provide steam heat between railway passenger cars, and experimented with wireless telegraphy.

A good number of his innovations, of course, were related to various aspects of building heating systems, including boilers (steam generators), compressors to run his pneumatic controls, thermostats and the like.

Johnson also was interested in entering into the automotive industry, producing steam and gas powered trucks and automobiles.

After Warren Johnson's death in 1911, the company decided to focus solely on its temperature control business for nonresidential buildings.

Johnson Multi-Zone Pneumatic Control System

Warren Johnson developed an entirely mechanical pneumatic control system that economically automated the process of maintaining the temperature in more than one zone for building comfort. His system employed a thermostat, which he patented that used a bimetallic element to control airflow through a nozzle and thereby operate a pilot regulator. The amplified air signal from the regulator was conveyed through a single tube to a steam or hot water valve on a heat exchanger, or alternatively the amplified pneumatic signal could be used to control the damper of a forced air system. It is believed this system was the first such innovation that allowed automatic temperature control for large, multi-room buildings.[hh]

The earliest thermostats, in use before the pneumatic invention of Johnson, were annunciators used to alert a boiler operator to conditions elsewhere in a building. When methods of direct fluid control were developed, they often employed a combination of electric and pneumatic/hydraulic devices. An electric switch closure, for example, might control a pneumatic valve through a solenoid relay, or an expanding fluid might directly actuate a valve through a capillary tube. The complexity or bulkiness of such methods limited the range of application of temperature control.

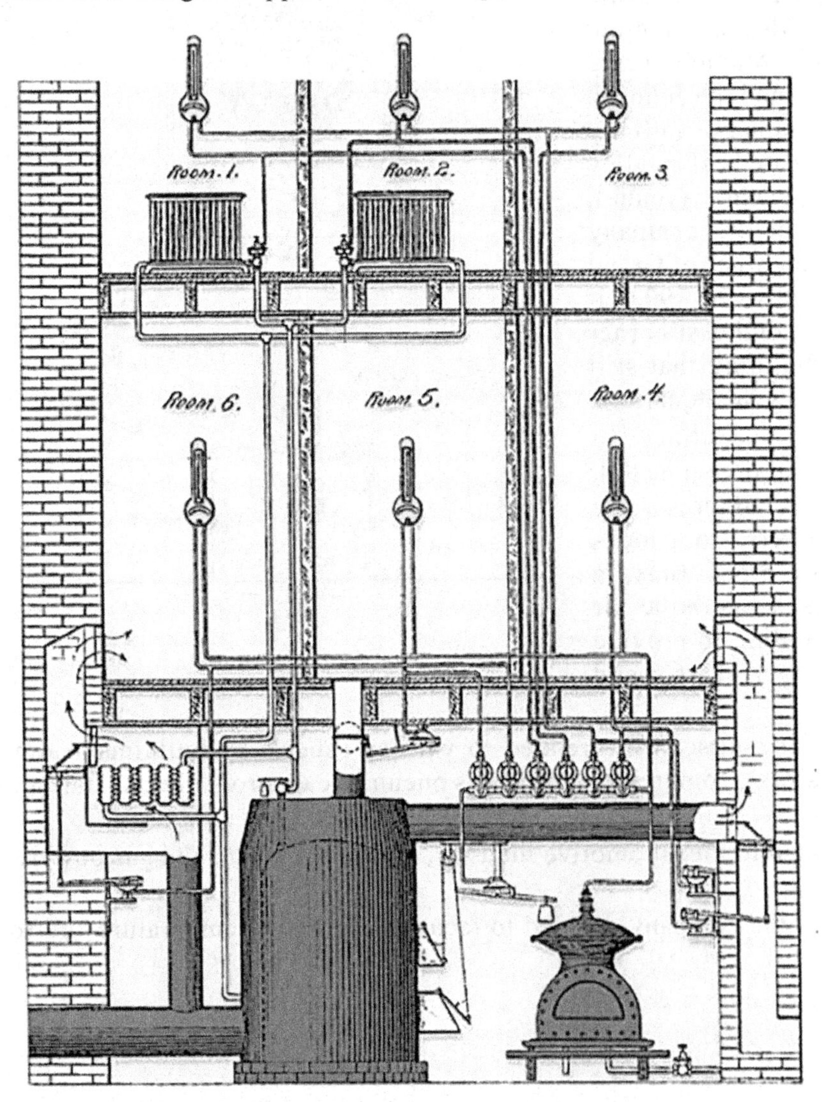

The innovations associated with Johnson's control system allowed the automation of multi-zone building heating systems for the first time. His pneumatic control system was economic and practical and was immediately adopted throughout the United States and eventually the rest of the world.

The adjacent drawing is an adapted illustration from the original patent drawing (US Patent No. 542,733) of the Johnson "Heat Regulating Apparatus," showing a conceptual multi-zone heating system where the Johnson pneumatic control system is regulating the temperature of six rooms. Two of those rooms (shown as Rooms 1 and 2 in the diagram) use the input signal from individual room thermostats to control the input valves to the radiators located in each room. Four other rooms (shown as Rooms 3 through 6) use individual thermostats to control dampers from a forced air heating system.

The pneumatic air for the system is supplied from a mechanical air pump (the device in the lower right corner of this illustration), powered by building service water.

[hh] The novelty of this approach was contested in court. On February 26, 1906, the Circuit Court of the District of Massachusetts declared for the inventor.

The system covered by the 1895 patent was the first complete automatic temperature control system that could be economically manufactured, installed and maintained. The secret was a unique three-way valve, which controlled the flow of compressed air via the pneumatic thermostat. The way this valve operated delivered a combination of reliability and precision that had been unattainable before. The Johnson System of Temperature Regulation became the standard for pneumatic control systems in commercial buildings.

Competing systems were eventually invented that used other means to achieve control, but the Johnson System was the dominant pneumatic control system for decades. Thousands of buildings throughout the world continue to use pneumatic building temperature control systems based on the Johnson System.

One of the early businesses to benefit from this innovation was Milwaukee's downtown Pfister Hotel.

Johnson subsequently invented the "Humidostat," a word coined by the inventor. It was the first humidity-sensing element used in a building air control system. It was patented in 1909. The expansion of a maple dowel as it absorbed moisture from the surrounding air was used to control the leakage of air from a pneumatic port. The amplified signal was then used to control an apparatus for the addition of moisture to the air.

Milwaukee's Pfister Hotel, built in 1893, was the first hotel to incorporate Johnson's automatic multi-zone temperature control system.

Johnson Building Clock Mechanisms

Warren Johnson also developed a pneumatic system for powering building clocks. Clocks of the period for large offices and schools relied upon a master pendulum clock, which provided a signal that controlled satellite clocks. Johnson developed a system that used air pressure to control the remote clocks. He soon employed the system to also power large tower clocks, building his first large tower clock in 1895 for the Minneapolis courthouse. That same year the company bid on and was successful in obtaining the order to provide the clock for Milwaukee's City Hall tower.

Milwaukee City Hall's four-sided clock tower. It originally had clocks designed and manufactured by Johnson Electric Service Company in 1895. When first occupied, Milwaukee's City Hall was the tallest habitable building in the United States.

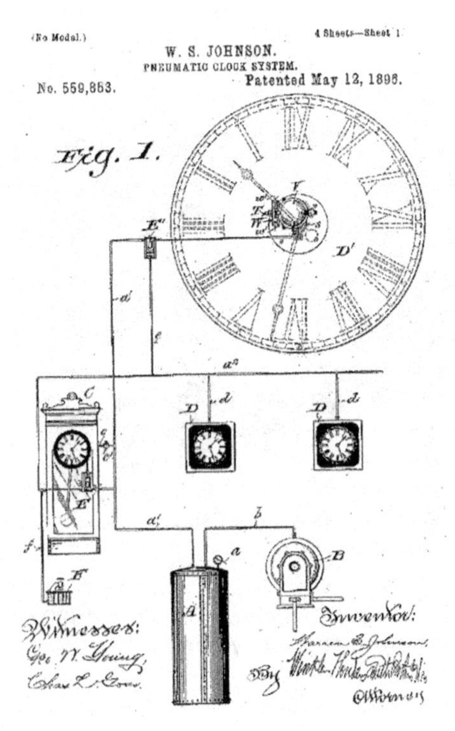

Patent illustration for Johnson's pneumatic system for running clock towers. It used a master clock, which then triggered the mechanism for the tower clocks as well as other satellite clocks, using compressed air. A large compressed air tank was used to power the system in case of loss of power to run the compressor.

The company's largest tower clock was installed in Philadelphia City Hall. It gained the greatest acclaim, however, for a giant floral clock installed at the Saint Louis World's Fair in 1904. The massive clock was located on the front lawn of the Agriculture Building and had a clock face that was over 100 feet in diameter and had 13,000 flowers covering its surface.

Stereoscopic images of the Floral Clock designed and manufactured by Johnson Electric Service Company at the 1904 St. Louis World's Fair. The images were observed with a viewer that produced the effect of looking at a 3D scene. Library of Congress Photograph.

Johnson Electric Service Company entrusted a young engineer, Paul J. Ostrowski, to the highly visible task of designing the system and overseeing its construction and operation. Ostrowski proved equal to the task. He designed the clock mechanism, which used compressed air, to power the 75-foot long minute hand and a shorter hour hand. The minute hand weighed 2,700 pounds and moved five feet every sixty seconds. A clock bell was included which weighed 5,000 pounds and was struck on the hour and half hour.

The clock mechanism was located in a small pavilion located just above the face of the clock.

Johnson's Trucks and Automobiles

Warren Johnson was enthusiastic about entry into the automobile business. After the company designed a steam-powered truck for Johnson's personal use, it followed with several other truck models. A fleet of eight steam-powered trucks was built for Milwaukee's United States Postmaster. These were some of the first specialty-built mail vehicles in the US. Only Baltimore may have used a non-horse-drawn vehicle before Milwaukee—using a modified Columbia automobile.

This postcard shows a steam-powered vehicle built by Johnson Electric Service Company for use by the US Postmaster in Milwaukee. The inset on the upper right of the face of the postcard shows how the interior of the vehicle could be used for sorting mail. The message on the back reads, which was written in 1908, states in part, *"This is the way they take the mail up around over town here. They also carry mail to neighboring towns."* From author's collection

This 1912 US postage stamp, showing a Johnson mail truck, commemorated the United States Post Office's use of automobiles to collect mail.

In 1905, Johnson began building steam-powered automobiles and, in 1907, switched to internal combustion engines with 30, 40 and 50 horsepower. They were large, expensive vehicles and were often custom built. A Johnson *Landaulet* listed for $3,500 in 1909—equivalent to $90,000 today. Ten different models of automobiles were offered in 1911—all of which were dubbed the *Silent Johnson*, along with a wide-range of fire trucks, ambulances, limousines and a sightseeing bus. A thirty horsepower, five-passenger special touring car was offered at a starting price of $1,500.[127]

The vehicles were sold nationally. It is estimated that the company manufactured a thousand vehicles between 1901 and 1912. Johnson Electric Service Company's entries in the automobile business ended with Warren Johnson's death in December 1911. The company's board of directors decided to focus on its core business of automatic temperature control products.

A Johnson fire truck is shown at right.

Johnson-Fortier Wireless Telegraphy

As noted earlier, Warren Johnson had a creative mind and pursued numerous interests. One of the most interesting was his involvement with the development of wireless technology. At the turn of the century, a competition was underway to develop a workable means of wireless communication—the transmission of telegraph messages without connecting wires. Notables such as Nikola Tesla, Guglielmo Marconi, and Thomas Edison were involved in the pursuit, as well as some lesser-known inventors.

Most investigators were working with a spark-gap transmitter for creating radio waves, and a coherer to detect them. The coherer was initially developed by Édouard Branly of France. The device consisted of an insulated tube that contained iron filings. Electrodes on either side of the device were used to conduct an electric current across the filings. The thin layer of oxide present on the surface of the iron filings caused the assembly to have a high resistance to current. Branly discovered, however, that the filings would lose their resistance to current in the presence of electric oscillations. The oscillations, or electromagnetic waves, were initially referred to as *Hertzian* waves (after Heinrich Hertz who demonstrated electromagnetic waves in his laboratory) and eventually became known as radio waves. According to the common explanation at the time, when electric waves are set up in the neighborhood of this circuit, electromotive forces are generated in it, which appear to bring the filings more closely together, that is, "to cohere," and thus the electrical resistance across the iron filings decreases.[128]

The device would stay in this low-resistance state until the voltage was removed and the coherer was physically tapped. Tapping the device caused the filings to reconfigure and be ready to respond to radio waves again. Warren Johnson's innovations were largely directed at automating this process of breaking up, or "de-cohering," the iron filings. Like many of Johnson's innovations, he used pneumatics to accomplish the task.

Working with Charles L. Fortier, a telegraph operator by occupation, Johnson developed his system. It was first demonstrated on November 22, 1899, during a talk at the Plankinton Hotel. Messages were sent between a private second-floor dining room and the Club Room, which was located about four-hundred feet away.

The two men formed the American Wireless Telegraph Company (AWTC) in early 1900, with funding from a number of prominent Milwaukeeans, to further refine their apparatus and design a marketable product.

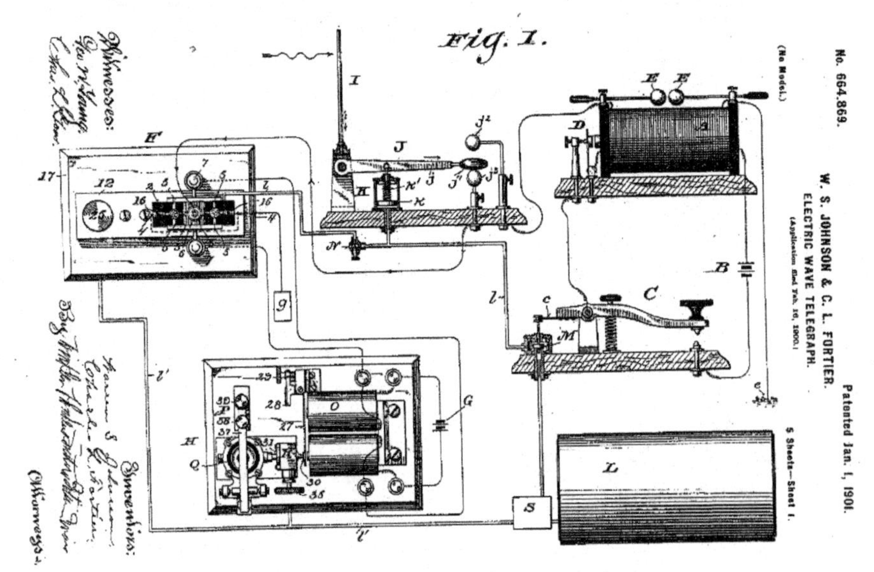

The drawing at left is from Warren Johnson and Charles Fortier's patent for an "Electric Wave Telegraph." The apparatus includes a Morse Key (Denoted C) and a spark-gap transmitter (E-E, which the inventors referred to as an oscillator) to send signals and a coherer (F) for receiving signals. The compressed air tank (L at the bottom right) provided an impulse of air at the appropriate time to "de-cohere" the filings in the coherer tube.

The inventors set about testing their system. Paul F. and Carl F. Johnson, Warren's sons, also participated in the tests. Eight wireless stations were located in the Milwaukee area. In addition to one on the fifth floor of the Johnson Controls building on East Michigan Street, stations were located as far north as the top of the North Point Water Tower and as far south as a 150-foot, free-standing tower the company built at Cudahy in what today is Sheridan Park. The most useful of these stations, however, was likely a station on the end of the pier on the harbor break wall near the mouth of the Milwaukee River. The advantage of the harbor pier site was that it also afforded line-of-sight between it and the station on the Johnson Controls building and allowed telegraph operators to use flag signals to help communicate the status of experiments.

The interesting photograph above is of the wireless station on the fifth floor of the Johnson Controls building at 507 East Michigan Street. The operator could not be identified, but is thought to be one of Warren Johnson's sons. The boxes in the foreground apparently house batteries. The spark-gap transmitter can be seen at the top of the box in front of the operator.

In October 1899, Warren Johnson delivered a lecture before the Western Society of Engineers in Chicago and demonstrated the current version of the company's decoherer. By this time, the apparatus connected two coherers in parallel. Each was fitted with pneumatic ports to allow impulses of air to alternately go from one cohere receiver or the other, which apparently increased the overall speed of reception. It was reported that the air engine was "exceedingly small and would run for hours from a tank of air which was filled by hand in a minute." Johnson referred to his activities employing compressed air as the field of "micro-pneumatics." It was reported that the demonstration "showed a very satisfactory working."[129]

Charles Fortier traveled to Europe in the summer of 1900 to exhibit the Johnson-Fortier system at the Paris Exposition, in competition with three other exhibitors, one of which was Guglielmo Marconi. A German competitor was awarded the first place prize in wireless apparatus. The Johnson-Fortier system took second place—besting Marconi's system, which came in third.[130]

During Fortier's four-month absence, Johnson decided to hire Dr. Lee de Forest to assist the team. De Forest had recently attained his doctorate in electrical engineering and was eager to work in the field of wireless telegraphy. He had applied previously to work for both Nikola Tesla and Guglielmo Marconi and saw the opportunity to work with Johnson and Fortier as "an opportunity of a lifetime to enter first at the start of a scientific art all new, all to be developed, of a boundless, ever-widening, ever-enriching future."[131]

De Forest had been working on his own wireless receiver, a device he called a *responder.* It consisted of tinfoil fastened to a glass plate. A gap was created in the tinfoil with a razor blade and was bridged with a glycerin fluid solution. De Forest's device was referred to as an anti-coherer receiver since it interrupted the circuit when a radio signal was received.

De Forest moved to Milwaukee and started working with Johnson's team in May of 1900. He was put to work testing the Johnson-Fortier system, working out of the station in Cudahy. De Forest wasn't impressed with the company's coherer, calling the device "insensitive and quite discouraging to me." Indeed, the team was having difficulty. They were only able to receive signals relatively short distances, on the order of five to seven miles, and somewhat inconsistently. The testers were frustrated that the good results on one day were not reproducible on the next. It appears that moisture introduced by the pneumatic pulses may have played a role in changing the characteristics of the coherer. Different compositions of metals were tested, along with a device to remove the moisture from the air used in the pneumatics.

Meanwhile, de Forest tested and worked on his own device; apparently testing it alongside the Johnson-Fortier coherer. He believed his responder to be superior. Johnson soon found out about de Forest's clandestine tests on company time and fired de Forest on the spot, after having only worked at the American Wireless Telegraph Company about four months.[132]

The picture at left is an image of the AWTC's 150-foot tower in what is today's Sheridan Park in Cudahy. Lee de Forest worked at this station for a time, testing the Johnson-Fortier coherer receiver, as well as a receiver of his own design.

De Forest's firing was unfortunate because it appears that his receiver was a more reliable and sensitive device. While it appears Johnson was justified in firing de Forest, given his transgressions, the AWTC lost an opportunity to capitalize on his technology and Milwaukee lost an engineer who went on to international prominence.

The AWTC team continued to work on refining the Johnson-Fortier system, with limited success. At the end of 1901, it was reported that Guglielmo Marconi had succeeded in receiving a wireless telegraph signal across the Atlantic. Upon this news, Johnson directed his attention elsewhere, although Charles Fortier continued to work on refinements to his electric wave receiver.[133]

De Forest went on to obtain a patent for his responder. More importantly, he eventually designed the audio vacuum tube that provided the breakthrough for radio. In 1906 he invented a three-element (triode) vacuum tube, which he called the "Audion." It was the first practical electronic amplification device. While it appears that de Forest had only a limited understanding of how it worked, it was the foundation of the field of electronics and was used in numerous applications. He had over 180 patents during his career.[134]

Image of Warren S. Johnson, courtesy of Johnson Controls Corporate Archives

WARREN SEYMOUR JOHNSON was a college professor who was frustrated by his inability to regulate individual classroom temperatures. His multi-zone pneumatic control system solved the problem. Johnson established a company to manufacture, market and service the product—a company that has grown into a large international concern that continues to provide building energy control services.

Johnson was born in Rutland County, Vermont, in 1847. His family moved to Wisconsin three years later, eventually settling in Menomonie. It appears that he had only limited formal educational training—but supplemented his knowledge with self-study of scientific subjects. He worked for a time as a printer, surveyor, schoolteacher, principal and school superintendent. In 1876 he obtained a teaching position at the State Normal School in Whitewater—now known as the University of Wisconsin—Whitewater. Five years later, he was named professor of natural science.[135]

He had an inquisitive mind and was particularly interested in electricity. In 1883, Johnson developed a thermostat, which he deployed at the State Normal School. He coined the instrument an 'electric tele-thermoscope' in the patent application. It was a bi-metal coiled thermostat with a mercury switch, which could be used to ring a bell to alert the fireman to open or close the heating damper. While not the first bi-metal thermostat, Johnson received a patent for the device and interested William Plankinton, heir to the Plankinton Packing Company, to provide financial backing to manufacture the device.

The Johnson Electric Service Company was established in 1885. Johnson's most notable contribution to temperature control was the automatic multi-zone temperature control system—a pneumatic system that used a bi-metal thermostat to control air flow through a nozzle and thereby operate a pilot regulator. The amplified air signal from the regulator was then used to control a steam or hot water valve on a heat exchanger, or to control a damper of a forced air system. He received a patent for the system in 1895.

Johnson continued to invent additional control devices, as well as products such as chandeliers, springless door locks, puncture-proof tires, thermometers, and a hose coupling for providing steam heat to passenger railcars. He also designed pneumatic tower clocks—one of which was built for Milwaukee's City Hall tower.136

He experimented for a time with wireless communications, forming the American Wireless Telegraph Company. The company's exhibit at the Paris Exposition Universelle of 1900 won second prize, beating out Guglielmo Marconi. A test tower was built several miles south of Milwaukee, but the tests were unsuccessful. For about three months Lee de Forest, who eventually went on to design the audio vacuum tube that provided the breakthrough for the development of the radio, worked on the project with Johnson.

Johnson also sought to form an automobile company, introducing first a steam-powered truck and then a line of automobiles using gasoline-powered engines. The company was among the first to receive a contract to deliver mail with a horseless carriage.

He is credited with more than 50 patents. He died in 1911.

BRIGGS & STRATTON

The story of the formation of the Briggs & Stratton Company starts with an engineering student from South Dakota State College of Agricultural and Mechanic Arts, a football coach from that same college, and a grain merchant from Milwaukee. It provides an interesting illustration of the convoluted way companies sometimes come together, as well as the circuitous path that innovation can take.

Let's start with that engineering student. Stephen Foster Briggs was born in Watertown, South Dakota in 1885. While taking an upper-level engineering course at South Dakota State, he developed a six-cylinder, two-cycle engine. Following his junior year, he took a summer job working in Milwaukee for the A.O. Smith Company as a machinist. Upon graduation in 1907, he was eager to enter into the burgeoning automotive industry, but lacked both the financial resources as well as a facility to work on his endeavor. Briggs must have impressed his coach, Bill Juneau, because Juneau intervened on his behalf.

Coach William Juneau, about 1918. From Wikipedia Commons

William (Bill) Juneau was the grandnephew of Solomon Juneau, one of the founders of the city of Milwaukee, Wisconsin. His family lived on a farm in Troy Center—just northwest of East Troy Wisconsin. Bill attended college at the University of Wisconsin, where he played football as an end and halfback from 1899 to 1902. He captained the football team in 1902 and began coaching at Fort Atkinson High School in Wisconsin. His coaching career advanced rapidly and Juneau took head coaching assignments at Colorado College in 1904, before becoming the head coach of the football, basketball and baseball teams at South Dakota State in 1906.[ii] Stephen Briggs played on Juneau's basketball team, and became acquainted with the coach. In the process, Juneau became aware of Briggs' engineering abilities, as well as his need for a financial backer.

Fortuitously, Bill Juneau's family farm in Troy Center adjoined that of Harold M. Stratton. By that time, Stratton was a successful grain merchant and was president of the Stratton Grain Company. Juneau developed a friendship with Stratton and was aware of his entrepreneurial interests. He introduced the two and Briggs and Stratton established an informal partnership in 1908. Stratton invested $50,000 in the venture.

Briggs moved to Milwaukee, where Stratton rented factory space in the City's Third Ward. His six-cylinder engine proved too expensive to produce for the automotive industry at the time. Henry Ford had introduced his Model T in 1908, which was priced for the mass market. All other automotive companies were attempting to compete. In this environment, they found it was commercially impractical to introduce the new engine to the market.

Briggs & Stratton then attempted to enter into the automobile assembly business, producing automobiles from components produced by others. They assembled three vehicles under the name "Superior," but almost went bankrupt in the process—the cars were too costly to produce. Interestingly, A.O. Smith supplied the frames for those three vehicles, which was the start of a relationship that would eventually be extremely important to Briggs & Stratton.

[ii] Bill Juneau went on to head coaching positions at Marquette (1908-1911), Wisconsin (1912-1915), Texas (1917-1919) and Kentucky (1920-1922). His football teams had an overall record of 86 wins, 39 losses and 12 ties.

Initially, Briggs & Stratton operated out of rented space at 258 Milwaukee, located directly across the river from Walker's Point. The map of the Milwaukee Lithographing & Engraving Company was published by Holzapfel & Eskuche Stationers and Book Sellers. Map courtesy of the Library of Congress.

Fortunately, Stephen Briggs had other innovative ideas. In 1910, Briggs received a patent for an automotive engine igniter. The mechanism was designed to provide “an igniter for gas engines and the like of a construction which will be neat and compact, combining the induction coil and the timer and the distributor in a single mechanism which is simple and efficient its operation, inexpensive to construct and easily maintained in operative condition.”[137] While the new product didn’t meet with immediate success, it provided sufficient business to keep the company solvent. It also set the company on the path as a producer of electrical specialty products for the automotive industry.

By 1920, Briggs & Stratton was the largest producer of electrical specialty products in the United States. Its product line included ignitions, regulators, starting switches and specialty lighting, and its customers included all of the major US automobile producers. As the sales increased, expansion became critical. In 1920, Briggs & Stratton moved into their “East Plant,” located at 13th and Center streets. The plant employed fourteen hundred workers. The company produced automotive parts out of this facility until 1973, when it moved its automotive lock division to a new factory in Glendale.

In 1924, a new die-cast automobile lock cylinder was introduced, which outsold competing brass models. By the end of the decade Briggs & Stratton was the largest US producer of automotive locks. The company ventured into other automotive products, including body hardware such as door handles, inside knobs and levers, compartment locks, door locks, and hinges. It also introduced a “Cushion Action” starter drive in 1938 and it became standard equipment on the Ford V8, as well as the Mercury and Lincoln Zephyr.

The company also searched for other markets to enter. For a time it produced refrigerators, radio tuners, and a device called a battery eliminator. It also manufactured a variety of stamped metal products, including soap containers, calendar banks, display stands and a coin operated paper towel dispensing cabinet. Most were met with limited success.[138]

Then, in 1919, Briggs & Stratton acquired the rights to produce the “Motor Wheel” from the A.O. Smith Corporation. The decision was a seminal one for the company, causing them to enter into the small internal-combustion engine business.

The significance of the Motor Wheel is discussed in the next section.

This gasoline engine ignitor was the first product produced by Briggs & Stratton for the automotive industry.

THE MOTOR WHEEL

In 1909, the Wall "Auto Wheel" made its debut in England. Designed by Arthur William Wall, it was intended as an attachment to a bicycle that served as a power booster. The single cylinder, air-cooled engine was mounted on a twenty-inch bicycle wheel and was rated at one horsepower.

Ohio industrialist Arthur Garford traveled to England in the fall of 1913 and negotiated and secured manufacturing rights for the device in the United States. Garford Manufacturing first exhibited the device at a motorcycle exhibition held in Atlantic City in August 1914. It is reported that two 'experts' were involved to provide support for the motor wheel—Reuben Stanley Smith of the A.O. Smith Corporation and W.V.C. Jackson. Both were employees of A.O. Smith.

The Smith family was familiar with Garford Manufacturing from its bicycle days—both A.O. Smith and Garford were both manufacturers of early bicycles. It is likely that Smith bicycle parts were used on Garford bicycles.

It is not known when A.O. Smith's involvement with the motor wheel began, but it is possible that the unit exhibited in Atlantic City was an early prototype manufactured by A.O. Smith for Garford.

It appears that A.O. smith quickly obtained the rights to manufacture the motor wheel from Garford, because by October of that year the Smith Motor Wheel was shown at a motorcycle exhibition in Chicago.[139] Reuben Stanley Smith is credited with several patents for improvements to the motor wheel. W.V.C. Jackson went on to become Smith's motor wheel sales manager.

A.O. Smith introduced at least four separate models of their bright red 'Smith Motor Wheel.'

An early advertisement for the Smith Motor Wheel. The original Wall Motor wheel was a two-cycle engine, whereas Smith's Motor Wheel ran on the 4-cycle principal.

In 1919, Briggs & Stratton purchased the manufacturing rights to the motor wheel from A.O. Smith. Smith produced approximately twenty-five thousand motor wheels between 1914 and 1919. It is not known why A.O. Smith was willing to sell the rights to the device, although it can be speculated that their motor wheel was considered as only a minor adjunct to their business. Their automotive frame operations were expanding greatly and the company was in the process of building an expensive new factory to expand production to meet demand.

It appears that Stephen Briggs was very familiar with the motor wheel. On several occasions, he provided A.O. Smith Corporation with blueprints for an internal flywheel magneto that could be used with the device.

Having obtained manufacturing rights, Briggs & Stratton set about to improve the motor wheel further. They increased the output to two horsepower and added their internal flywheel magneto. They called their new version the Model D. They sold it for about $90 and it was advertised to go one hundred miles on a gallon of gasoline.

In addition to offering the motor wheel as a bicycle attachment, Briggs & Stratton introduced a scooter with their motor powering the rear wheel.

This 1919 advertisement suggests that young boys ask Santa for a Briggs & Stratton scooter for Christmas. In the inset at the upper left of the photograph, a lad is seen writing a letter to Santa for that purpose

Briggs & Stratton also began manufacturing the motor-powered cart that was originally introduced by A.O. Smith. The 'Flyer' was essentially a wooden buckboard-style two-passenger cart, with a motor wheel mounted on the back. The Briggs & Stratton 'Flyer' was advertised as the country's lowest-priced car, selling for about $150 at the time. Briggs & Stratton sold about 2,000 Flyers between 1920 and 1923.

The Flyer, or 'Red Bug,' was a buckboard style vehicle that seated two. While the vehicle was inexpensive, it was also very rudimentary and offered little protection from the elements. Sales were disappointing.

While the company expected to market the Motor Wheel to Mexico and South America, and to Asia where the device could be used with rickshaws, sales provided to be disappointing. In 1924, Briggs & Stratton sold the rights to manufacture the Motor Wheel to Automotive Electric Service Corporation of New Jersey.

The company didn't give up on the small single-cylinder engine, however. Briggs & Stratton introduced a stationary version of the engine in 1920, called the Model P. This four-stroke gasoline-powered engine proved to be a popular power source for garden tractors and lawn mowers. However, the company's largest market for the engine at the time was for washing machines. Gasoline powered engines were sold in rural areas to allow customers without access to electricity to use automatic washing machines. The washing machine engine, denoted a model WM, was typically fitted with a foot starter and a hose to direct the exhaust out of the laundry room window, if used inside.

The Briggs & Stratton Type P engine was introduced in 1920. Its portability and reliability led to its use in many applications, including washing machines, garden tractors and generators. Eventually, the company became the world's largest manufacturer of air-cooled gasoline engines. Photo courtesy of Briggs & Stratton Corporation.

Briggs & Stratton Aluminum Engine

During the Second World War, Briggs & Stratton received a contract to produce generators for the war effort. This led the company to the use of aluminum for various engine components, and provided them with needed expertise in the design and manufacture of engines incorporating aluminum parts.

In 1948, Briggs & Stratton initiated a program to introduce a line of air-cooled aluminum engines, to replace the heavier cast iron engine blocks previously used for all of its engines. The company understood that in order for its four-stroke engines to compete with two-cycle engines that were gaining market share, they had to significantly reduce the weight of their engines. After extensive research and development, its lighter-weight engine was introduced in 1953. It was targeted initially for lawn mower applications, although eventually it was used in numerous other applications. The new engines, produced in both a horizontal and a vertical shaft format, were an immediate hit.

In 1954, a patent was granted to the company for its aluminum die casting process, one of many patents issued for various aspects of its new technology. In 1958, Briggs introduced the Sleeve-Bore version of the aluminum block engine, which contained a cast iron liner cast in the aluminum block. At that point, forward, the aluminum block engine then became known as the Kool-Bore engine due to the excellent heat transfer characteristics of aluminum resulting in lower engine operating temperatures.[140]

The company has manufactured well over four hundred and fifty million small gas engines. It continues to dominate the market for lawn mower engines and related applications.

A restored 1953 Briggs & Stratton aluminum engine is shown in the image above, mounted on a rototiller. The inset shows the operating instructions for the engine.

STEPHEN FOSTER BRIGGS was born in 1888 on a farm near Watertown, South Dakota. He attended grade and high school in Watertown and enrolled at South Dakota State College at Brookings in 1903. Reminiscing about his time there, Briggs remarked "I wired most of the town for electricity and fixed most of the first autos from 1903-1907." At South Dakota State, Briggs also designed and built a six-cylinder, two-cycle, air-cooled automobile engine. Much of the machining for the engine was done at the college's shop.

After his junior year in college, Briggs accepted a summer position as a machinist at A.O. Smith in Milwaukee. He met Harry Stratton that summer through William Juneau, then Athletic Director at Brookings. Stratton, a successful grain merchant, was impressed with Briggs' engineering abilities and the two established an informal partnership.

Briggs graduated from college in 1907 with a degree in Electrical Engineering and moved permanently to Milwaukee to work with Stratton.

With Harold Stratton's financial help, they began manufacturing Brigg's six-cylinder engine. However, it provided too costly to produce. They then turned their attention to the construction of automobiles. In 1908, they built their "Superior" automobile by assembling cars from parts that were manufactured by others. This also proved to be an unprofitable venture. However, along the way, Briggs designed an ignition switch for automobiles.

Stephen F. Briggs (driving) and Harold Stratton (L) are pictured riding in a Briggs & Stratton flyer.

In 1909, they launched the Briggs & Stratton Company to build ignition devices for gasoline engines. Later they branched into the manufacture of other switches, lights and starters and eventually expanded into the manufacture of ignition locks, door locks, and other automotive parts.

A decision to acquire the manufacturing rights to A.O. Smith's "motor wheel" in 1919 launched Briggs & Stratton into the small internal combustion engine business. The company set about improving the engine and then adapting it to other uses.

Stephen Briggs was also involved in a number of other corporate ventures. During the First World War, he formed the Briggs Loading Company to make rifle grenades for the US Army. Shortly afterward, he formed the Lemke-Briggs Electric Company to manufacture ignition coils, which were used by Ole Evinrude on his outboard engine.

In 1929, Briggs helped organize the Outboard Motor Corporation, which manufactured Evinrude outboard engines. The company later merged with Johnson Motor Company to form Outboard Marine Corporation, with Briggs serving as president.

Briggs was named on at least thirty patents for various automotive accessories, including automobile locks and a locking switch mechanism, which would allow the owner of a vehicle to secure it from being started without a key.

Stephen F. Briggs married Beatrice Branch in 1910 at the chapel at the University of South Dakota. He established a substantial foundation at the college that continues to award scholarships to deserving students.[141]

Stephen Briggs is shown in this 1906 picture taken at South Dakota State with his six-cylinder engine

METROPOLITAN MILWAUKEE SEWERAGE DISTRICT

The first sewers installed in Milwaukee were built more than one hundred and thirty years ago. They carried untreated sewage, along with stormwater directly to the rivers and Lake Michigan. As the City increased in population, this created an almost unbearable situation. By the 1920s, combined sewerage from a population of 500,000 was discharged to the Milwaukee, Menomonee, and Kinnickinnic rivers, which converge and flow together through a single outlet into Lake Michigan. Although fresh Lake Michigan water was pumped through two flushing tunnels into the Milwaukee and Kinnickinnic rivers, to dilute the level of pollution and to increase flow, public health problems remained unsolved.

In 1913, the City of Milwaukee created a sewerage commission to build and develop an intercepting sewer system and a sewage treatment facility. After much experimentation and testing, the commission chose the then-new technology of activated sludge for the treatment method of the treatment facility. In an activated sludge plant, microorganisms convert organic material into simple elements like carbon, hydrogen, and oxygen.

As property development and industrial growth increased, the Metropolitan Sewerage Commission was formed in 1921. At the same time, the Metropolitan Sewerage District of the County of Milwaukee was created to address sewerage needs outside the City of Milwaukee. Together, the Sewerage Commission of the City of Milwaukee and the Metropolitan Sewerage Commission of the County of Milwaukee ran the Metropolitan Sewerage District of the County of Milwaukee.[jj]

JONES ISLAND WASTEWATER TREATMENT PLANT

The Jones Island Wastewater Treatment Plant began operation in 1925 as an activated sludge plant with a sewage treatment capacity of 85 million gallons per day (MGD). This was America's first large-scale activated sludge plant. The waste sludge was used to produce a fertilizer product because of the high organic nutrient value of the dried sludge. This fertilizer is a commercially sold product called *Milorganite*, and is familiar to many, and is still produced today at the Jones Island Plant.

The production of Milorganite involves the dewatering of the waste sludge solids by various chemical and mechanical steps. The waste sludge solids are conditioned with chemicals to aid in filtration. Filter cake generated by vacuum filters is then fed to large, rotary kiln-type dryers. The dryers reduce the moisture to produce a granular product that has the appearance of coffee grounds after selective screening.

When the plant was originally designed prior to 1925, a coal-fired power plant was provided to meet the processing plant's needs for low-pressure process air, electric power, and process steam. The conventional coal-fired steam power plant included steam turbine-driven centrifugal compressors and electric generators to meet these basic requirements. Steam energy was also provided for large steam-driven reciprocating vacuum pumps, plant heating and other plant processes.

The wastewater treatment plant was expanded to meet the growing population and service area, first in 1935, and again after the Second World War in 1952. These two major expansions increased capability to two hundred million gallons per day and fertilizer production to 70,000 tons per year.

The energy source for both power generation and heat drying has always been dictated by cost. The original source—coal—was converted to coke oven gas in 1950, and in 1960 to natural gas with fuel oil for standby.

A complete study of the projected power requirements of the treatment facility was completed in 1967 and concluded that the antiquated power generating and electrical distribution facilities should be replaced.

[jj] In 1982 the Wisconsin legislature established the current Milwaukee Metropolitan Sewerage District (known as MMSD), which is governed by a single Commission of eleven members—seven of which are appointed by the City of Milwaukee.

In view of the continuous heat requirements in the sludge drying plant, the study concluded that the best power facility would be gas turbine-driven electric power generators with waste heat utilization in the existing sludge dryers. The decision was influenced by the availability of low-cost natural gas, the desire to assure the plant of an uninterruptable power supply, and the obvious need for a completely new electrical power system in the plant.

The close matching of the gas turbine exhaust to the dryer furnace air requirements was the key to the successful merger of these two systems.

The turbine exhaust also provides energy to heat-recovery boilers. These boilers provide a portion of the process steam required for heating and other miscellaneous plant services.

The performance of the waste heat recovery system has reduced dryer plant energy requirements approximately 72% and total plant energy requirements approximately 25% when compared on an identical plant production basis.

The successful operation of Milwaukee's sewage treatment plant led the way for many other American municipalities to adopt its methods of efficient environmental recycling.

Utilization of gas turbine exhaust heat in these waste heat boilers helped reduce overall plant fuel requirements by 20% in 1974. Photograph courtesy of MMSD.

BRADLEY CORPORATION

As manufacturing operations expanded at Allen-Bradley Corporation, Harry Bradley—brother of the founder of the company—came up with an innovation that would not only help workers wash up quickly, it would also reduce the floor space required for bathroom sinks. By 1918, Harry introduced prototypes of his "wash fountain," which would allow multiple users to wash up using the same sink. He applied for a patent on the device the following year.

The Allen Bradley Corporation was not interested in going into the plumbing fixture business, so Harry Bradley sought buyers to manufacture his innovation. He eventually sold the patent rights to some friends in 1921 and they formed a Wisconsin plumbing supply company that became known as the Bradley Corporation.

The company built a factory at 22nd and Michigan Street on Milwaukee's near west side.

DEVELOPMENT OF THE MULTI-USER WASH FOUNTAIN

Bradley's circular wash fountain was constructed as a single freestanding bowl, with water spraying upward from the bottom. It allowed a number of people to use the sink simultaneously. Sales were targeted toward factories, as well as other venues in which a large number of people were likely to need to wash up during a short period—such as at sporting events.

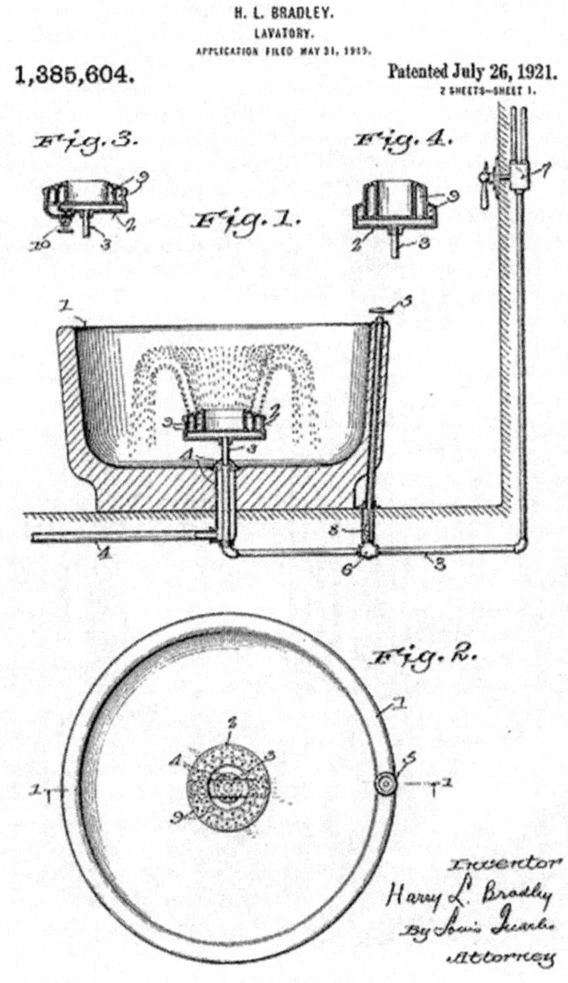

H. L. BRADLEY.
LAVATORY.
APPLICATION FILED MAY 31, 1919.
1,385,604. Patented July 26, 1921.
2 SHEETS—SHEET 1.

Fig. 3. Fig. 1. Fig. 4. Fig. 2.

Inventor
Harry L. Bradley
By Louis Tesche
Attorney

Right: Patent illustration for the lavatory designed for multiple use by Harry L. Bradley. The original 'wash fountain' sprayed up water from a sprinkler mounted in the bottom center of the sink.

In 1922, the company sold ninety-four units—demand increased rapidly in successive years and the company was soon selling thousands of units annually. In 1925, foot-operated models were introduced, as well as a semi-circular wash fountain that could be installed against a wall. Also, the company redesigned the water spray system. Users were complaining about occasionally getting soaked by water surges. In the 1930s, a downward spray head was introduced. A two-piece bowl was also designed to make the fountains easier to ship and install.

Over the succeeding years, Bradley Corporation has introduced numerous variations of its wash fountain product line, to meet the requirements of the marketplace. It has also launched a line of group showers, and various related plumbing accessories.

Workers are seen using the Bradley wash fountain in this promotional photograph

DIVING EQUIPMENT AND SALVAGE COMPANY (DESCO)

The Diving Equipment and Salvage Company was established in 1937—it is better known by its acronym, DESCO. The company was founded by two Milwaukeeans—divers Max Eugene Nohl and Jack Browne—with the financial backing of Norman Kuehn and the technical assistance of Edgar End, M.D. of the Marquette University School of Medicine.

Max Nohl was known nationally as a skilled diver. Born in Milwaukee in 1910, he obtained his initial diving skills in Lake Michigan. Supplementing these skills with a degree in engineering from the Massachusetts Institute of Technology, Nohl became the principal diver for an exploration of the South Sea Islands, which was chronicled weekly on a popular national radio show. He also served for a time on a team assembled by Hollywood producer, Colonel John D. Craig, who was interested in a deep-dive salvage of the Lusitania, the Cunard liner that had been torpedoed by Germany during the First World War. In exploring the possibility of a dive at that depth, Nohl teamed up fellow diver Jack Brown of Milwaukee and Dr. Edgar End of the Marquette University School of Medicine, who was a pioneer in studying hyperbaric physiology and medicine.[142]

Nohl had already been working on a suit design based on his MIT thesis. Browne and Nohl collaborated on the design of a new lightweight, self-contained diving suit. They contacted Edgar End about his pioneer work with a helium-oxygen mixture (dubbed 'heliox') to minimize the problems associated with nitrogen narcosis and decompression sickness (the 'bends' or caisson disease). End had developed decompression tables for the heliox gas.

Nohl and Browne established the new company to manufacture the newly designed diving equipment. Norman L. Kuehn, a Milwaukee businessman, provided financing. The company initially operated out of the rear of Kuehn's Rubber Company's building at 1053 North 4th Street but moved several times before locating at its current address on North Milwaukee Street.

In a demonstration dive on December 1, 1937, in Lake Michigan, Max Nohl used the newly designed equipment and a helium-oxygen mixture to dive to a depth of four hundred and twenty feet, breaking a record that had previously been set in 1915 by US Navy diver Frank Crilley.

The military's needs during the Second World War greatly contributed to the demand for diving equipment. US Navy contracts resulted in DESCO becoming the largest diving equipment manufacturer in the world.[143] The company was called on to produce over three thousand of the standard US Navy Mark V helmets during the war, to include Navy helium helmets. The company was also asked to design and build a compact oxygen rebreather, which resulted in the DESCO "B-Lung" and permitted divers to swim underwater without producing bubbles, which could disclose their position. The company also produced the Browne US Navy Diving Mask, the Browne Lightweight Suit and the Buie Mixed Gas Helmet.[kk] Additionally, DESCO produced ancillary diving equipment including weight-belts, shoes, knives, tools, and miscellaneous items including recompression chambers.

Toward the end of the war, DESCO installed its own pressurized wet tank for research and development. In April of 1945, Jack Browne used this tank to simulate a dive to a new record depth of five hundred and fifty feet of seawater, using a US Navy Lightweight Diving Suit—known as a "Bunny Suit" and breathing a heliox mixture under the supervision of Dr. Edgar End. The dive is considered a milestone in the development of modern diving techniques.

Following the war, Norman Kuehn and Jack Browne sold the company to Milwaukee businessman Alfred Dorst. In 1967, Thomas and Marilyn Fifield purchased DESCO and moved the company to its present address at 240 North Milwaukee Street in Milwaukee.

[kk] Commander E.D. Buie, a US Navy diver, worked with DESCO to design a low volume mixed-gas helmet for the US Navy to use clearing mines and for harbor clearance duties.

Over three thousand diving helmets were manufactured by DESCO during the Second World War for the US Navy. The large photograph shows helmets awaiting shipment. The superimposed image shows the helmet details. Photographs courtesy of DESCO.

Women are shown working the helmet soldering line at DESCO during the Second World War

Developments in Deep Water Diving Equipment

DESCO developed numerous innovations in deep water diving equipment. These developments resulted in world record dives of four hundred and twenty feet in 1937, and five hundred and fifty feet in 1945. Working collaboratively with Dr. Edgar End, the company developed equipment and expertise for helium-air breathing mixtures that permitted deep dives while minimizing the risks of nitrogen narcosis and decompression sickness.

During the Second World War, DESCO was contracted by the US Navy to design a compact oxygen rebreather that permitted divers to travel freely underwater without producing bubbles, which could reveal their location. During the war, the company also designed and produced the Browne US Nancy Diving Mask, the Browne Lightweight Suit—also known as the "bunny suit"—and the Buie Mixed Gas Helmet.

The development of the self-contained breathing lung—which eventually became referred to as SCUBA gear for "self-contained underwater breathing apparatus"—is perhaps the most important contribution to diving by DESCO during this period. It was developed five years before the famed Aqua-Lung was developed in France by engineer Émile Gagnan and Naval Lieutenant Jacques Cousteau. The Cousteau/Gagnon open circuit SCUBA technology was eventually universally adopted and the DESCO rebreather fell out of favor.

The Journal of Diving History describes Max Gene Nohl and Jack Browne as among "American's most innovative divers. Their working relationship during the 1930s created equipment, broke records, furthered deep diving technology and created DESCO. (Browne's) development of scuba diving equipment ... was one of his greatest contributions."[144]

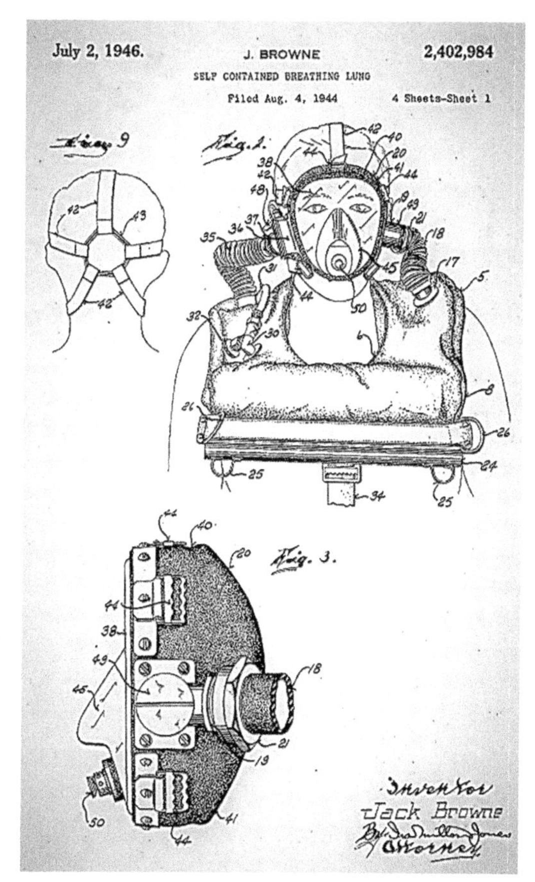

Illustration from Jack Browne's patent filing for his self-contained breathing lung, which eventually became described as SCUBA gear.

In the 1960s, the company developed the DESCO Air Hat to meet the requirements of the commercial diving industry. The Air Hat was designed by Thomas Fifield, owner of the company. It is a "free flow" helmet of contemporary design while employing time proven techniques and materials in its construction similar to the heavy gear helmets of the past. While very rugged, it can be disassembled and reassembled using only a screwdriver and wrench. It can also easily be donned or removed by the diver without assistance from the tender. Using an adjustable exhaust valve the Air Hat can be adjusted to either side of neutral buoyancy underwater. With its low center of gravity and excellent fore and aft balance, it rests comfortably on the divers head. The interior headliner is adjustable for size and foam pads covered in leather are provided for comfort. The DESCO Air Hat is often used for working in hazardous environments where contamination of the diver is a concern.

The DESCO Air Hat. Photograph courtesy of DESCO.

MAXIMILIAN EUGENE NOHL was born in Milwaukee in 1910. As a young child, he developed a fascination with underwater exploration and diving. He dove using a five-gallon paint can over his head and a garden hose supplied with compressed air—a dangerous practice which he fortunately survived. He went on to attend the Massachusetts Institute of Technology where he earned his degree in engineering—and designed several diving-related devices. For his thesis he explored the design of a self-contained diving suit for deep diving. Nohl took time out of his studies to join the Phillips Lord expedition.[ll] Lord hired Eugene Nohl for undersea exploration in the South Pacific using his "Hell Below" diving shell. The promotional materials for the expedition stated that Nohl would photograph "the sunken civilizations of the South Sea islands of its deep marine life and formations," and "search for sunken treasure and bring back film of shipwrecks."

Max Gene Nohl

The expedition was sponsored by the Frigidaire appliance company and set sail in December 1933 for the South Pacific via the Panama Canal.[145] The radio series attracted a good audience and was very popular. For Nohl, the radio show provided a good deal of publicity for his diving knowledge and skills.

It appears that Nohl left the expedition in Haiti, however, and returned to the United States.[146] He completed coursework at MIT and then set his sights on salvaging cargo and artifacts from various sunken ships to include the Prohibition-era rumrunner, John Dwight. He also participated in planning for an effort to dive and explore the RMS Liner *Lusitania*, working with photographer John D. Crag who owned film rights to the site. While this effort was eventually abandoned, Nohl provided diving support equipment that would be needed for the project, including the diving suit he had originally designed for his MIT thesis.

In 1937, Nohl, along with Milwaukeeans Jack Browne and Dr. Edgar End, formed the Diving Equipment and Salvage Company to manufacture and sell newly designed diving equipment. Norman Kuehn, a Milwaukee businessman, largely financed the new corporation. In December of that year, Nohl established a world record for diving to a depth of 420 feet in Lake Michigan, using DESCO's new diving equipment and breathing a heliox mixture under the direction of Dr. End. This dive broke a record that had been held since 1915.

The following year Nohl filed for a patent (granted in 1943) for improvements in respiratory apparatus for diving. That year he also designed, in collaboration with Jack Browne, a SCUBA rig—five years before the famed Aqua-Lung was developed in France by engineer Émile Gagnan and Naval Lieutenant Jacques Cousteau. After testing the equipment in the chamber at the Milwaukee County Emergency Hospital and in Lake Michigan, Dr. End took the equipment to the West Indies to search for Spanish treasure ships in 1938-39. From 1938 to 1940, Max Nohl worked on the Tarzan films starring Johnny Weissmuller.

While returning from a vacation in Mexico in 1960, Nohl and his wife were killed in an automobile accident.

[ll] Phillips Lord was a radio personality who had created the character "Seth Parker," a clergyman and backwoods philosopher. The show was syndicated on NBC Radio. In 1933, Lord came up with the idea of sailing a ship to exotic places around the world and broadcasting show from the ship via short-wave radio.

JACK BROWNE was also a survivor of tin can diving as a boy. His father was an executive with the Goodrich Transportation Company in Milwaukee. Jack became interested in diving while a freshman in high school. He would take on jobs for pay or just dive for fun. He also had a knack for invention and his first diving helmet was literally a tin can. In 1934, he and fellow Milwaukeean Max Nohl dove on local wrecks. A captioned photo appeared in The Milwaukee Journal on April 4, 1935, of Jack and his friends Paul Gallun, Fred Lange, and Bob Wescott testing a homemade diving helmet in a Shorewood swimming pool.

When Max Nohl and John Craig began work on the equipment for diving to the RMS *Lusitania*, Jack was ready to pitch in. The custom diving dress they needed was stitched together from canvas and taken to the N.L Kuehn Rubber Company to be made watertight.

Norman Kuehn became a mentor to Jack and an angel to the new company—Nohl, Browne, and Craig collaborated in 1937 to form the Diving Equipment and Supply Company. Jack experimented with new designs for breathing tanks and lighter suits.

Browne served as president of DESCO during the Second World War, during which time he is credited with the development of the Browne US Nancy Diving Mask, the Browne Lightweight Suit, the Browne Utility Mixed Gas Helmet and compact oxygen rebreathers for US Navy covert operations. Browne had at least a half-dozen patents for various diving equipment.

In 1945, Browne used DESCO's pressurized wet tank to simulate a dive to a new record depth of 550 feet of seawater. As in the case of Nohl's earlier dive, he breathed a heliox mixture under the supervision of Dr. Edgar End. Both dives were milestones in the development of modern techniques of mixed-gas diving. Max Nohl was asked how he felt about Jack breaking his record. Max replied, "Records are made to be broken."

In 1946, Jack left DESCO to help run the family automobile dealership, Browne Motors Chrysler-Plymouth in Milwaukee. In June 1949, he became president of the dealership after his father George passed away. He left Milwaukee in 1950 to start a commercial fishing business in Florida.

DESCO's website states that in 1958 Browne was captured by the Baptista Cuban Air Force and imprisoned for flying guns to Fidel Castro's Cuban rebels. He apparently managed to escape from prison and steal back his plane, but ran out of fuel on his flight to Florida and was rescued by the US Coast Guard. Newspaper accounts report that in 1963 his newly outfitted boat, the 115-foot *Shrub*, sank about a hundred miles north of Cuba and that he and the other divers on board spent several days in a lifeboat that eventually drifted to Cuba, where the party was arrested by the Cuban government and held for several months before being released.

Jack retired to the Virgin Islands. He died in 1998.

MILWAUKEE COUNTY EMERGENCY HOSPITAL

The Milwaukee County Emergency Hospital, more formerly called the Milwaukee County Dispensary and Emergency Hospital, was located at 24th and Wisconsin Avenue. It was built between 1927 and 1929 to serve the medical needs of the near west side of the City.

The building was designed in the Neo-Classical style by architectural firm Van Ryn & DeGelleke. The hospital became well known, in part, because it housed the practice and laboratories of Dr. Edgar End, a pioneer in hyperbaric chamber work which aided in deep sea diving.

After serving as a medical facility for many years, the hospital closed and the building became vacant for a number of years. More recently, it was rehabilitated to serve as a Chinese language elementary school and a project-based high school for the Milwaukee School system. It has also served other social service needs of the community.

Recent photograph of the former Milwaukee County Emergency Hospital, which was later converted into an elementary school.

WALK-IN HYPERBARIC CHAMBER

The first walk-in Hyperbaric Chamber in the United States was designed by Joseph Charles Fischer and located at the Milwaukee County Emergency Hospital. It was used for medical treatment, as well as for pioneering research, for 50 years.

Fischer, who was chief engineer at Milwaukee County Institutions, designed and built the hyperbaric chamber in 1928. Prior to this achievement, the only such chambers in use were designed like typical pressure vessels with small hatches for entry.

Through the years, the chamber was used to revive hundreds of individuals with carbon monoxide poisoning, and was used in the healing of patients with gangrene and senility, of patients who had suffered strokes, and of divers and deep tunnel workers stricken with the "bends."

Starting in 1936, Edgar End, M.D., of the Marquette University School of Medicine began using the chamber for his studies in helium-oxygen compressors. His research led to a new diving record of 420 feet on December 1, 1937, by Max Gene Nohl. The earlier record had been set by Frank Crilly, who in 1914 dove to 306 feet for the Navy at Pearl Harbor. The chamber was also used for other pioneer work in hyperbaric medicine, including open-heart surgery in an oxygen atmosphere.

On December 22, 1938, Dr. End and Max E. Nohl made the first intentional saturation dive using the device. Nohl spent twenty-seven hours breathing air at a 101 feet (30.8 m) diving depth. He then went through decompression in the chamber, leaving Nohl with what was reported as a mild case of decompression sickness that resolved with recompression.[147]

In 1942, the first tables for diver decompression were published based on work in the Milwaukee chamber. The data greatly shortened the time required for recovery after a dive.

The chamber was also used during the Second World War after conversion as a vacuum chamber for the study of the bends brought on by high altitude flying, and for Navy tests that led to the manufacture of the Browne Lightweight suit and of frogman gear.

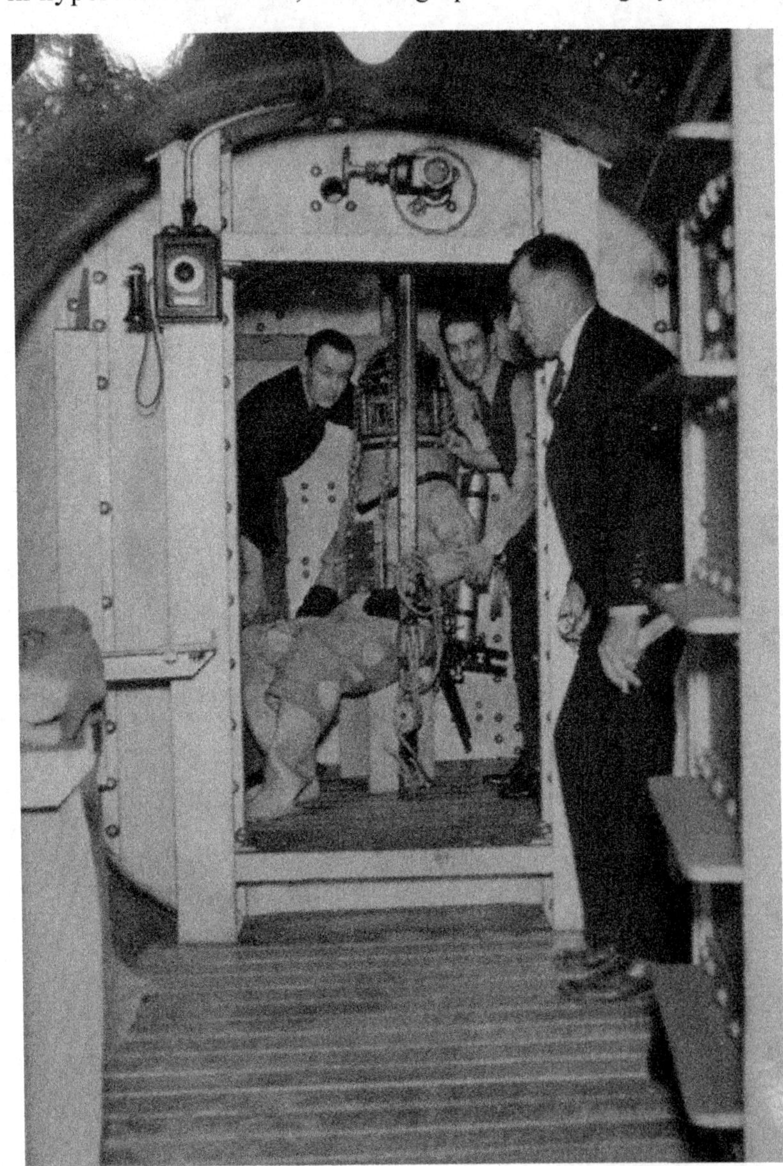

Picture of the chamber being used for testing of diving gear. Dr. End is shown outside the chamber.

In 1947, the chamber was used to improve the design of SCUBA gear,[mm] five years before Jacques-Yves Cousteau and Émile Gagnan Cousteau introduced their "Aqualung" in France.

The chamber was 18 feet long, seven feet in diameter and tested at 150 psi. It was built for an installed cost of only $5,000. The chamber's service pressure was gradually reduced to 30 psi before being taken out of service in 1976 and dismantled in 1979. It was used longer than any such chamber in the world.

EDGAR END, M.D attended the Marquette University School of Medicine, graduating in 1936. During his career, he served as Assistant Clinical Professor of Environmental Medicine at Marquette, an intern at the Milwaukee County General Hospital and eventually as Director of the Milwaukee County Hyperbaric Chamber. He specialized in hyperbaric oxygen therapy—medical use of oxygen at a pressure above atmospheric pressure.

End began tests using a helium-oxygen mixture, which he called "HELIOX," believing that helium would be less soluble in the body than nitrogen. He developed new decompression tables for the mixture and tested them on himself in the Milwaukee County's hyperbaric chamber. The tables specify the maximum rate of accent (or the maximum rate at which pressure is reduced) required to safely avoiding entrainment of gas bubbles in the diver's tissues. Typically maximum ascent rates are in the order of ten meters (33 feet) per minute for dives deeper than six meters (20 feet).[148]

He and diver Max Nohl used the mixture to set the then world's record dive of 420 feet. The dive took place in Lake Michigan and was broadcast live on NBC radio. Dr. End and Nohl designed and built the first self-contained underwater breathing apparatus (SCUBA), years before well-known French diver Jacques Cousteau invented his "Aqua-Lung" in the 1940s.

During the Second World War, Dr. End performed further experiments and broke new diving records. He also helped to design diving apparatus for the US Navy, including frogman gear, suits for mine disposal experts and underwater demolition team diving outfits.

Dr. End continued as Director of the Milwaukee County Hyperbaric Unit and treated nearly all local cases of the bends. He also used the chamber for the treatment of carbon monoxide poisoning and other injuries. He had a practice of staying in the hyperbaric chamber with his patients during their treatment—a practice that generated a good deal of praise.

He died in 1981 at the age of 70.[149]

[mm] SCUBA is an acronym for "self-contained underwater breathing apparatus."

CHAPTER 5: BAY VIEW, TOWN OF LAKE AND SOUTH

On January 2, 1838, the Wisconsin territorial legislature divided Milwaukee County into two townships—the Town of Milwaukee which encompassed everything north of the present Greenfield Avenue, and the Town of Lake which encompassed the area to the south of Greenfield Avenue. The Town of Lake was sparsely settled, with just over four hundred residents at the time of the 1840 census.[150]

One of the more populated areas of the Town of Lake was located just south of the present Greenfield Avenue and adjacent to Lake Michigan, an area that came to be known as Bay View, because of the scenic views of Lake Michigan from its bluffs. In 1868, the Milwaukee Iron Company opened a plant in the area, as described below, and Bay View was established as a company town.[nn]

This drawing looks north toward the City of Milwaukee, from the Village of Bay View in the Town of Lake.

In 1879, Bay View incorporated as a village with a population of 2,592. It was annexed into the City of Milwaukee in 1887.

Located south of Bay View, the community once known only as the settlement adjacent to the Buckhorn railroad station, became Cudahy in the late 1800s when Patrick Cudahy purchased seven hundred acres of property in the Town of Lake to build a meatpacking plant. Located along Lake Michigan, his new property was adjacent to the railroad. Cudahy considered railroad access essential for future shipping goods and receiving materials for industry, and providing transportation to the area.[151] This pattern will be also seen in future chapters—Milwaukee's industrial companies tended to follow the railroad lines when looking for property for expansion.

The Cudahy Brothers Company started to build the new plant in 1892. As a result, the Chicago and Northwestern Railway station was rebuilt. It was renamed the Cudahy Depot.

The new community of Cudahy did indeed attract industry to the area. It also attracted workers, who purchased home sites near the factories, and merchants, who wished to meet the needs of the young community. The little settlement grew into the Village of Cudahy in 1895 and became a city in 1906.

[nn] A company town is a community where practically all housing and stores are owned by a company that is the principal employer. The company provides infrastructure (housing, stores, transportation, sewage and water) to enable workers to move there and work in its factories.

Following several expansions, the City of Cudahy now borders Lake Michigan on the east, General Mitchell Airport on the west, East Lunham Avenue on the north and College Avenue on the south.

The area that now encompasses the cities of South Milwaukee and Oak Creek was first settled in 1835 as the Town of Oak Creek, which in 1841 became the seventh and final of Milwaukee County's townships. The area was largely rural, but a small community built up along the old Chicago Road that traversed the area. The town was also crossed by the Chicago & Northwestern Railway, which was essential for its industrial development.

In 1891, two local businessmen, along with some Milwaukee-area capitalists, formed the South Milwaukee Company—a land development company that began securing land options in the northeast corner of the township. In the following year, the Village of South Milwaukee was incorporated.

The South Milwaukee Company provided enticements to attract manufacturers to the area. The Chicago & Northwestern Railroad helped this effort by agreeing to run a spur line to each manufacturing site.

The effort was quite successful. Among the early companies enticed to the new village was the Bucyrus Steam Shovel & Dredge Company, which relocated from Ohio. The factories, in turn, attracted thousands of workers to the area. As a result, the village transformed quickly. A downtown shopping district was formed, along with a hotel and a new railway station.

Meanwhile, the balance of the Town of Oak Creek remained largely rural in nature. Farming was the prominent occupation well into the 1950s. Then, in the early 1950s, the Wisconsin Electric Power Company started to build a new power plant in the Town of Oak Creek on the shores of Lake Michigan. The City of Milwaukee was covetous of the utility taxes the plant would generate and sought to annex the area. The residents of the area struggled to remain independent. The situation was resolved in Oak Creek's favor with the passage in 1955 of a Wisconsin statute known as the "Oak Creek Law."[152] Crafted by the town's attorney Tony Basile, it was shepherded through the Wisconsin legislature with the help of Wisconsin state senator Leland McParland, who represented the 7th district, which included the Town of Oak Creek.[153]

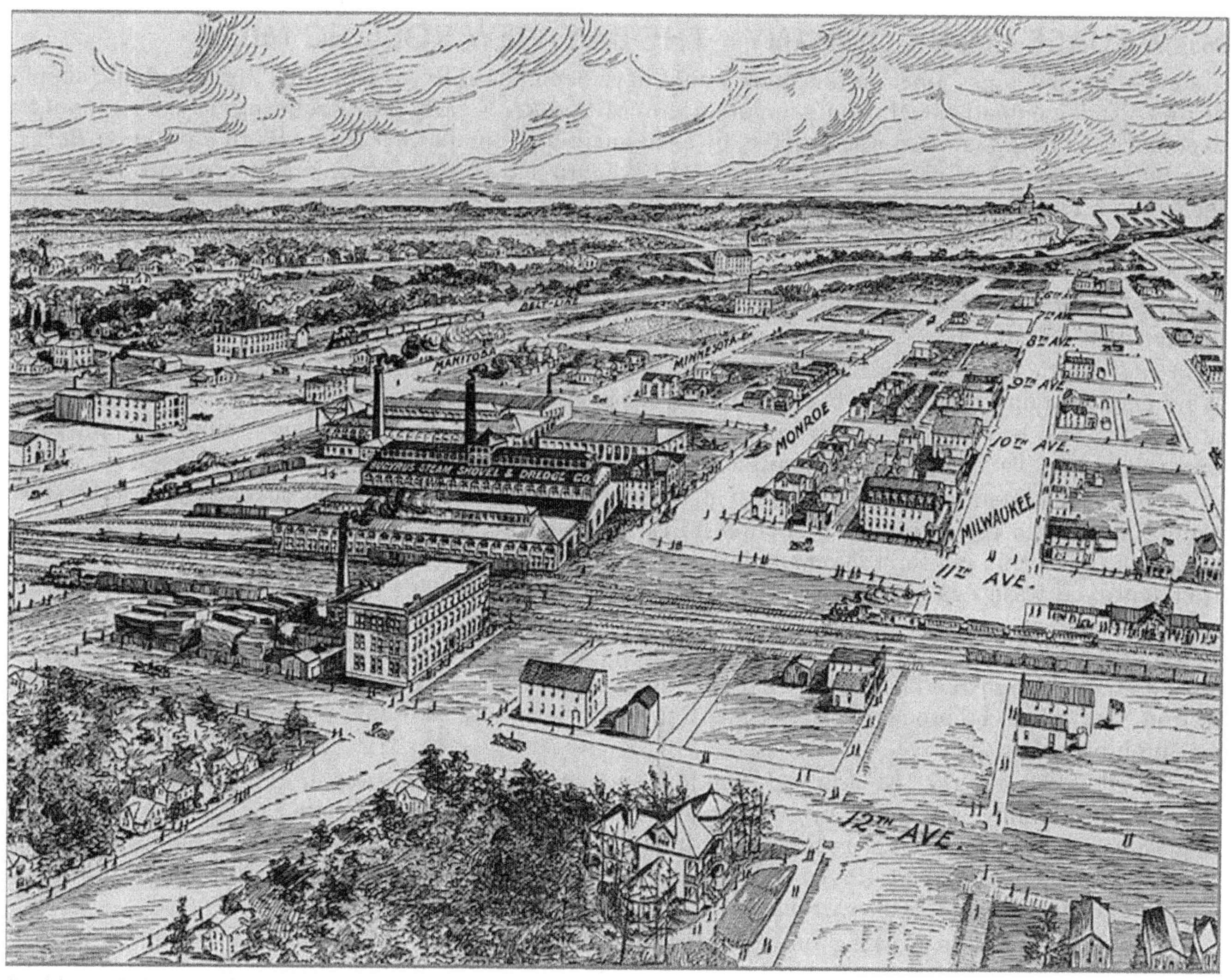

As this early image of South Milwaukee suggests, the railroad was an important asset to the Bucyrus Steam Shovel & Dredge Company. It allowed the company's products to be shipped anywhere.

MILWAUKEE IRON COMPANY—THE BAY VIEW ROLLING MILLS

Few of the machines noted in this book would have been possible without a readily affordable supply of iron—and eventually steel. Iron's importance to Milwaukee's early manufacturing industry cannot be overstated. In 1845 that need was satisfied by a shipment by Captain Eber Brock Ward's freighter *Baltic*. The *Baltic* unloaded its cargo at a warehouse at Clybourn and Chase streets. Ward apparently also used the trip to establish contacts with industry in Milwaukee—realizing that Milwaukee was a large potential market for iron.

Eber Ward's shipping business, which carried freight between Chicago, Milwaukee, Detroit and other ports, expanded significantly between 1845 and 1854. He came to own one of the largest shipping fleets on the Great Lakes.

In addition to shipping flour, lumber and other goods from Chicago and Milwaukee to eastern ports, Ward expanded into the iron ore business in Michigan's Upper Peninsula—as well as into the processing of iron. In 1853, he built the Eureka Iron Works at Wyandotte, near Detroit, using iron ore from the Lake Superior region. By 1847, he built a mill an iron mill in Chicago.

Ward concentrated on the need for providing rails for the booming railroad business. His mills were rolling mills, also referred to as reduction mills, in which hot iron bars are passed through rollers to reduce the thickness and form the metal into long strips of the desired shape—in this case railroad rails. Iron, however, wasn't an ideal metal for railroad rails. Iron rails exhibit an excessive wear rate, requiring replacement in as little as two years. The rails had to be removed, returned to the mill, heated up and re-rolled to achievc the desired shape. While the need for re-rolling was certainly good for Ward's business, he and others recognized the need for a better, more durable material. By controlling the carbon content in iron, steel could be produced. But at that time, steel was only available in very small quantities and was prohibitively expensive.

That all changed in the 1850s. It had been long recognized that impurities in iron caused it to be more brittle than desirable. An American, William Kelley of Pittsburgh, came up with a solution to remove the impurities. By blowing oxygen through molten pig iron, he discovered that he could remove most impurities, including carbon. However, experiments revealed that removing all of the carbon resulted in a product that, while better than wrought iron, was not as durable as desired.

Henry Bessemer was working on a similar solution in England. He discovered that forcing high-pressure air through molten iron ignited the silicon and carbon impurities in the iron. As the iron became hotter, more impurities burned off, further reinforcing the conversion. He patented his process in 1855, and licensed it to four ironmasters, who tried the process but failed to achieve a durable steel.

It fell to Robert Mushet of Scotland. He developed a process to burn off the impurities and carbon, and then reintroduced the desired amounts of carbon and manganese—resulting in the required durability. The puzzle had been solved.

Back in the United States, Eber Ward heard of the Bessemer process and sent Zoheth S. Durfee to England to investigate. While Zoheth Durfee was traveling, Ward also proceeded to gain control of the Kelley-Mushet patents and erected an experimental steel plant next to his Wyandotte Iron Company. Upon Zohelth's return, he and his cousin, William, succeeded in making Bessemer steel—the first produced in the United States.

Alexander Holley, who had secured the Bessemer patents in the United States, sued Ward. In order to resolve the litigation and obtain clear rights to produce steel, Ward brought Holley into his company as a shareholder.

After successful tests of using the steel produced in Wyandotte, Eber Ward greatly expanded his entry into the steel business. In addition to the Eureka mill in Wyandotte, Michigan, Ward built a steel mill in Chicago called the Illinois Steel Company, and the Milwaukee Iron Company in Bay View.

When Ward first visited Bay View, the area was dotted with a few cabins and farms of early Yankee settlers—and not much else. It was, however, conveniently close to the mouth of the harbor,[oo] which provided ready access to Lake Michigan shipping. The Bay View site was only forty miles from the iron mining town of Iron Ridge in Dodge County.[pp] The site was also close to the more established pre-Milwaukee communities of Kilbourntown, Juneautown and Walker's Point. In Ward's view, this later point made the area a reasonable location to establish an iron mill.

In order to raise capital for the venture, Ward approached several influential Milwaukeeans into the venture. Once capital was secured, the enterprise purchased an 114-acre site in Bay View for the mill.[154]

Construction began in 1867. In addition to the initial mill for producing railroad rails, the group built an office, machine shop, a blacksmith shop, boiler shops, a storehouse and a pier on Lake Michigan. They also constructed a large boarding house, known as the Palmer House, along with 24 company houses for the workmen. The first rail was rolled in April of 1868.

Ward quickly made plans to produce steel in Bay View, using the Bessemer process. He assembled skilled workers for the effort—many of which were from Staffordshire England and were housed in the recently completed Palmer House. The new blast furnace was placed into service in 1870, along with the equipment to manufacture and reroll up to 30,000 tons of rail annually.

It was soon discovered that the pier on Lake Michigan did not provide adequate protection for ore ships under storm conditions. A decision was made to abandon the pier and to dredge the Kinnickinnic River to accommodate the ore traffic. The improvements to the Kinnickinnic not only provided excellent protection for the boats serving the mill, it also helped to open up the Milwaukee harbor basin for later development.

The mill prospered for a time, as did the community of Bay View. However, an economic depression in 1873 reduced demand for iron and steel. The railroad companies halted nearly all new construction, greatly reducing the demand for new rails. The impact upon the Milwaukee Iron Company was severe. The situation was made even worse by Eber Ward's death in 1875. By 1876, the company was forced to close.

The Milwaukee Iron Company was acquired by the North Chicago Rolling Mills Company, which in 1889 was folded into the Illinois Steel Company and became known as their "Milwaukee Works." Soon, however, Illinois Steel was acquired by Federal Steel, which in 1901 became part of the United States Steel Corporation.

The Milwaukee Works continued to operate until 1929—the buildings were demolished about a decade later.

[oo] The Milwaukee and Menomonee rivers combined and then flowed south toward the Kinnickinnic, where the three rivers then exited to Lake Michigan. The original outlet was considerably south of the current harbor near today's Jones Island—essentially on the doorstep of Bay View.

[pp] Iron ore was discovered near Mayville in 1845 by William Foster, Chester May and E.P. May. A large sample of the ore was sent to Milwaukee, tested and found to be good quality iron ore. Nearby Iron Ridge became the site of the Iron Ridge Mine which operated from 1849 to 1892 and from 1896 to 1914. The ore deposits were acquired by Eber Ward's three mills in 1869.

The Milwaukee Iron Company's rolling mill was the first major heavy industry in the region and an important producer of iron and steel for the Midwest. The mill, which opened in 1868, transformed ore from Dodge County and Lake Superior area mines into iron products including thousands of tons of rail for the region's growing railroads. It was eventually acquired by the Illinois Steel Company.

WEISEL & VILTER MANUFACTURING COMPANY

In the year 1867, Peter Weisel equipped a machine shop in Milwaukee for general jobbing and the manufacture of slide-valve steam engines. Shortly after starting in business, Weisel formed a partnership with Ernest Vilter and the firm did business under the name Weisel and Vilter.

The firm was located on Chicago Street between Water Street and what is now known as Broadway. During the first 20 years, the firm broadened its scope from a general machine shop to include the manufacture of Corliss engines, refrigerating equipment, and brewery equipment.

Weisel & Vilter built its first refrigerating machine in 1882. It was a horizontal double-acting compressor, driven by a slow speed Corliss engine. Its advantages over the vertical single-acting compressors manufactured by nearly all of its competitors were immediately recognized.

On October 28, 1892, a fire that started in the Union Oil Company's plant destroyed almost the entire third ward of the city, including the Weisel & Vilter plant. It was decided that a new plant should be built on a larger site and the firm located on Clinton Street, which is now known as South 1st Street, near Beecher Street in Bay View. The firm also shortened its name to the Vilter Manufacturing Company with Theodore O. Vilter, son of Ernest, serving as its president.

Prior to 1890, the refrigeration industry was small since the applications for refrigeration were few. In that winter, however, seasonal conditions were not favorable for harvesting ice from the rivers. Milwaukee breweries and other ice users began turning to mechanical refrigeration. As a result, Vilter experienced a decade of rapidly increasing business.

Vilter's Corliss engine-driven refrigeration equipment produced six tons of ice per ton of coal. Probably the outstanding improvement during the decade was in the automatic oiling and lubrication devices added, designed by an engineer with the Company—William Nugent. Nugent eventually left Vilter and in 1897 founded, in Chicago, the Wm. W. Nugent Company to manufacture devices for central oiling systems.[155]

In 1910, Vilter furnished the equipment for the first large railcar pre-cooling plant. It was for the Santa Fe Railroad and was installed in San Bernardino, California. It had the capacity of pre-cooling 150 carloads of fruit per day. The cars were spotted along the dock and large air ducts ran along the dock for supplying and returning refrigerated air. The products were cooled in four hours, after which the railroad cars were iced in the usual manner.[156]

Prior to 1915, almost all low-temperature cooling applications used absorption refrigeration machines. Late that year, Vilter bid on and received the order for equipment for the Chicago Central Cold Storage Company. In order to achieve the -20° F brine temperature required for the job, Vilter designed and patented a two-stage compression machine with the two-stage compressors driven by a newly designed poppet valve steam engine. Vilter was able to furnish the chilled brine at a large savings in power over that achievable with absorption refrigeration machinery. Because of the success of the installation, absorption machinery was quickly displaced from the refrigeration field.[157]

Theodore O. Vilter served as the president of the American Society of Refrigeration Engineers in 1915-1916.

This photograph of the Vilter shop floor shows castings for refrigeration compressors lined up in the foreground. Photograph from the Historic Photo Collection of the Milwaukee Public Library.

Vilter's VMC 440 reciprocating compressor, developed in 1945, is still considered the refrigeration industry standard.

An early Vilter refrigeration compressor.

During the Second World War, Vilter, like most Milwaukee manufacturers, dedicated its factory to the wartime production effort. Vilter Manufacturing received a contract of over a million dollars to manufacture 105-millimeter howitzers, replacing the 75MM guns used during the First World War. Tooling was finished in early 1941 and the first gun was accepted by the US Ordnance Department on May 7, 1941. The guns started to roll off Vilter's production line and additional orders were received.

Vilter's howitzers were the first 105-millimeter guns turned out by private industry in the United States. The eight-foot long, twelve-hundred-pound barrel could hurl a thirty-three pound shell twelve thousand yards with a high trajectory.

This is a war production worker at the Vilter Manufacturing Company making parts for M5 and M7 guns for the U.S. Army. Her name is not known—she was described as an "ex-housewife, age 24, filing small parts." Her husband and brother were in the armed service at the time. US Farm Security Administration Photo.

In 2005, Vilter was acquired by a private equity group, which sold the company to Emerson Climate Technologies four years later. Vilter continues to operate as part of Emerson's refrigeration division—providing commercial refrigeration products and solutions. Their manufacturing plant continues to be located in Cudahy.

WHITEHILL MANUFACTURING COMPANY

Scottish immigrant Robert Whitehill was born in June 1845 and eventually immigrated to the United States, settling in Boston. In 1882, he established the Whitehill Manufacturing Company in Milwaukee for the principal purpose of manufacturing sewing machines of his design. With the financial support of Milwaukee-area businessmen, including Guido Pfister and Fred Vogel, the company constructed a large brick factory adjacent to the Kinnickinnic River on Becher Street.

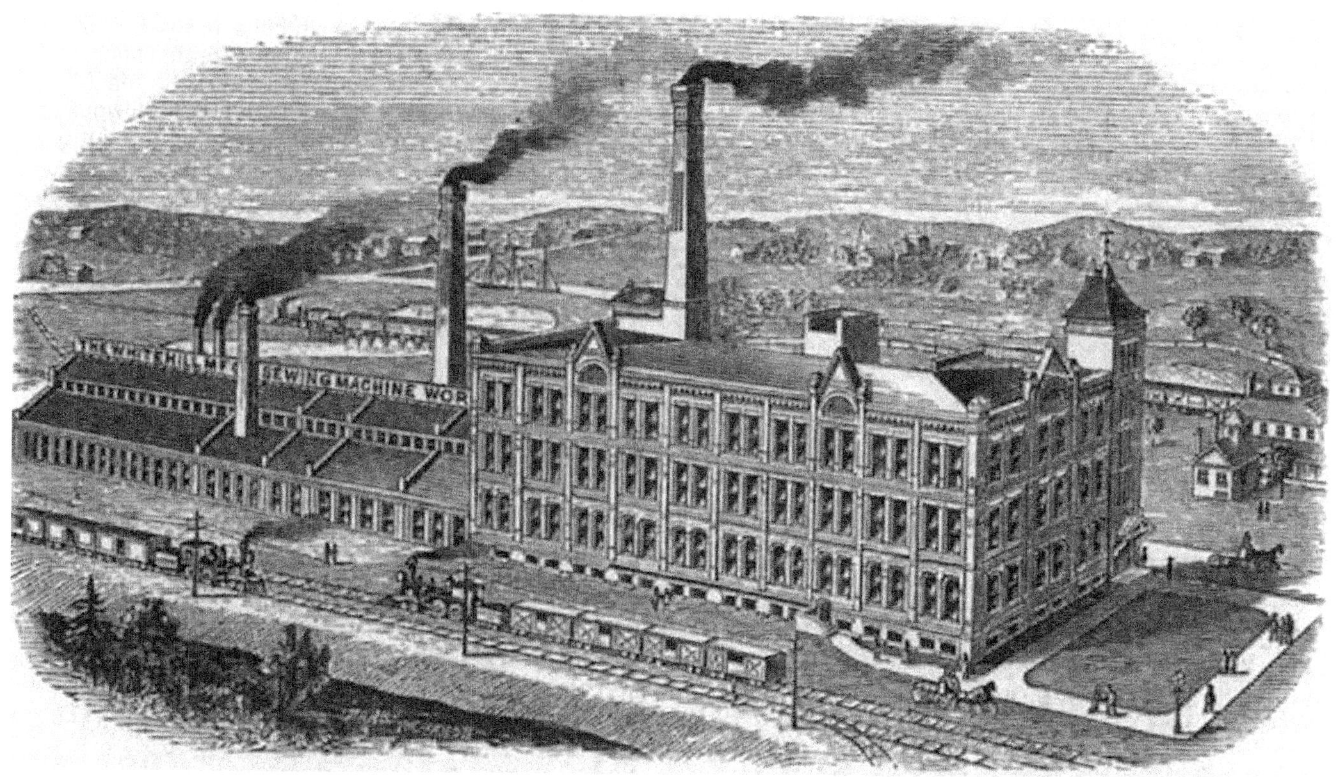

Illustration of the Whitehill Manufacturing Company's factory located in Bay View on Becher Street. From Industrial History of Milwaukee, 1886.

The company quickly established production in Milwaukee, hiring as many as three hundred employees to manufacture Whitehill's patented sewing machine. The staff included Henry Harnischfeger, who left his position at the Singer Sewing Machine Company and moved to Milwaukee to work as a foreman in the Whitehill milling department, and Alonzo Pawling, who worked at Whitehill as a patternmaker.

The patented sewing machine was recognized as the "most rapidly running sewing machine" by the centennial commission. It had a novel shuttle mechanism that featured a cylindrical shuttle, which operated with a crank motion, at a time when shuttles typically employed noisy and irregular cam motion. As a result, Whitehill's machine could operate at high speed and provided even stitches across a range of thread sizes, without the need for adjusting tension.[158]

In spite of the features of the new sewing machine, the market failed to materialize. Within a short time employment was reduced to half, and after struggling for an additional year the company was forced to shut down by 1887. The local investors in the enterprise lost a substantial amount of money. The building and its machinery passed into the hands of the Wilkin Manufacturing Company.

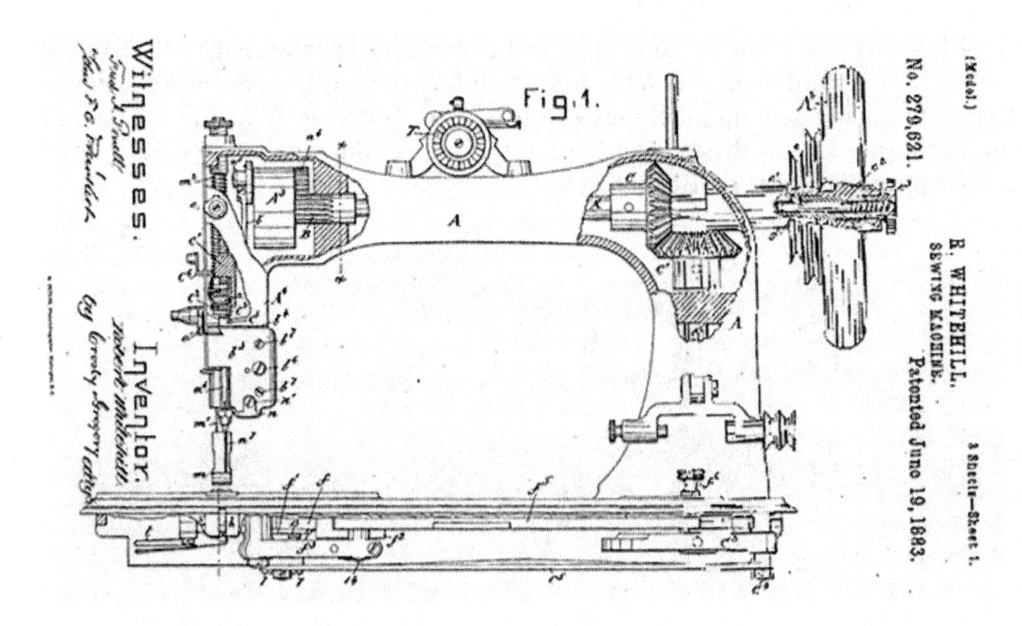

Drawing from Robert Whitehill's 1883 sewing machine patent. This appears to be the machine manufactured in Milwaukee. Drawing rotated for clarity.

Three employees of the company, Henry Harnischfeger, Alonzo Pawling and Maurice Weiss, left the company in 1883 to go into business for themselves, starting the Milwaukee Tool and Pattern Shop—which eventually became Pawling and Harnischfeger and is today the Joy Global Corporation.

Robert Whitehill went on to refine his sewing machine mechanism further. He obtained a patent for a new mechanism that employed a bullet-shaped, vibrating shuttle and approached the Singer Sewing Machine Company with the new design. He showed them a working model and demonstrated its superiority over Singer's best models. Singer promptly acquired the rights to the new mechanism, paying Whitehill about $8,000 (equivalent to about $200,000 in 2016).[159]

NORDBERG MANUFACTURING

When Bruno Nordberg outgrew his factory in the Menomonee Valley, he selected an area near present day Oklahoma and Chase Avenues for expansion—an area just west of the Bay View neighborhood in the Town of Lake, which is now in the City of South Milwaukee. The site had convenient access to the Chicago, Milwaukee and St. Paul Railway.

The company began moving into the new buildings in 1901. It is reported that the machine shop that Nordberg built was the first in the world to incorporate separate electric motors for each machine tool, instead of the universally accepted practice of installing overhead line-shafts and separate belt take-offs for the various machine tools.[160]

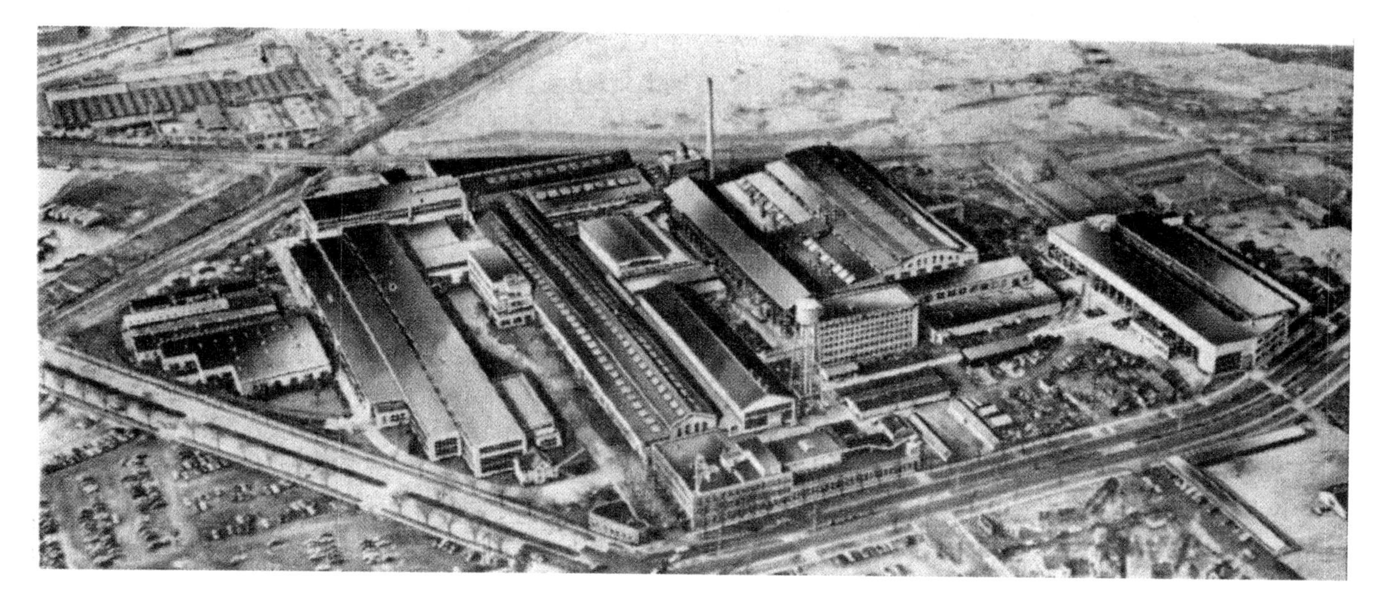

Nordberg initially manufactured steam engines, air compressors, gas compressors, mine hoists and blowing engines. Eventually, the company extended its product line to build mineral and rock crushing equipment, screens, grinding mills, railroad maintenance equipment, hydraulic valves, presses and heavy-duty diesels

In 1944, Nordberg designed and built the largest diesel engine ever built in the United States. It was destined for a Victory ship for the United States Maritime Commission.

The company was acquired by Rex Chainbelt Incorporated (formerly the Chain Belt Company) in 1970 and became a division of that company.

Nordberg produced some remarkable machines at its plant in South Milwaukee, several of which are discussed in this section.

Nordberg's Quincy Mine Hoist

When the Quincy Copper Mining Company was planning a new and deeper mine shaft—known as the Quincy No. 2—it solicited bids for the mine hoist. Bruno Nordberg went to visit the mine in the Houghton-Hancock area of Michigan's Upper Michigan. When pressed about the ability of his firm to deliver the enormous mine hoist required for the task, it is reported that Nordberg asked the firm to pull out the drawings for the hoist for Quincy's first mine. He pointed out that the drawings bore his signature and promised to design and build a better one. On the strength of this, he succeeded in obtaining the order.

In 1919, the Nordberg Machinery Company installed what was to become the world's largest steam engine-driven hoist on the No. 2 shaft at the Quincy Mine. The hoist was the crowning achievement in the design and manufacture of almost 40 smaller hoists up to that time for Michigan's "Copper Country" development.

The hoist was a balanced design of conico-cylindrical cast construction, designed to lift 20,000 pounds of ore at a speed of 3,200 feet per minute from a vertical depth of 6,300 feet. The hoist drum itself was thirty feet in diameter. The overall installation, including the steam engine, was sixty feet high by 54 feet long and weighed 1,800,000 pounds. It carried over 10,000 feet of cable and was driven by a 2,500 horsepower steam engine.

The large diameter made it possible to couple the double cross-compound Corliss-type steam engine directly to the hoist, thereby eliminating the need for gear reduction. The entire engine, hoist, foundation, building and erection cost the company the magnificent sum of $370,000. The building was the first built of concrete block in the Keweenaw Peninsula.

The hoist operated from 1921 to 1931, through the depths of the Great Depression, and contributed to the general prosperity of the mine, which was important to the financial health of the area. The design and manufacture of this great hoist consolidated Nordberg's position as the major builder of large mine hoists throughout the world.

The world's largest steam engine-driven mine hoist was installed at the Quincy mine by Nordberg Mfg. Go. Currently, the Quincy Mine Hoist Association maintains the hoist as a tourist attraction. The hoist was designated an ASME landmark in 1984.

Nordberg Self-Powered Adzing Machine

From the beginning of railroads, it has been necessary to replace the rails periodically. This is required because of wear, or due to an increase in the loads it must carry. In replacing rail, procedure and practice required that it be done one rail at a time. This permitted the rail remaining on the track to be used for reference. Since the new rail, being placed, was gaged to the old rail still in track the straight line or curve was retained in the proper location. To provide support for the new rail, a smooth surface on the tie must be obtained. For many years, this rail seat preparation was done using hand adzes.[qq] Due to the wood grain, knots and variance in workman skill, the rail seat surfaces were almost never smooth, flat or at the proper and uniform cant. The imperfect seat produced by hand tended to create torsional stresses and gage variations. It also contributed to rapid initial wear, shortening the life of the railroad rails and ties. This condition could also cause rail bending and an uneven running surface for the trains. It was necessary to correct gage and surface variations with additional work forces when these unwanted conditions appeared after trains had operated over the new rail for a period.

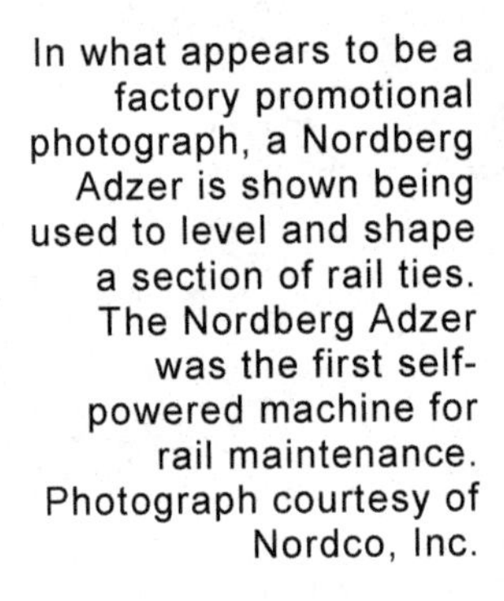

In what appears to be a factory promotional photograph, a Nordberg Adzer is shown being used to level and shape a section of rail ties. The Nordberg Adzer was the first self-powered machine for rail maintenance. Photograph courtesy of Nordco, Inc.

Observing the above conditions, Henry Talboys, Vice President of the Nordberg Manufacturing Company, set out to mechanize the adzing of railroad ties. He, along with Helmer Erickson, Chief Engineer, and Rudolph Buettner, Shop Superintendent, designed and built a "mobile milling machine." Their broad knowledge of machine tools, mobile machinery and railroads permitted them to develop and patent the Nordberg Adzing Machine. In 1928, they assembled this novel machine, which utilized the one rail in track for support and as a guide, while milling a properly located surface to permit the second rail to be placed in track correctly positioned and fully supported by a smooth surface on the tie.

The Nordberg Adzing Machine was essentially a milling head with a vertical shaft that was belt-driven from a small gasoline engine and cleverly mounted on a frame supported by flanged wheels, positioned

[qq] An 'adze' is a tool similar to an ax with an arched blade at right angles to the handle, used for cutting or shaping large pieces of wood.

by guide rollers on each side of the railhead. It was quickly adopted by the railroad industry as the standard adzing method since it provided greater quality, production rate and safety while at the same time improving the economy of a laborious task. This machine was the first machine to be used in railway "maintenance of way" work to operate on only one rail in track. It was also the first self-powered machine for roadway maintenance operation.

This machine established Milwaukee as a pioneer in the railway maintenance equipment field. It was the first of many machines used by the railroads in the maintenance of their roadways. This business line was eventually sold and transferred to Milwaukee's Nordco, Inc., which continues to be a recognized leader in manufacturing such equipment.

Here, five Nordberg Adzers are shown working on a rail job in the Midwest. It is reported that these five machines replaced forty hand adzers that would typically have been used for the task. Photograph courtesy of Nordberg Corporation.

NORDBERG GAS-FIRED DIESEL ENGINES

In 1911, Bruno Nordberg studied the developments of Rudolf Diesel who had designed a unique oil-burning engine. Nordberg started work on a small horizontal two-cycle engine, which developed 50 horse-power. In 1914, Nordberg secured a license agreement with Carels Freres of Ghent, Belgium to build diesels of the Carels design in the United States.[rr] One year later the first large diesel engine built in this country was completed by Nordberg; a five-cylinder, 1,250 HP two-cycle engine. This beginning led to Nordberg's eventual position as the number one builder of heavy-duty diesels in the Western Hemisphere.

Some of the significant Nordberg contributions to diesel engine progress in America include:

- the largest single acting, two-cycle diesel engine
- the first turbocharged four-cycle diesel engine
- the first uniflow scavenging, two-cycle diesel engine
- the first port scavenging two-cycle, vertical diesel engine
- the first diesel geared drive for propelling large cargo vessels
- the first inherently balanced, heavy-duty industrial radial engine

However, the company's most significant diesel development may have been the design and manufacture of the first engine based on the diesel-cycle principle using natural gas as a fuel. This innovation can be attributed to Bruno Nordberg's son, Bruno Victor Edward Nordberg (often referred to as Bruno V.E. Nordberg, to distinguish him from Bruno V. Nordberg). Bruno V.E. spent five years experimenting with the technology before introducing a natural-gas fired diesel engine to the market. The diesel engine was particularly suited to diesel-electric generating stations, because natural gas was often significantly less expensive than fuel oil. As a result, Nordberg's natural-gas fired diesels had wide acceptance by municipal electric utilities.

[rr] While Nordberg was able to secure an order from Phelps-Dodge for this 1,250 horsepower diesel engine, he did not have drawings of the Carel engines. The First World War started a month after Nordberg signed the agreement with Carels and Belgium and the company was not only occupied by the Germans, but the area was near the middle of the front line. When Nordberg was about to tell Phelps-Dodge he couldn't execute the order, he received a packet of drawings via the Netherlands that had been smuggled out. While it was not a full set of drawings, it provided enough detailed information to allow Nordberg and his team to build their first diesel engine.

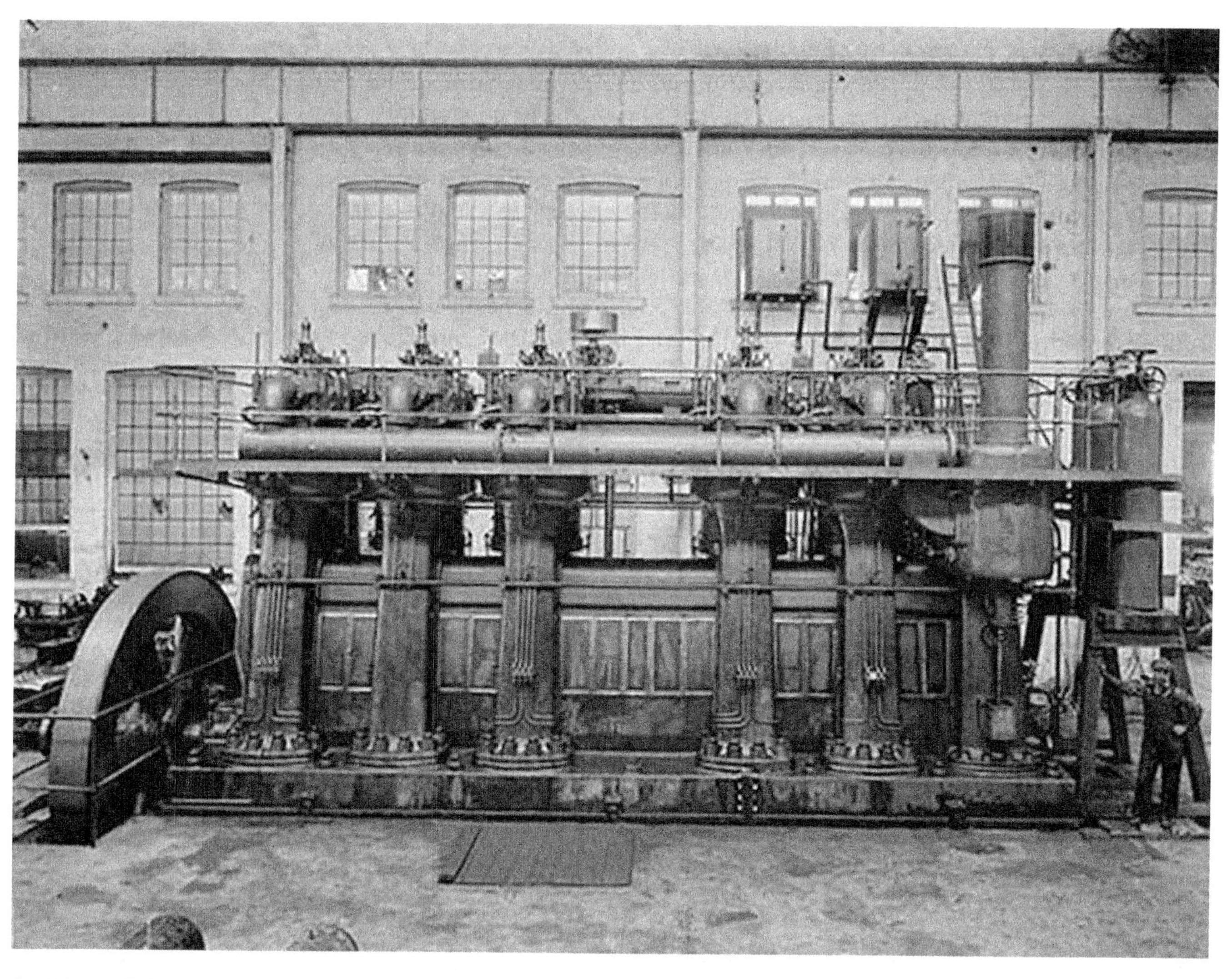

A picture of the first Nordberg diesel engine produced in the United States. Notice the size of this engine compared to the two men in the picture—one of which is on top of the engine on the right side of the cylinder head assemblies.

Nordberg 'Victory Ship' Engine

In 1944, Nordberg produced the largest diesel engine ever built in the United States at that time—a six-thousand horsepower engine for the United States Maritime Commission. The nine-cylinder engine was designed for direct-drive marine service. The crankshaft alone weighed eighty-thousand pounds.

The engine used advanced "mechano-pneumatic" controls that governed starting, fuel and speed regulation, with interlocks to prevent damage to the engine. It was powerful enough that a single engine could power a ship, compared to two normally used.

The engine was completed in 1944 and installed in the *M.S. Emory Victory.* The *Emory* was a VC-M-AP4 Victory ship with an overall length of 455 feet and a beam of 62 feet. She was launched at the Bethlehem Fairchild Yards in Baltimore, Maryland in April 1945 and moved to Sun SB & DD Co. for the installation of her machinery. The ship was finally delivered in October 1945, shortly after the conclusion of the Second World War.

This is a factory photograph of the assembled Nordberg nine-cylinder diesel engine, before shipment for installation in the *M.S. Emory* Victory Ship.

NORDBERG RADIAL ENGINE

In the 1940s, the Nordberg Manufacturing Company developed a compact, lightweight radial engine. Over the next two decades, a large number of these engines were installed for electricity production at several aluminum reduction plants. The engines were also suitable for driving pumps, compressor, blowers and similar rotating equipment.

A total of 475,000 horsepower was produced by 220 normally scavenged and 22 supercharged Nordberg Radial Engines at the Point Comfort reduction works of the Aluminum Company of America at Port Lavaca, Texas. Shown here are forty of the radial engines in one of seven similar powerhouses.

Two hundred and forty-two Nordberg radials were installed at Port Lavaca, Texas by ALCOA. Producing 475,000 horsepower, the Point Comfort Reduction Works was the largest internal combustion engine plant in the world. Other such plants included the Kaiser Aluminum and Chemical Corporation's Chalmette, Louisiana, reduction plant.

Nordberg Radial Engines were installed with the axis of the crankshaft in a vertical plane with a single crank at the top. Cylinders, equally spaced radially about the shaft in a horizontal plane, were bolted to the frame. Simple parts provided for scavenging and exhaust and eliminated the need for intake and exhaust valves and attendant valve mechanism. Scavenging air was normally supplied by a separate blower, usually motor drives for electric generating units.

Nordberg Radial Engines were shipped as complete units, thus substantially reducing the time and expense of installation work. The engines were balanced to a very high degree, achieved by actual convergence of combustion pressures and inertia forces at one focal point on the crankshaft axis. Because of this, they could be mounted on simple, inexpensive foundations, containing less than 25% of the mass required for in-line cylinder engines of equal capacity and speed.

Nordberg Symons 'Cone' Crushers

In 1927, Nordberg was granted exclusive rights to manufacture and sell the Symons mechanical crushing machine. The machine was conceived by Loren Symons who was an early pioneer of mechanical crushing machines. The design was further refined by Nordberg personnel.

Symons Cone Crushers were important to the development of the United States steel industry and the increasing utilization of lower grades (lesser percentage of iron content) of harder ore such as taconite. Such ores are dependent largely on dependable means for economical reduction. Similarly, the Copper Country in Northern Michigan and in the Southwest involves treatment of huge amounts of ore and waste due to the low percent of copper content in the ore.

Eighteen Symons cone crushers are installed in this large copper factory.

The crushers were also used in building the United States Interstate road system, due to the need to produce huge amounts of rock, coarse and fine aggregate, and sand to close specifications.

The "cone" crusher is a special case of a group of crusher designs that have the purpose of reducing (comminuting) large particles into smaller ones for various extractive and other industrial processes. The cone crusher consists of three main assemblies of moving parts arranged within the interior of a frame containing an inverted bowl assembly, which is adjustable in position to set the gap, or, fineness of crushing, but which, during crushing, is retained against any movement.

The assembly opposed to the bowl is referred to as the head assembly. The head is supported on a spherical bearing in the frame and actuated by means of an eccentric assembly driven by a drive assembly to achieve a conical rolling action so that feed material in the cavity is crushed successively in the peripheral gap between the head and bowl and is discharged by gravity.

There are many styles of mechanical arrangement and many sizes of machines as there are manufacturers. The Symons cone, however, was the exclusive design of the Nordberg group and was recognized as the most durable and productive of all cone crushers in aggregate, mining, and processing industries.

The general acceptance of this machine made it possible to process rock and ore at high production rates, close size specifications for the product, and with dependable rugged trouble-free performance of the equipment. The Symons principle is the basis for almost all secondary and finer crushing operations in ore dressing applications where the ore is hard, abrasive or otherwise difficult to crush. Almost one-hundred percent of all the crushers being used on Mesabi Iron Range in Minnesota are Symons cone type manufactured by Nordberg. In over fifty years of continuing development, the basic principle remains the same. Sizes range from two-foot head diameters to a monstrous ten-foot diameter crusher installed in an iron mine in South Africa.

The features, which set the Symons principle apart from other methods of comminution, are the rapid hammer-like compressive impacts that reduce the material, which meanwhile moves by gravity down through the cavity. While five or six blows are experienced by almost every particle, the last one or two blows are delivered in a so-called parallel zone which sizes the material to the pre-set gap, assuring close regulation of particle top size, an essential requirement for handling, conveying and processing.

Largest Aluminum Extrusion Press

In 1954, Nordberg Manufacturing Company built the largest and most versatile aluminum extrusion press ever constructed. It was manufactured for the Kaiser Aluminum and Chemical Corporation to be utilized in making the heavy cross section shapes used in aircraft construction, particularly in military and subsonic commercial aircraft such as B-52s, the 707 and others.

The 8,000-ton extrusion press is essentially a horizontal axis hydraulic machine consisting of forged cylinders and tie rods and cast steel crosshead construction arranged on a foundation and support structure of welded steel fabrications. In principal, the machine extrudes shapes by applying very high (fifty- to one-hundred-thousand psi) pressure to a confined aluminum billet at a temperature in excess of 1,000°F in much the same manner as toothpaste is squeezed from a tube. The press is powered by a high-pressure water accumulator station at a pressure of 4,500 psi. The shapes, which may be extruded, are limited only by practical considerations in the die design. Shapes include tubes, structurals, flats and irregular contours.

The versatility of the machine and the size of the billet that it could extrude made it possible to have long enough structural members in aircraft without having to splice, thus avoiding potential points of high stress and fatigue failure. The independent mandrel feature made it possible to produce thin wall tubular members or integrally stiffened panels for maximum strength and minimum weight.

The machine revolutionized aircraft design and manufacture by freeing the designer from limitations in plate and stringer construction which was the rule in aircraft construction until then. It resulted in greater economy in manufacture as well making possible most sophisticated aircraft.

This press was the largest machine ever built at Nordberg. It was followed by two other similar machines. All machines were built under the Air Force Heavy Press Program, Air Materiel Command and were leased to the operators such as Kaiser, which supply material for military needs. The concept and design of this press were developed by personnel of Loewy-Hydropress of New York

Eight thousand ton Loewy Extrusion Press built by Nordberg Manufacturing Company and installed at Kaiser Aluminum. Photograph courtesy of Rexnord Corporation.

BRUNO VICTOR NORDBERG was born in 1858 in Finland and educated at the University of Helsinki, where he graduated in 1878. He came to the United States in 1880. His educational background was primarily in machine design and thermodynamics. Because the E.P. Allis Company was an outstanding builder of various kinds of machinery, including steam engines, he came to Milwaukee and secured a position as a draftsman at the company. In a very short time, Nordberg was able to demonstrate his abilities and became the private designer for Irving Reynolds, who was then chief engineer at Allis.

In 1886, Nordberg began his own company in rented quarters on the second floor of a building just north of the old Allis plant, which also contained facilities for Pauling & Harnischfeger as well as for Christopher LeValley who later started the Chain Belt Company.

The new company's first product was a revolutionary type of cut-off governor invented by Nordberg for use on steam engines. Soon the company began building simple poppet valve steam engines, condensers, steam pumping engines, and mine hoists.

Nordberg quickly demonstrated his broad engineering expertise, designing high-efficiency steam engines using a regenerative cycle, hoists, compound steam traps, gas compressors, vacuum pumps, etc.

Nordberg was awarded a medal and honors by the French Academy in 1891 for his outstanding inventions. The degree of doctor of engineering was conferred upon him in 1923 by the University of Michigan, in recognition for his achievements in the field of mechanics. He was a member of several technical societies and published many technical papers.[161] Nordberg died on October 30, 1924.

BRUNO VICTOR EDWARD NORDBERG was born in Milwaukee in 1884, the son of Bruno Victor Nordberg. He attended South Division High School and the University of Wisconsin's College of Engineering. After his graduation in 1907, he went to work for his father's company as a draftsman. He was named manager of the company's oil-engine department in 1914 and concentrated on diesel engine design. He was the first to deploy natural gas in an engine operating on the diesel principle successfully. He had sixteen patents for diesel, steam and compressed air engines and hoists.

He was appointed to the City of Milwaukee's sewerage commission in 1926 and named vice chairman in 1931. He served on the commission until his death in 1946. Bruno V.E. Nordberg served on many professional and engineering organizations, including the American Society of Mechanical Engineers and the Engineers and Scientists of Milwaukee, of which he was past president.

BUCYRUS COMPANY/BUCYRUS-ERIE

Daniel P. Eells of Ohio acquired the Bucyrus Machine Company of Bucyrus Ohio in 1880, founding the Bucyrus Foundry and Manufacturing Company. The new company set about manufacturing various components for the railroad industry. In 1882, the company received orders from two railroads for steam-powered shovels. Within a year, eighty percent of the company's business was in the manufacture of steam shovels. As the company developed a reputation for designing and building specialized excavating equipment for mining and construction projects, it changed its name to the Bucyrus Steam Shovel and Dredge Company.

In 1891 the company was enticed to relocate to the City of South Milwaukee by an attractive relocation package—a fifteen-acre site and $50,000 toward construction of a new factory. The company moved into its newly built South Milwaukee facility in 1893.[162] The company used its new steel foundry to develop special-allow steels for its line of heavy-duty excavating equipment. It also adopted new heat-treating techniques to improve material properties for excavating tools. The company's equipment established a good record for quality and reliability. As a result, orders increased three-fold between 1897 and 1901.

In 1904, the Isthmus Canal Commission began ordering Bucyrus shovels for excavating the Panama Canal. Eventually, over seventy steam shovels were ordered for the project. During the peak years of the project, almost all earth and rock was moved by Bucyrus steam shovels (see the following article for additional information).

Bucyrus established several early innovations in excavation equipment. Among them were the steam shovel that could revolve 180-degrees, and the first "back-acting" shovel that could dig below its carriage. In 1910-11, Bucyrus introduced its first dragline machines, its first "crawler" shovels and draglines, and its first excavators powered with internal combustion engines supplied by Wisconsin Motor.[163] It also introduced a line of pile drivers.

During the First World War, Bucyrus manufactured high-explosive shell blanks for the British government. It also created the Wisconsin Gun Company, along with three other Milwaukee-area firms, to produce artillery barrel and breech mechanisms. In addition, the war precipitated a boom in non-military construction orders.

In 1929, Bucyrus acquired the Erie Steam Shovel Company, which was then the largest American producer of small excavating equipment. With the acquisition, it changed its name to the Bucyrus-Eric Company and quickly consolidated operations.

The Second World War dramatically increased orders for excavating equipment for mining activities, as well as for export to US war allies and use by the armed forces. Bucyrus intentionally limited its involvement in munitions production in order to meet the demand for excavation equipment, which was in critical need. However, the company did manufacture equipment for gun carriages and mounts.

Orders continued at a high level during the post-war years and, in 1947, the president of Bucyrus-Erie admitted that "demand [was] beyond its capacity to produce." Bucyrus expanded its plant, increasing its capacity by twenty-five percent. Bucyrus also expanded its product lines—producing rotary blast-hole drills. By 1955, Bucyrus-Erie employed five thousand people.

Bucyrus-Erie was acquired in 2011 by Caterpillar Inc., and run as its mining equipment division. In mid-2016, the company announced that it was moving the engineering and technology jobs from its mining equipment division in South Milwaukee to Tucson, Arizona.

Bucyrus Panama Canal Steam Shovels

On April 14, 1904, the ownership of the New French Panama Canal Company and of the Panama Railroad passed to the United States government. In August of that year, the Isthmus Canal Commission placed orders with the Bucyrus Company of South Milwaukee for a 70-ton shovel and two 95-ton shovels for test purposes at list prices and without competitive bids. The machines quickly proved their merit. The commission then asked Bucyrus and other shovel manufacturers to submit bids by October 26, 1904, on eleven additional machines, five of the 70-ton size and six of the 95-ton variety. The Bucyrus bids were low on both sizes of shovels and the company was awarded the entire order. In January 1905, the commission asked Bucyrus to supply three more 70-ton shovels. In April of 1905 the commission was again in the market for shovels, this time by the dozen. Bucyrus got the order for all twelve shovels, and a month later, the commission took up its option for a second dozen of the machines. Bucyrus was also awarded several other bids, bringing the commission's order of the new Bucyrus equipment to $625,000 and encompassing 77 steam shovels.

Digging the Panama Canal was a gigantic undertaking. By comparison in cubic yards excavated, the Suez Canal required 80 million cubic yards; Manchester (Great Britain), 54 million, Albert (Belgium), 100 million; Panama, 225 million. The Culebra Cut (later renamed Gaillard Cut) carried the canal through the high hills constituting the backbone of the Isthmus and required over 102 million yards of dry excavation—about the same as digging a ditch 10 feet deep and as wide as a two-lane highway from New York to San Francisco. This one nine-mile stretch of the canal required about ninety percent of all the steam shovel excavation performed in the channel up to 1914.

An early estimate of the amount of earth and rock to be removed from the Culebra Cut was 54 million yards, but this was continually revised upward as the work progressed.

At times, slides played fantastic tricks. Once a 95-ton steam shovel was picked up and carried halfway across the cut. On another occasion, 75 acres of the town of Culebra broke away from the hillside location and moved foot by foot toward the canal, carrying hotels and clubhouses. Another time a slide kept moving downward as fast as the steam shovel worked, so that the shovel was able to make 103 trips across the toe of the slide without moving its track an inch.

In addition to the slides, material moved up from the bottom of the cut. This unusual occurrence resulted when the downward pressure of the small mountains on either side of the cut distorted the underlying strata and caused material underneath to swell up in the channels that the shovels had created.

In spite of the many difficulties, work proceeded at a rate unprecedented in the history of construction. At the peak of the activity around 1908, a total of over sixteen million cubic yards of earth and rock was excavated in one year, practically all of it by Bucyrus 95-ton and 70-ton machines. The Bucyrus Company, and its equipment, established an excellent reputation for performance under adverse conditions. In 1915, it was reported that Bucyrus machines held "all records of output at the canal proper, where the most severe digging conditions were encountered."

Theodore Roosevelt climbed aboard a Bucyrus 95-ton shovel on an inspection trip of the Panama Canal.

"Big Muskie"

In 1912, Bucyrus Company of South Milwaukee introduced the first crawler-mounted dragline. It was available with a steam engine, gasoline engine, or electric motor drive. Previous draglines had been mounted on railroad wheels that ran on short, movable sections of track, or roller and skid arrangements. Crawlers gave the dragline new mobility. They were accepted widely for the jobs they could perform on soft ground.

Although an improvement over earlier methods, large machines equipped with crawlers still occasionally became mired in wet or sandy material. Also, these early crawlers had numerous mechanical parts and needed constant maintenance.

A solution to these problems was developed in 1913 by a young engineer named Oscar Martinson working for Monighan Machine Company. The Martinson invention consisted of a pair of pontoon-like "shoes" mounted on cams attached to a sturdy shaft running through the revolving part of the excavator. When the cams revolve, the shoes come down, lifting one end of the machine and sliding it along several feet.

Recognizing the advantages of this machine, Bucyrus-Erie acquired controlling interest in the Monighan Company in 1931.

In 1966, Bucyrus-Erie began production in South Milwaukee of what was to become the largest mobile land machine in the world. Central Ohio Coal, a subsidiary of American Electric Power, ordered a mammoth walking dragline, dubbed "Big Muskie," to excavate overburden from a strip-mine in Southeastern Ohio. Massive in dimensions, it was designed to remove three million cubic yards of overburden a month. By the end of 1980, Big Muskie, by itself, removed more earth in about ten years than was taken out of the Panama Canal by more than 42,000 men and hundreds of machines—more than 270 million cubic yards.

The massive machine can hoist a load as high as a thirty-story building and drop the load two city blocks away. Its 220-cubic-yard bucket loads 325 tons of material with each pass. A hydraulic system had to be developed for propelling the machine, since the conventional cam on crank-operated mechanism had size limitations. The system employed has four lifting cylinders, two per shoe, and four shoe cylinders, two on each shoe. After the four lifting cylinders raise the machine, the four push cylinders move it forward a variable distance up to 14 feet. With four points of lift, the walking mechanism raises the base completely off the ground, thus eliminating sliding friction of the base on the ground and reducing wear and stress on the base.

Assembled on location at the Muskingum mine (hence the "Muskie" name), parts were shipped in 340 railcars and 260 truckloads. Special wire ropes for hoist and drag service of five-inch diameter, the world's largest, were developed for the machine.

This giant walking dragline, built by Bucyrus-Erie in 1968, is the largest mobile land machine in the world. It has a bucket capacity of 220 cubic yards. Shown here in the final stages of assembly at a site in Southeastern Ohio. Notice the people adjacent to one of the feet, which provides an idea of the scale of this massive machine.

"Big Brutus" Coal Mining Shovel

Bucyrus-Erie's Coal Shovel Model 1850B was placed in service by the Pittsburg and Midway Coal Mining Company in 1962. Shop fabrication of the shovel was completed in Bucyrus-Erie's South Milwaukee factory. The coal shovel was shipped from Milwaukee in 150 railroad cars to Hallowell, Kansas. Subsequent field assembly occupied fifty-two Pittsburg and Midway employees for eleven months.

When the shovel began operation at the Pittsburg and Midway Mine No. 19 in May 1963, it was the second largest operating coal shovel in the world. Emil Sandeen, then superintendent of the mine, called the machine "Big Brutus" and the name stuck.

Big Brutus was designed to make possible the recovery of relatively thin seams of bituminous coal. The machine could remove overburden from approximately one square mile of surface per year. As overburden removal was completed, other smaller-scale equipment was employed to extract the coal. The over-burden that Big Brutus had removed was then reshaped and re-vegetated, and the land reclaimed.

During its lifetime, the machine removed the overburden to allow recovery of nine million tons of bituminous coal. The bucket scooped out 90 cubic yards, or 135 tons of earth, with each bite. The mined coal was used locally for electrical power generation.

Retired in 1975, Big Brutus is currently the centerpiece of a mining museum in West Mineral, Kansas. It was designated a landmark by the American Society of Mechanical Engineers in 1987.

Big Brutus is the nickname of the Bucyrus-Erie model 1850B electric shovel, which was the second largest of its type in operation in the 1960s and 1970s. It was designated a landmark by the American Society of Mechanical Engineers.

FILER & STOWELL

Filer & Stowell moved to the Town of Lake in 1902, acquiring land near the relocated Nordberg factory. The company continued to specialize making sawmill equipment, along with other machinery and equipment.

During the Second World War, Filer & Stowell made steam engines for EC-2 cargo ships—more commonly known as the Liberty Ships. By May 1943, the company had produced over fifty of the standard 2,500 horsepower, triple-expansion marine steam engines; each weighing 135 tons. The company was awarded the "M" pennant by the Maritime Commission—the first company in the state of Wisconsin to receive this designation for outstanding production achievement.[164]

The original purpose of the Liberty ship-building program was to provide cargo space across the Atlantic faster than Nazi U-boats could sink the supply ships plying that route. By autumn of 1940, the British Isles were experiencing severe shortages, and it was in September of that year that the British Merchant Shipbuilding Mission visited the United States and contracted for the construction of sixty "Ocean class" vessels to be built in the United States for the trans-Atlantic route. Ultimately, 2,751 ships were made, which allowed the United States to meet the needs of the Allied forces during the war. Filer & Stowell was one of fourteen American manufacturers that produced steam engines for the Liberty Ships.

Filer & Stowell continues in business as a producer of high quality sawmill machinery for the lumber industry—a remarkable 150 years after being established in 1865. The company is now located at 3939 West McKinley Avenue.

A typical Liberty Ship steam engine is shown as right. It is not known if the pictured steam engine was one built by Filer & Stowell. Photograph from the United States Maritime Commission.

OBENBERGER/LADISH DROP FORGE

As noted in Chapter 2, a decision by Obenberger Drop Forge to concentrate on forging axles for the automotive industry was a good one. The business expanded dramatically. In order to meet the demand, a decision was made to build a new factory in Cudahy. The company acquired five acres of land on Packard Avenue. The need to attract investors for the expansion interested Herman Ladish of Ladish Malting. By 1915, Herman Ladish owned seventy-six percent of the business and changed the name of the company to Ladish Drop Forge the following year.[165]

In 1917, the company had three drop-hammers in operation. By 1927, Ladish had thirty steam hammers and called itself "the axle forger to the auto industry." It was also doing business for tractor and railroad companies. In the 1930s, the company introduced forged industrial pipe flanges and started the production of brake drums for the aircraft industry and aircraft parts.

Business greatly expanded during the Second World War. Among other things, Ladish forged parts for aircraft landing gear, propellers, crankcases for aircraft engines, tank treads, tractor gears, diesel drive shafts, and nose cones for bombs. Toward the end of the war, the company began forging parts for jet engines using heat-resistant alloys.

In more recent years, the company made significant contributions to NASA's space program, including the production of the casings use for the solid rocket motors used for the space shuttle.

WORLD'S LARGEST, MOST-POWERFUL FORGES

As the Second World War was ending on the Eastern Front, Ladish company president Victor Braun was invited to participate with a delegation of American industrialists to tour German industrial facilities. He discovered that German engineer, Hans Beche, had invented a counterblow forge hammer,[166] which could apply significantly more force than anything the Allies had in their industrial arsenal. In a counterblow forge hammer, both halves of the die are driven together, with the workpiece between them. Since the forces of the two halves largely off-set one another, the foundation requirements are reduced. By 1949, Ladish took delivery of the largest Beche hammer at the time, rated at 80,000 meter-kilograms.

This massive hammer, and others that were to follow, gave Ladish unprecedented forging capabilities—capabilities that have resulted in the company forging large parts requiring massive forces, including gas turbine wheels and nuclear reactor vessels.

In 1949, Victor Braun made another key acquisition. He hired Otto F. Widera, a German engineer, to oversee design activities for Ladish. Within a short time after accepting the position, Widera designed a small ring-rolling forge to produce seamless rings and cylinders. In 1951, Widera was asked to design a much larger ring roll, designated #125. It could turn out seamless rings weighing up to 30,000 pounds.[167]

In 1957, Widera designed and built what continues today to be the world's largest and most powerful hammer, designated Hammer #85. It is rated at 125,000 meter-kilograms and can handle forgings up to 50,000 pounds. The massive, five-story machine was completed in 1959. It provides Ladish with unique capabilities to handle large workpieces for the most demanding applications.

In 2007, the company overhauled the forge, replacing the top and bottom rams—each one weighing 375,000 pounds. The forge remains in service and is still ranked as the world's largest. As a result, Ladish has become a major supplier of steam chests for some of the largest steam turbines, some of the largest closed-die forgings in the world.[168]

Over the years, Ladish has continued to increase its forging capabilities. In 1965 it added a ring roll (#202) measuring 28' in diameter, in 1970 a press (#112) with 4,500 ton isothermal capabilities,[ss] and in 1981 a press (#116) with a ten-thousand-ton isothermal rating, the worlds' largest. In 2008, Ladish installed the World's largest isothermal press. It is used to forge aircraft engine parts for companies such as General Electric and Rolls-Royce.

[ss] Ladish's isothermal press #112 was also designed by Otto Widera and was the largest in the world at time of its construction.

Ladish can make ring forgings weighing up to 350,000 pounds with 28-foot diameters, and can fabricate larger seamless aluminum rings than any other company.

The company employs a sophisticated computer model to predict accurately the behavior of the workpieces it is forging. The model is particularly helpful in addressing the one-of-a-kind work that Ladish is often called upon to perform, where a mistake could result in costly wasted material and delivery delays.

Ladish's Hammer #85 is the world's largest and most powerful. It is a counterblow hammer with a baseline rating of 125,000 meter-kilograms and a nominal operating energy of 15,000 meter-kilograms. It was designed by Otto Widera and completed in 1959. Photograph from *Forge Work*, by Michael Schultz. Image copyrighted by Michael Schultz Photography—used with permission.

Ladish's unique capabilities and reputation have resulted in the manufacture of some of the most critical forgings ever created. For example, Ladish has made the forgings for the Apollo lunar excursion module and for the rocket motors used on NASA space shuttles. It has also made components for the US's intercontinental ballistic missiles.

In 2011, Ladish was purchased by Allegheny Technologies Incorporated, one of the largest and most diversified specialty metals producers in the world. Today, the company is known as ATI Ladish Forging. It continues forging for the space, airframe and aircraft engine markets.

Casings for the Space Shuttle's Solid Rocket Boosters

Ladish fabricated the casings for the Solid Rocket Boosters (SRBs), which operated in parallel with the main engines for the first two minutes of flight of NASA's space shuttles. The casings were made of a high-grade D6AC alloy and were shaped on a Ladish cold shear forge designed by Otto Widera. The casings were critical to making the space shuttle program possible.

The SRBs were designed to separate from the space shuttle assembly at an altitude of approximately 24 nautical miles, descend on parachutes and land in the ocean. They were recovered by ships, returned to land, and refurbished for reuse. The thrust of both boosters was over five million pounds.

The SRBs were the largest solid propellant motors ever developed for space flight and the first built to be used on a manned craft. The huge motors were comprised of a segmented motor case loaded with solid propellants, an ignition system, a movable nozzle and the necessary instrumentation and integration hardware.

In a vehicle assembly high bay building, a segment of a Solid Rocket Booster is lowered toward a segment already in place. Photo Credit: NASA

A segment of the Solid Booster Rocket is assembled for use to launch a Space Shuttle. The casings for each segment were fabricated by Milwaukee's Ladish Corporation. Photo Credit: NASA

The joint between each segment of the casing that made up the SBRs was sealed with a high-performance O-ring. The failure of one of these O-ring joints caused the space shuttle Challenger to break up, seventy-three seconds into its flight on January 28, 1986. The incident resulted in the tragic deaths of its seven crewmembers. Investigations into the cause of the failure revealed that both NASA and Morton Thiokol (manufacturer of the SBRs) were aware that the performance of the O-ring seal could be too easily compromised by a number of factors, including low temperature at the time of launch. The SRB casing joint was redesigned subsequent to the Challenger accident, using an additional interlocking mortise and tang with a third O-ring.[169]

OTTO F. WIDERA was born in 1906 in Tomice Posen, Germany, which is now located in Poland. After the First World War his family moved to Cologne. He was one of eleven children and his father was a school principle. Otto put himself through college by working as an apprentice electrician with the Automotive Electrical Company Scheele in Cologne. He graduated summa cum laude with a mechanical engineering degree from the Cologne Technical College. After a series of jobs, he secured employment at Wagner Forging and Steel in Dortmund where he designed machine tools, principally hammers and presses.

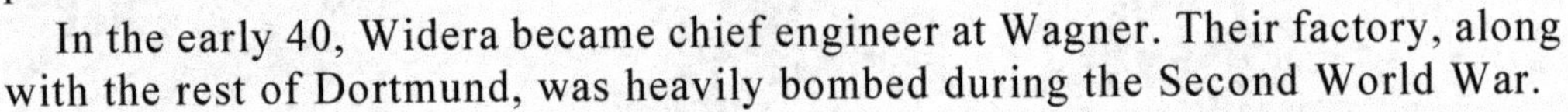

In the early 40, Widera became chief engineer at Wagner. Their factory, along with the rest of Dortmund, was heavily bombed during the Second World War.

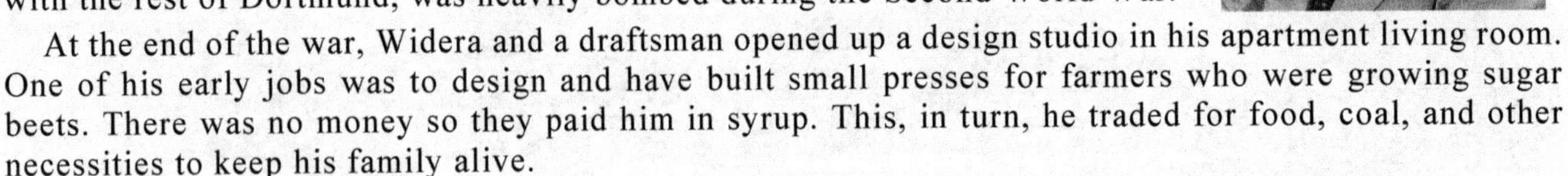

At the end of the war, Widera and a draftsman opened up a design studio in his apartment living room. One of his early jobs was to design and have built small presses for farmers who were growing sugar beets. There was no money so they paid him in syrup. This, in turn, he traded for food, coal, and other necessities to keep his family alive.

After the war, the allies dismantled Germany's heavy industry and, as a result, a counterblow hammer from the Krupp Works was shipped to Ladish in Milwaukee. Ladish had no one who knew how to assemble it, so Victor Braun, the president, came to Germany to look for a knowledgeable engineer to oversee its assembly. Everyone he contacted referred him to Otto Widera. Since Germany was still in an iffy recovery state, Widera relocated to the United States in 1949. A year later, he brought his family over.

At Ladish, in addition to setting up the hammer relocated from the Krupp Works in Germany, which was designated as Hammer No. 80 at Ladish, Widera also designed a larger counterblow hammer, which is the world's largest. He also designed the world's largest ring roll, as well as other first-of-a-kind equipment.

Gene Janikowski of ASME-Milwaukee's Section reported an interesting tale about Otto Widera: "The Ladish ring roll was making noise and Otto Widera was called in to diagnose the problem. Widera walked onto the shop floor in a white shirt and tie and then proceeded to roll up his shirtsleeves. He plunged his hands into the grease and fished out some metal pieces. A quick examination showed that the metal pieces came from one of the bearings, confirming that the bearing had failed. An on-the-spot assessment of the problem. Things like this caused the employees at Ladish to revere Widera for his skills as an engineer.

At an Apollo rocket launch in Florida, an Air Force general told Otto Widera that if it weren't for him, the country wouldn't be able to go to the moon. Widera designed a cold shear spinning machine that produced the solid rocket motor cases made of a high-strength D6AC steel alloy, which was critical to making the space shuttle program possible. NASA had attempted to develop a carbon-fiber composite casing replacement instead, but it was not successful. Over the duration of the space shuttle program, Ladish supplied all casings for the solid rocket boosters.

Widera had several patents, including one for "the art of roll-reducing ring-wall thickness" and one for "the art of manufacturing ball valves." Due to his accomplishments at Ladish, Otto F. Widera became Chief Engineer, vice president of engineering and eventually a member of the Ladish board of directors.

In 1972, he semi-retired from full-time employment and worked from Florida during the winter months. He died in Milwaukee on October 24, 1993.

THE HEIL COMPANY

As noted in Chapter 3, early in his career Julies Peter Heil worked as an apprentice for Falk. When he was 22, Falk sent him to South America to introduce a new process for welding streetcar and railroad rails. After two years, he was a recognized expert in the technology, working on rail projects in New York and St. Louis. In 1901, he left Falk to form his own company, the Heil Rail Joint Welding Company.

In a small rented building at 412-15 Poplar Street (just south of today's Vliet Street between 4th and 5th Streets) Heil applied welding technology to the assembly and repair of streetcar rails. He found, however, that the field was limited. He eventually decided to use his welding skills to build wagon tanks.

Business was good. The superiority of welded tanks, over riveted ones, was recognized. Local breweries ordered tanks for the washing of bottles, other companies used Heil tanks for oil storage, and the City of Milwaukee ordered galvanized iron garbage tanks.

In 1909, Heil moved into his own building at Montana and 26th Street in what was then the Town of Lake. Heil obtained an order from the United States Army for welded steel water tanks. This government purchase helped to bring in other orders for welded steel tanks.

Heil's first garbage collection trucks, ordered by the City of Milwaukee in 1914, were horse-drawn.

Sterling Motor Truck approached the company in 1912, asking it if it could build a steel motor truck body. Heil's first truck bodies were successful. When the United States entered the First World War, the US ordinance department sent Heil an order for ninety- seven ammunition truck bodies—for delivery in 20 days! Heil met the tight delivery schedule, which led to many more orders. The Army truck body orders had a side benefit when the war ended. The trucks were distributed throughout the United States, leading many to see the advantage of steel truck bodies over wooden bodies that were typically used.

Heil's first refuse collection truck bodies were built for the city of Milwaukee beginning in 1914. In 1929, Heil built the first fully enclosed van-type refuse truck body. By the 1930s, Heil garbage trucks were collecting solid waste in hundreds of American cities of all sizes.

Heil built the first fully enclosed "van-type" refuse truck in 1929.

But Heil's most significant innovation in refuse collection trucks occurred in 1945 when the company introduced the first refuse body capable of packing its own load. As a result, Heil became the world's leading producer of refuse collection vehicles.

The Colecto-Pak truck body was introduced in 1945. It was the first refuse body capable of packing its own load.

JULIUS PETER HEIL was born in Dussmund-an-der-Mosel, Germany in 1876. He immigrated to the United States when he was five. At age fourteen, Heil started work at the Milwaukee Harvester Company as a drill-press operator. He later took a job with the Falk Corporation where he worked installing welded steel track for street railways throughout South America. When he turned twenty-five, Heil started up his own company to weld railway tracks—it was called the Heil Rail Joint Welding Company. As other product lines were introduced, the name was shortened to the Heil Co. The company eventually grew into a major manufacturing concern, producing road machinery, tanks for storage and transportation, dump-truck truck bodies, and furnaces for home heating.

Heil had over a dozen patents issued in his name, most of which related to various innovative truck bodies.

In 1933, Julies Heil was appointed by President Franklin Roosevelt to head the Wisconsin advisory board for the National Recovery Administration. His taste of politics was strong. Heil successfully ran for governor of Wisconsin in 1938—defeating Philip F. La Follette.

While governor, Heil created the Department of Motor Vehicles out of five other agencies. He also consolidated welfare and institutional programs into a single Department of Public Welfare. Referred to as "Julius the Just," he enjoyed his terms in office. The New York Times featured him on its cover in 1939, reporting in part that Heil was known for clowning and silly antics.[170] Julius Heil toured the country to promote Wisconsin's dairy products.

After losing a third term as governor, Heil returned to his company as president and later chairman of the board, until his death in 1949.

OILGEAR COMPANY

In 1908, William E. Magie, an engineer who felt industry needed a machine to transfer controlled power smoothly at a variable speed, brought his ideas and two experimental fluid power transmissions to the attention the Bucyrus Company for further development. Walter Ferris and E. K. Swigart of the Bucyrus Company were impressed with the idea and provided financial assistance. In the following years, the three developed several compact fluid power devices to transmit power. These new devices were applied experimentally to excavators and a variety of machine tools.

After many developmental years, the Oilgear Company was formally organized in 1921 to manufacture and market multiple radial piston variable volumetric fluid power pumps, motors and transmissions. The company's attention was focused on the strong demand for an improved means of traversing and feeding machine tool carriages. Small volume, radial piston, variable displacement pumps with built-in large volume low-pressure pumps and a line of cylinders were designed for this specific purpose. Providing readily adjustable feeds and quick traverse in either direction, these fluid power "feed controllers" received favorable acceptance from the automotive and machine tool industries. Their successful performance on various types of lathes, drilling, boring and milling machines and presses provided a much-needed "boost" for fluid power by industry.

Hand in hand with the production of the feed controller was the development of multiple, radial piston, constant displacement motors for rotary drives on machine tools. The company also introduced larger rotary fluid power transmissions and successfully applied them to various types of lathes and boring machines.

Further progress and added interest in fluid power was realized when the first simple and effective method of putting a mass production machine through its entire cycle by the mere push of a starting lever was perfected by Oilgear in 1924. Applied to the feed controller, this "load and fire" control device was used on a large multiple spindle drilling machine. This was the beginning of semi-automatic and fully automatic hydraulic feeds used on production machine tools. From then on, the development of more compact feed controllers for mounting integral with machines, for larger and higher pressure pumps with built-in controls, for new fluid power motors, and for additional types of vertical hydraulic presses and improved hydraulic broaching machines by The Oilgear Company proved increasingly important to industry.

To meet the demands of industry for more and improved fluid power, Oilgear discontinued the manufacture of broaching machines and presses in 1952. Customer acceptance had influenced other manufacturers to incorporate Oilgear fluid power and control systems in their broaching machines and presses.

In the ensuing years, the company's entire product line has been devoted to fluid power.

Oilgear Variable Displacement Piston Pump

Oilgear invented the radial rolling piston pump in the 1930s, despite the depressed economic conditions caused by the "Great Depression." Small, one piece, rolling pistons replaced crosshead piston assemblies, each involving fourteen parts.

The input shaft drives the cylinder, piston and rotor assembly. The cylinder rotates on a lubrication film on the stationary "pintle." Centrifugal force keeps the pistons against the thrust rings to rotate the rotor with the cylinder. During the lower half revolution, the pistons move outward to suck fluid through the pintle and fill the radial bores in the cylinder. During the upper half revolution, the pistons move inward to discharge the high-pressure fluid from the radial bores to pintle and the pressure line.

Oilgear developed the variable displacement, radial, rolling piston pump.

Being small, light, pressure lubricated and free of mechanical attachments, the pistons move smoothly and quietly at high speeds.

New integral stroke changing devices simplified application of manual, hydraulic, electric, pneumatic and electrohydraulic system controls. Smaller unit size standards, higher input speeds, quieter operation, added durability and a wider offering of unit sizes, pressures and controls enabled Oilgear's line of radial rolling piston pumps, motors and transmissions to open up new uses on machine tools, presses of all types and other equipment in the chemical, paper, lumber, rubber, food, textile, machine tool, steel, printing and other industries.

FROEMMING BROTHERS INCORPORATED

Froemming Brothers was founded in 1919 by Ben Froemming and his brothers Walter and Herbert. Ben, who was born Bernhard Arthur Froemming, started the firm to specialize on city sidewalk replacement and similar masonry jobs. He was only eighteen at the time, but the small firm's construction expertise grew quickly. By the early 1940s, the company had grown into an engineering construction company, building roads in Texas, bridges in Pennsylvania and Mississippi, a new airport for the Panama Canal project, and various tunnels in the Midwest.

In 1941, Ben Froemming answered an appeal from Milwaukee's mayor Carl Zeidler to bring back shipbuilding to the City. While the firm had no shipbuilding experience, it bid for and obtained millions of dollars of contracts from the United States Maritime Commission. To establish its shipyard, the company acquired twelve acres along the Kinnickinnic River, just north of Beecher Street and 1st Street. The shipyard was readied for operation in sixty days. The initial order was for eight seagoing tugs, the first of which was launched on July 21, 1942.

Wisconsin Governor Julius Heil attended the launch ceremony and told the guests, "Ben Froemming didn't have to build ships to make a living, but he is building them because his country needs them. He had the guts to go ahead and build a shipyard even when they told him it was impossible."

All eight tugs were completed by 1943. Each displaced 1,117 gross tons and was 194 feet long.

While the tugs were still under construction, the company was given a contract to build four anti-submarine frigates. Named *Allentown, Bath, Machias*, and *Sandusky*, the frigates were completed in 1943 and 1944. Each displaced 1,430 tons and measured 304 feet in length. They were armed with three 3-inch guns, four 40-mm anti-aircraft guns and four depth charge throwers.[171]

Froemming received one more contract during the war—this one for 14 cargo ships. These "*Alamosa*-class" cargo ships were designed by the Maritime Administration and were intended for rapid construction. They were 339 feet in length with a fifty-foot beam and were powered by single Nordberg diesel engines rated at 1,750 horsepower. The last ship under this contract was launched on November 1, 1945. All were named after counties that started with the letter 'C' in the United States: The *USS Claiborne, Chicot, Clarion, Chatham, Codington, Charlevoix, Craighead,* and the *Colquitt.* Most had only brief service during the war, when ended on September 2, 1945. However, the *USS Charlevoix* provided aviation fuel for the New Zealand Air Force based on New Britain, supporting their twice-daily raids on the Japanese.

Most of the cargo ships saw service following the war. As an example, the *USS Colquitt* (AK-174) was transferred to the US Coast Guard, where she was renamed the *USCGC Kykui* (WAK-186). She was sold to the Republic of the Philippines in 1972 and delisted by their navy in 2001. The *USS Clarion* was sold to Norway after the war and operated until 1970 when she was wrecked off the coast of Peru.

The *USS Claiborne* (AK-171) was launched on September 3, 1944, by Froemming Brothers, Inc. Note the number of Milwaukeeans that witnessed the launch ceremony. Photograph from the Historic Photo Collection of the Milwaukee Public Library.

MMSD'S SOUTH SHORE WATER RECLAMATION FACILITY

Milwaukee's first waste treatment plant began operating on Jones Island in 1925—an activated sludge plant with a sewage treatment capacity of 85 million gallons per day. By the 1960s, planning began for additional capacity to meet the growing needs of southern Milwaukee County. A decision was eventually made to locate a new facility in the City of Oak Creek. The new plant, called the South Shore Water Reclamation plant, was started in 1968. It was expanded in 1974.

The facility uses anaerobic digesters, which turn organic material into methane gas. The gas is collected and burned to produce electricity.

In 1982, the Wisconsin legislature established the current Milwaukee Metropolitan Sewerage District (known as MMSD). It provides wastewater services for twenty-eight municipalities within Milwaukee County and portions of the surrounding counties. It is the largest municipal wastewater treatment provider in the state.

DIGESTER GAS USE FOR SEWERAGE TREATMENT

When the Milwaukee Metropolitan Sewerage District's South Shore Plant was designed, anaerobic digesters were included to produce a high heat content gas, which in turn is used to supply the energy requirements of the facility. Because of this far-sightedness, this plant has realized enormous savings in the cost of energy for the facility.

The anaerobic digester process was first practiced on a commercial basis in Birmingham, England and in Baltimore, Maryland, both in 1912. Anaerobic digestion is a biochemical process that breaks down organic wastes into a granular material and a usable methane gas. It takes place in the absence of oxygen, using microorganisms to break down the organic wastes. The purpose of this process is to stabilize these solids from the standpoint of odors and pathogenic organisms and to produce digester gas.

The early anaerobic-digestion wastewater treatment plants used a batch process. Open tanks, heated with steam during the winter months, led to odor and scum nuisances. As a result, the process fell into disfavor.

In 1921, a pilot project demonstrated that a continuous digestion process could result in a facility that was essentially odor free. A plant at Brownsville Texas used anaerobic digestion successfully, although the temperature requirements of the digestion process caused concerns for employing such facilities in northern climates. However, in 1926 at Antigo, Wisconsin, a successful anaerobic digestion plant was placed into service. The plant has served as the forerunner of all modern day anaerobic digestion facilities. The Antigo facility used a heated, covered, mechanically mixed digester that was continually fed, with a digester gas-collection system.

MMSD's South Shore Wastewater Treatment facility in Oak Creek is a good example of a modern anaerobic digestion plant. The digester gas produced by the process is used to generate much of the electric power needs of the plant, producing the compressed air used for the secondary treatment processes, providing heat to the digesters, and for the incineration of grit, screenings and grease. Digester gas has a heating value between 600 and 650 BTU per cubic foot.

The solid by-product is a thickened sludge, which MMSD conveys to its treatment plant at Jones Island to produce Milorganite. (See Chapter 4 for a discussion of the Jones Island Milorganite facility.)

The South Shore Waste Water Treatment plant used digester gas in spark-ignited Nordberg six-cylinder turbocharged gas engines, driving Allis-Chalmers electric generators. Photograph courtesy of MMSD. Note, these diesel engines were subsequently replaced.

To generate the plant's electrical power requirements, three six-cylinder, 638 horsepower, dual-fuel (either natural or digester gas) engines were used. Each engine drove a 350 kW generator and consumed 7,000 cubic feet of digester gas in the process.

The air used in the secondary treatment process is provided by four 12-cylinder, 1,375 horsepower, dual-fuel engines. Each of these engines drove a blower capable of producing 35,000 cubic feet of air per minute.

In addition to electrical power and air production, digester gas is utilized in make-up heat boilers for digester heating. The plant's two boilers used for this purpose each consumes 8,700 cubic feet of digester gas per hour. The plant also uses digester gas in its incinerator in place of natural gas.

The digester gas is used to produce at least half of the plant's electric power requirements, at considerable savings.

WISCONSIN NATURAL GAS COMPANY

In 1912, the North American holding company formed the Wisconsin Gas and Electric Company. Initially it provided light, power and transportation services to Kenosha, and manufactured gas to the City of Racine. Racine had a substantial gashouse on the lakefront, providing gas to a variety of customers for heating, cooking and for factories, and selling coke. Soon the company extended gas lines as far north as South Milwaukee, and south to Kenosha.

The company also expanded its electric operations, acquiring a number of small electric companies. Following the passage of the Public Utility Holding Company Act in 1935, Wisconsin Electric acquired Wisconsin Gas and Electric. Electric operations were consolidated in 1950.

In late 1947, the Michigan Wisconsin Pipeline Company began work on a natural gas transmission pipeline from gas fields in Texas, Kansas and Oklahoma to existing utilities in Michigan and Wisconsin. Every mile of the new pipeline was furnished with pipe from Milwaukee's A.O. Smith Company.

Wisconsin Gas and Electric connected to this new source of gas in early 1950 and converted its system to use natural gas. To symbolize the event, it changed its name to Wisconsin Natural Gas Company. Its Racine manufactured gas plant was demolished a few years later.

LIQUEFIED NATURAL GAS STORAGE PLANT

In 1965, Wisconsin Natural Gas Company became the first in the United States to place a liquefied natural gas (LNG) storage plant into commercial operation. The storage plant is used by the company to reduce the peak annual demand for natural gas from its pipeline supplier. The plant is located in the City of Oak Creek. Construction began December 15, 1964, and the plant was on stream in September 1965.

Chicago Bridge and Iron, the primary contractor, was awarded a "turnkey" contract for the design and construction of the plant. George Bomier, president of the company at the time, is credited with the feasibility analysis and the development of the economic considerations that justified the use of a liquefied natural gas storage plant to help the company meet the peak demands of its customers.

The LNG storage tank is an above ground, double-walled, metal tank with a storage capacity of 72,000 barrels, equivalent to 250 million cubic feet of gas. The flat bottom, dome roof vessel has an overall diameter of 81 feet 6 inches and a straight side height of about 105 feet. The inner tank is constructed of tapered aluminum plate, providing a tank wall closely approaching ideal design. The space between the inner and outer tank is insulated with non-flammable, inorganic Perlite, limiting boiloff of LNG under worst anticipated ambient conditions to 0.7 percent per day of total storage volume.

Natural gas is liquefied by a cascade refrigeration-cycle system utilizing propane, ethylene and natural gas as refrigerants, which reduces the temperature of the natural gas to 260 degrees below zero. The liquefaction capacity is three-quarters of a million cubic feet per day. The cascade cycle permits simple extraction of a portion of the heavy hydrocarbons and nitrogen. The result is a liquid product which, when vaporized, has a controlled heating value and specific gravity consistent and interchangeable with the pipeline supply. The liquefaction plant design and process vessel arrangement facilitate ease of operation and maintenance.

An aerial photograph of the liquefied natural gas storage plant of Wisconsin Natural Gas Company. It is located at 4375 East Elm Road in Oak Creek.

During periods of high demand, LNG is pumped from the tank, vaporized and superheated to enter the pipeline at the proper temperature. The vaporizers are plate fin heat exchangers heated by isopentane, which, in turn, is heated by hot water. This system combines safety, reliability and simple control for a wide regasification range. Redundant vaporization systems provide 100% standby capability. Operated together, they provide additional capacity in the remote possibility that the normal pipeline service might be interrupted. Each redundant vaporization system has a capacity of thirty million cubic feet per day.

Wisconsin Natural Gas Company was merged into Wisconsin Electric, which operates under the name We Energies as part of WEC Energy Group, Inc.

CHAPTER 6: RIVERWEST AND 'RIVER-EAST'

A number of early Milwaukee companies moved north along the routes of the Chicago, Milwaukee, St. Paul and Pacific Railroad—better known as *the Milwaukee Road.* The route passes through an area just west of the Milwaukee River, now commonly referred to as *Riverwest.* A competing rail line of the Chicago and Northwestern Railway traveled east of the Milwaukee River and attracted industry seeking expansion.

The Milwaukee Road's rail line was initially constructed in 1854. In that year, the La Crosse and Milwaukee Railroad built a line along the west bank of Byron Kilbourn's canal, traversing the natural bluff. This rail line serviced a number of factories, as well as mills, tanneries and coal yards. From the downtown area it went north to the Humboldt Yards, under North Avenue and continued to today's Gordon Park, before turning northwest. The line was eventually absorbed into the Milwaukee Road's system. Because this railway line served the Blatz, Pabst and Schlitz breweries, the line was locally dubbed the *Beer Line.* The *Beer Line* name was appropriate. On some days more than 250 rail cars were loaded and traveled over the line. It is reported that this section of track generated more freight revenue per mile than any other part of the Milwaukee Road's system.[172]

Milwaukee industry fanned out along these routes, including a number of important manufacturing firms that located near Capitol Drive (formerly referred to as Lake Street). These companies, served by the Milwaukee Road, included Nash's Seaman Body Plant, Globe-Union, Milwaukee Electric Tool and Cleaver-Brooks. Nash's Seaman Body plant alone required thirty-five to fifty freight cars a day.[173]

National Brake & Electric and a Ford Motor assembly plant were located east of the Milwaukee River and were served by the Chicago and Northwestern Railway line.

The Beer Line Tracks at Schlitz Brewery, From *Trains and Travel Magazine*, Wallace W. Abbey, Used with permission of Kalmbach Publishing Co.

The map above, taken from an 1898 map of TMER&LCo, has been highlighted to show the railway routes of the Milwaukee Road (west of the Milwaukee River) and the Chicago and Northwestern Railway (east of the river). The location of the manufacturing facilities is approximate. Note: Lake Street was eventually renamed Capitol Drive.

NATIONAL BRAKE & ELECTRIC

As noted in Chapter 2, Niels Christensen formed Christensen Engineering as demand for his air braking system for streetcars and trolleys increased. It was a joint-stock company established to finance a new factory on Milwaukee's north side. The site was located on the east bank of the Milwaukee River, at the foot of Belleview Place and adjacent to the city's Riverside Park.

Frank Bigelow, Samuel Watkins and Henry Goll owned the majority of the stock in the company. The three renamed the enterprise the National Electric Company. When the owners pushed to diversify into other products, Christensen objected. To settle the matter, Christensen left the company in exchange for some preferred stock and a five percent royalty on his patents. National Electric soon ran into financial difficulties. One of the principal owners, Frank Bigelow who was the president of the First National Bank of Milwaukee, defaulted on $1.4 million, which he lost in speculation in wheat and stocks.[174]

Westinghouse Air Brake Company acquired National Electric out of bankruptcy, renamed the company the National Brake & Electric Company, and refused to pay royalties to Niels Christensen. In December 1906, Westinghouse sued Christensen for patent infringement. George Westinghouse had developed an air braking system for steam locomotives and railcars, and he didn't take lightly what he considered an illegal use of his patent rights.

Christensen promptly countersued Westinghouse, starting a 24-year legal battle that went before the US Supreme Court on three separate occasions over the rights to manufacture compressed air brake systems for streetcars.

National Brake & Electric prospered under Westinghouse's ownership and the factory on the banks of the Milwaukee River was expanded to meet demand for its streetcar braking system, as well as for a multitude of other products including air compressors, electric motors, gasoline-driven locomotives, tractors, and water pumping systems. During the First World War the company also produced four-wheel drive and steer tractors, lathes and gun mount castings. National Brake & Electric's factory eventually became one of the largest in the state of Wisconsin, occupying a fifteen-acre site. The facilities combined floor space totaled over 540,000 square feet and at its peak employed over 1,400 individuals.

In 1931, the company began transferring work to another facility in Pennsylvania. Most of the facility was demolished in 1937. The area today features the Urban Ecology Center and Milwaukee's Rotary Centennial Arboretum.

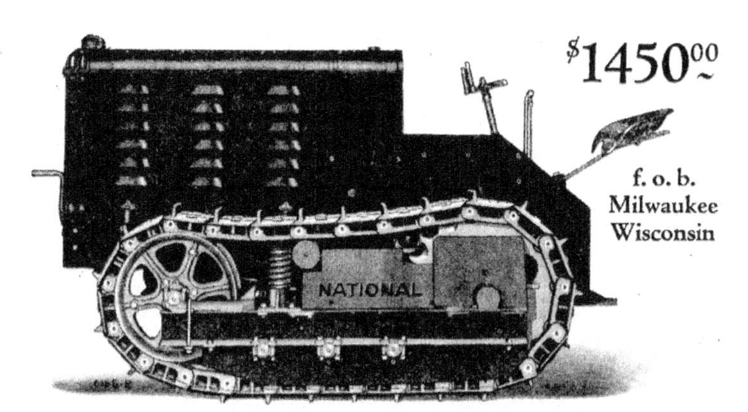

One of the company's products was this heavy-duty, tracked-tractor, advertised here.

Photograph of the National Brake & Electric Company, located on the east bank of the Milwaukee River—about four blocks north of North Avenue. Today the site houses Milwaukee's Urban Ecology Center and Riverside Park. Photograph from the Historic Photo Collection of the Milwaukee Public Library.

The Milwaukee Locomotive

The Milwaukee Locomotive Manufacturing Company was organized in Wisconsin in 1907. It's not known whether the company was formed by National Brake & Electric, or acquired by them, but by 1910 it was operating as a subsidiary. The company produced a patented, gasoline-powered locomotive principally for use in mines, manufacturing plants, mills, lumberyards and quarries, and for tunnel and canal construction.

Milwaukee Locomotives production grew from two-ton locomotives in the early days to twenty-ton in the later years. They were all four-wheel locomotives, and were available in gages from 18 to 56.5 inches.[175] Many of the Milwaukee locomotives had Modine sectional core radiators bearing the Milwaukee name cast across their tops.

A Milwaukee locomotive is shown in use at an unknown mine.

A patent for the locomotive design was filed in 1924 by W. A. Beimling, a mechanical engineer for the company.[176]

Milwaukee Locomotive manufactured locomotives in Milwaukee until 1932, when it was acquired by competitor Whitcomb Locomotive Company, a subsidiary of Baldwin Locomotive. The company continued to operate in Rochelle Illinois for three years until it was fully absorbed into Whitcomb's operations. It is believed the Milwaukee Locomotive Company manufactured over nine-hundred locomotives—the highest record number assigned was 981, although some earlier locomotives may have been rebuilt and assigned new numbers.

Pictured is one of Milwaukee Locomotives largest locomotives. It is believed it features a radiator built by Modine of Racine

GLOBE-UNION INCORPORATED

Globe Electric Company was founded in Milwaukee in 1912 by G. W. Youngs, D. Decker and Julius H. Gugler. However, the real genesis of the enterprise was the formation of the Milwaukee Electric Construction Company in about 1905 by Oscar Werwath. Werwath established his company to provide work for students at his newly formed School of Engineering, to help defray tuition expense and provide students with hands-on training. Milwaukee Electric Construction's principal product was electric storage batteries. The demand for these batteries eventually exceeded the available space for their production and Werwath sold the business to a Youngs, Decker and Gugler.[177]

Globe Electric initially produced electric storage batteries for streetcars, as well as for rural light plants and switchboards. The company expanded into the manufacture of batteries for automobiles and radios, and entered into the farm electric plant business.

Right: Early Globe automobile batteries were made of thick lead plates, which were enclosed in glass jars. The jars were then inserted into a lead-lined wooden box. Photograph courtesy of Johnson Controls.

In a series of transactions, the company merged with Western Engine and Dynamo and Western Utilities in 1919, retaining the Globe Electric Company name. The resulting company included John I. Beggs of the Milwaukee Electric Railway and Light Company as president, Julius Gugler as vice president, John Daniel Wanvig, Jr. as secretary and general manager and Chester Odin Wanvig as treasurer. It established its combined operations the following year at 900 East Keefe Avenue, just south of Capitol Drive and about a block west of the Milwaukee River.

In the early 1920s, Globe Electric introduced a battery-powered radio, which it produced for a number of years, before concentrating on the production of various radio components including a "low-loss" tuner, which was patented in 1924. The Globe tuners were featured in stories of commendation by every radio magazine of the era.

In 1925, Chester Wanvig, Globe Electric's treasurer, entered into an agreement with Sears, Roebuck and Company to produce automobile batteries for sale by Sears. However, Globe Electric's shareholders turned down the opportunity. Chester Wanvig decided to produce the batteries for Sears himself and established the Union Battery Company to manufacture the batteries. The venture was apparently successful, because in 1929 Globe Electric and Union Battery consolidated as the Globe-Union Manufacturing Company with Chester Wanvig as president.

Globe-Union was one of the first manufacturers to conduct experiments to make automotive batteries more reliable under extreme climatic conditions. The company installed refrigerating facilities to allow batteries to be tested in temperatures as low as -20° F, and to use the results to develop batteries specially designed for all-climate use.

Battery production was somewhat seasonal in nature. In order to help fill the manufacturing void that occurred in the summer months, the company manufactured clip-on, adjustable roller skates for many years. As a result, the Globe-Union name was well recognized by Milwaukee children of the era. The company also manufactured spark plugs and, during the early years of the Great Depression, golf clubs.

These Globe-Union roller skates were popular on Milwaukee sidewalks in the 1940s and 50s. Inexpensive, they clipped onto a pair of regular shoes. Their metal wheels made quite a chatter on concrete sidewalks. The company also made similar clip-on ice skates.

Globe-Union Thin-Wall Injection Molded Battery Case and Cover

In the 1920s, battery manufacturers began using cases and covers made of resin rubber mixtures, to replace the formerly used lead-lined wooden boxes. These cases and covers were made by compression molding highly filled resin rubber mixtures. Filler materials were coal dust, anthracite dust, clay, rags, waste paper, and other locally available material. The result was heavy, porous, brittle, and unreliable battery cases and covers.

In the late 1950s, Globe-Union developed the thin-wall, injection-molded polypropylene battery case and cover. This significant development reduced the weight and bulk of the battery case and cover, thus giving it the capability of greater energy content in a lighter weight battery. This was achieved after several years of development. The unique features of both the mold and final product were patented. This accomplishment continues to have significant value in view of the needs of lightweight fuel-efficient vehicles. Generally all United States automobiles today contain these polypropylene batteries.

In 1967, this technology was adopted by Sears in its Diehard batteries, which were made by Globe-Union.

The development of the thin wall, high strength polypropylene battery represented a major milestone within the history of the battery industry. In the words of one of their largest worldwide competitors, "Globe's development of the thin wall polypropylene battery dragged the battery industry kicking and screaming into the 20th century." This Globe-Union patented development is now a world standard. Licensees of their technology include most large battery manufacturers in the United States, as well as in several other countries in the world. The development resulted in a reduction of approximately seven pounds per unit in battery weight, and an increase in the electrical efficiency of batteries by better space utilization of ten to twenty percent.

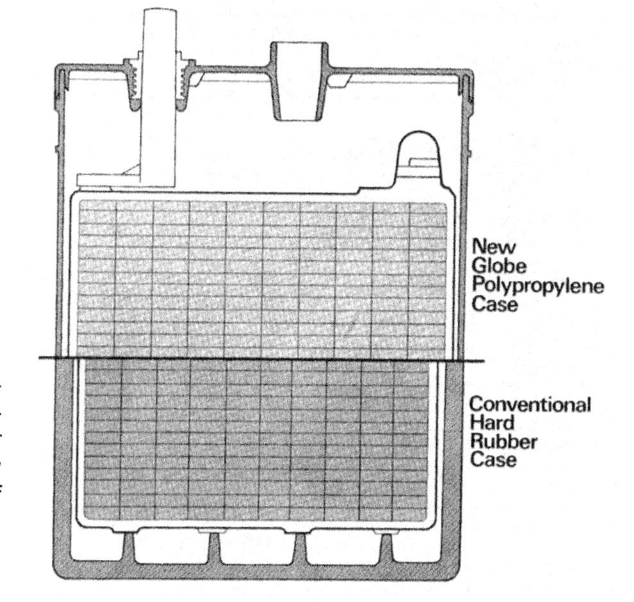

The thin wall injection molded polypropylene battery cases and covers resulted in a reduction of seven pounds per battery. It increased space utilization by 10-20%. This view shows a comparison of the new design (top) with the old (bottom)

In 1978, Globe-Union was acquired by Johnson Controls. Johnson Controls continues to be a major researcher, manufacturer and supplier of automotive and specialty batteries.

Centralab Division

During the Second World War, most of Globe-Union's battery production went into wartime efforts, including the production of batteries for Army Jeeps and other vehicles. The company was also called upon to expand its radio parts business, producing radio communication devices through its newly formed Centralab Division.

During the later stages of the war, Centralab was also asked to design a proximity fuse for cannon shells. These fuses used a special radar sensor to detect the proximity of the target and to explode just prior to impact. It was a difficult engineering feat because the circuits had to withstand the incredible force of being fired from a cannon. The fuses also needed to be produced quickly and in quantity. Centralab met the requirements by employing a ceramic plate that was screen-printed with metallic paint for conductors, and used a carbon material for the resistors. Ceramic disc capacitors and subminiature vacuum tubes were soldered in place. This circuit design, which eventually became known as a thick-film hybrid circuit, was successful. Armaments equipped with these proximity fuses were critical during the Battle of the Bulge, as well as during engagements on the Pacific Front.[178]

A patent for the technology was assigned to Globe-Union, but was classified by the US Army. In 1984, the Institute of Electric and Electronics Engineers (IEEE) awarded Harry W. Rubinstein, the former head of the Centralab Division, with its Cledo Brunetti Award for early key contributions in the development of printed components and conductors on a common insulating substrate.[179]

Following the war, Centralab's considerable experience with electronic circuitry miniaturization and early techniques for placing silver "ink" on a ceramic base to replace wiring led the division to pioneer in the production of various electronic devices, including printed circuits for hearing aids.

This advertisement for an early hearing aid that used Centralab's printed circuit notes the heritage of the technology—first developed for use as a proximity fuse for mortar shells.

In 1947, Centralab hired engineer Jack Kilby, who had just graduated from the University of Illinois-Urbana. Kilby was put to work designing various consumer electronic products using this ceramic base, silk-screen circuit technology. While there, Kilby had a number of patents. He was also exposed to a new innovation—the transistor. The Centralab folks must have had a great ability for scouting talent because Kilby went on to achieve phenomenal success in circuitry miniaturization.

Kilby eventually left Milwaukee and soon after invented one of the most significant devices in engineering history. That stage of Kilby's career is covered in Chapter 9.

FORD MOTOR COMPANY

Anyone familiar with Henry Ford's development of the Model T automobile, including his revolutionary manufacturing methods, recognizes that most of the innovation was centered on Detroit, Michigan. What may not be as readily known is that the Ford Motor Company manufactured its Model T in Milwaukee for a number of years.

Ford Motor Company introduced the Model T on October 1, 1908. It had some advanced features, including a four-cylinder engine with a removable cylinder head and a one-piece block. It also used a lightweight, high-strength vanadium alloy steel.[180] The vehicle had a high ground clearance, which gave it the ability to traverse rural roads which often contained numerous ruts and uneven surfaces.

Even more important to the success of the new automobile, however, was its relative affordability. When first introduced it sold for $825—an aggressive price but still more than the average worker's annual salary. Henry Ford, however, desired to "... build a car for the great multitude ... so low in price that no man making a good salary will be unable to own one." Toward the end of 1913, Ford and his engineers built a new factory in Detroit that featured a moving assembly line. Automation and volume production allowed the company to reduce the price to $440 the following year. By 1925, Ford was selling the Runabout version of the Model T for $260.[181]

The low selling price did exactly what Henry Ford predicted. The demand for the vehicle soared. In order to meet the demand for the Model T, Ford added a number of "branch" assembly plants. One was built in Milwaukee at the intersection of Prospect and Kenilworth streets. The five-story building with "deep basement" was completed in late 1915 and the plant went into production the following April. The building was enlarged in 1920, and operated as an automobile assembly plant until the fall of 1932, producing Model T Fords initially and eventually Model A's. At peak the factory employed 450 individuals. Employees were paid $6 for an eight-hour day in 1919, building 180 cars a day on a conveyor-based assembly system. Automobile assembly was interrupted during the First World War, when the factory was taken over the US government.

Between 1921 and 1927, the Milwaukee plant assembled 281,334 automobiles and 11,116 Model T-based trucks. Eventually, Ford located twenty-six branch assembly plants in cities throughout the United States and Canada. These branch plants assembled almost half of the over fifteen million Model T automobiles the company manufactured.[182]

Ford shuttered most of its branch assembly plants in 1932, because of the declining demands for automobiles during the depths of the US Depression.

The Milwaukee factory was sold by Ford in 1942 to the US Government's Defense Plant Corporation and leased to the A.O. Smith Company during the Second World War. A western addition to the building was completed in 1945. Following the war, the General Motors AC Spark Plug Division leased the building from 1947 to 1970. The building was acquired by the State of Wisconsin for the University of Wisconsin–Milwaukee in 1970 and renovated in 1975 for storage, shops, printing services, mail services and administrative offices. In 2004, the building was redeveloped for student housing, classrooms, studios, and retail space and is now referred to as UWM's Kenilworth Square.

This picture was taken in 1916 at a Ford assembly plant in Oklahoma City, believed similar to the one in Milwaukee. It shows Ford Model 'T' chassis moving along the assembly line on the distant left, while auto bodies are moved by an overhead rail from the adjacent bay. Image used with the permission of the Henry Ford Museum.

MILWAUKEE ELECTRIC TOOL

In 1918, Henry Ford approached a young manufacturer named Arno H. Petersen with a request to produce a lighter, smaller one-quarter inch capacity power drill. Ford was interested in a portable, easily handled power drill that could be used on his automobile assembly lines. Petersen's factory was located at 3522 North Fratney in what is now referred to as Milwaukee's Riverwest neighborhood.

At the time, Petersen's company had been producing tools and dies—some of which were for the Ford Motor Company. Peterson jumped at the chance to develop a drill for Ford Motor Company. His team developed the "Hole-Shooter," a small yet powerful drill powered by a Westinghouse electric motor. The drill weighed five pounds and was light enough to be used by assembly line workers in the automotive industry. Ford was enthusiastic about the new drill because the drills in use at the time were much heavier and only the strongest mechanics could use them productively.

The drills were prone to breakdowns. During its initial years, the company often kept itself in business by repairing the tools they had sold. As its employees repaired and refurbished the tools, they analyzed the failures and refined their drills to be more robust—yet designed to be manufactured at a reasonable cost. As reliability increased, so did sales.

In 1922, Petersen founded the A.H. Petersen Company to manufacture the company's newly engineered hand drill. Petersen hired Albert F. Siebert as sales manager. A fire destroyed the manufacturing facility the following year, causing a financial setback and forced the company into bankruptcy. Siebert acquired the company's assets at auction, along with Ray Beckwith, calling the company Siebert and Beckwith. Three years later the partnership dissolved and, in 1927, Siebert incorporated the company under the name Milwaukee Electric Tool Corporation. The company's operations at that time were located in a corner of the Kempsmith Milling Machine Plant in West Allis.

Based on the success of the 'Hole-Shooter,' Milwaukee Electric Tool introduced electric sanders, polishers, the electric hammer and portable hand grinders.

In addition to improving its tools, the company also expanded its facilities to manufacture its own fractional horsepower motors. Their new electric motors had improved overload capacity and increased performance—leading to lower maintenance and longer tool life.

The Second World War resulted in a large demand for the company's portable power tools. Hole-Shooters were used extensively for the manufacture of airplanes and the demand was significant. Many new product lines were also developed during the time.

Petersen's original "hole shooter." Image used with permission of Milwaukee Electric Tool

Development of the Lightweight Portable Electric Drill

The Hole-Shooter was the tool industry's first, lightweight, one-handed drill capable of handling heavy-duty workloads. The initial tool used a Westinghouse electric motor to power the mechanism. Over time, Milwaukee Electric Tool engineers spent hundreds of hours studying failure modes to refine the product. They also analyzed competitive electric, portable power tools.

A series of changes and improvements were made to the Hole-Shooter, making it more durable. Bronze bushings, spur-gear drives, cords and switches were replaced with stronger components. The result was a robust device that won acceptance in automotive and heavy metal working industries.

Here a work crew is seen using Milwaukee Electric Tool's Hole Shooters on an aircraft panel.

SEAMAN BODY DIVISION OF NASH-KELVINATOR CORPORATION

As noted in Chapter 2, Seaman reached an agreement with Nash Motors Company in 1919 under which the company would build bodies for Nash vehicles. Seaman decided to build a new factory on a fifteen-acre site on Milwaukee's north side adjacent to the tracks of the Chicago, Milwaukee and St. Paul Railway—the so-called "Beer Line" and just south of Capitol Drive. Nash Motors acquired a fifty percent stake in the company in order to provide Seaman with the capital to expand their facilities.

Plans were drawn up for a five-story factory that would eventually become one of the largest automotive body factories in the United States. Construction started in 1919 and the first portion of the building, about one-fifth its eventual size, was completed by the end of the following year. A recession delayed further progress, however, and the factory was not completed until 1938.

Early photograph of Seaman Body Division of Nash Motors on Capitol Drive and Richards Street. Photograph courtesy of the Milwaukee Public Library.

When the plant was completed, it was the world's largest auto body factory. It ultimately provided 1.6 million square feet of floor space, spread out over five floors.[183] The factory was entirely devoted to the manufacture of automobile bodies, almost all of which were shipped by truck to Kenosha to be mated with engines and other mechanical components. Since the new plant was designed to build all-steel auto bodies, the company instituted a program to retrain the Seaman employees that previously worked in its wood milling and framing departments.

In 1936, Nash combined with Kelvinator and became a diversified company and, in 1954, Nash-Kelvinator acquired the Hudson Motor Car Company of Detroit to form the American Motors Company.

At its peak, the Milwaukee body plant employed as many as nine thousand workers. In addition to the Nash automobile, the factory produced bodies for the Rambler, American, Ambassador, Matador, Marlin, Javelin, AMX, Hornet, Gremlin, Pacer and Eagle automobiles. American Motors was acquired by Chrysler in 1987 and AMC's Milwaukee body plant was shut down and eventually demolished.

Nash Automobile Body Firsts

Nash Motors Company had several engineering innovations in its automobile body design and features:

- In 1936, it introduced its "bed-in-a-car" feature. The initial design created an opening into the trunk such that two adults could sleep with their bodies on the rear seat cushions and their legs and feet in the trunk. This arrangement was eventually modified to create a bed by designing the front seat backs to fold back flat, which created a sleeping area entirely within the passenger compartment. Dealers offered inflatable air mattresses and window screens to complete the package. Nash referred to this seat design as "Airliner Reclining Seats."
- In 1938, Nash announced its "Conditioned Air System," the first heating system to take fresh outside air and pass it over a heater core. The system was refined the following year by adding a thermostat to make temperature control semi-automatic and renamed it the "Weather Eye" System, which was also a first for the industry (see article below).
- The 1941 Nash 600 was the first mass-produced automobile in the United States to employ unibody construction (see article below).
- In 1949, the Nash "Airflyte" was introduced—the first redesigned Nash vehicle following the Second World War. The car featured a scientifically designed aerodynamic body shape, incorporating theories of Nash's vice president of engineering, Nils Wahlberg on reducing automobile body drag coefficient. The shape was refined in a wind tunnel. It resulted in a controversial cutting-edge aerodynamic shape with enclosed front fenders. The design also incorporated a one-piece curved safety glass windshield. The design was wider and lower than the 1948 vehicle it replaced, and yet provided more interior room.[184] Nash also offered seat belts as an option in 1949—the first automaker to do so.
- The Nash Airflyte also featured a scientifically designed, ergonomic driver-oriented instrument panel, called the "Uniscope," which was a first for the industry. The Uniscope placed the speedometer, fuel gauge, ammeter, oil pressure gauge, and temperature gauge into one unitized module that was mounted in direct line-of-sight of the driver. Similar one-piece instrument panel assemblies were eventually adopted by the industry.

Nash Weather-Eye Conditioning System

The Nash "Conditioned Air System," introduced in 1938, was the first automotive heating system to take fresh outside air and pass it over a heater core that used hot engine coolant as its heating source. The Nash system used a fan to draw in the air and provide a slightly pressurized engine compartment to minimize infiltration of cold outside air. The system also had a disposable filter to clean incoming air.

This was a revolutionary system. Other automobiles of the era offered relatively crude heaters, which recirculated the air in the cabin and did not have ducts to distribute the heat. This resulted in a very hot area around the heater, and cold areas in the rest of the vehicle. There was no system to defog windshields, other than wiping the glass with towels and scraping ice away manually.[185]

In the following year, Nash further refined the system by adding a thermostat and rebadged it as the Nash "Weather-Eye" Heater. The thermostat control system was jointly designed by engineers from the company's Nash and Kelvinator divisions led by Nils Wahlberg.[186] The Weather-Eye was a first for the industry and was heavily promoted as eliminating drafts, steamed-up windows, and stale interior odors by continually drawing in outside air through the heater, which also allowed the air to be warmed to the desired temperature and even partially dehumidified. This was a revolutionary system at the time.

Also capitalizing on Kelvinator's experience in refrigeration, Nash was also the first to offer a modern automotive air conditioning system. Other automaker air-conditioning systems had a large condenser in the trunk, which reduced the available storage space and was heavy and costly. Calling its combination heating, ventilation and air-conditioning system the Nash "All-Weather Eye System," it was introduced in 1954. The system sold for $345 when introduced, which was considerably less expensive than other automotive air-conditioning systems at the time. As an example, the optional air-conditioning system offered by Oldsmobile cost $544 and weighed twice as much.[187]

The Nash system was entirely located in the engine compartment, and the conditioned air entered into the passenger compartment through dash-mounted vents—all similar to today's automotive air-conditioning design. It added only 133 pounds to the vehicle.

Controls from a Nash All-Weather Eye air-conditioning system from a 1956 Nash Ambassador.

Unibody Construction

The Nash 600 is generally credited as the first mass-produced American automobile that used unibody construction.[tt] The vehicle employed a unitized body/frame construction in which the body and frame were integrated into a single strong structure. Prior to that time, almost all vehicles used a body-on-frame design in which the car body is separately assembled and then bolted to the frame. Unibody construction allowed Nash to produce a car that was lighter—reducing the vehicle's weight by five hundred pounds. Nash also claimed that the design was quieter and more rigid than its competitors were.[188]

Nash's adoption of unibody construction caused significant changes to the industry. Not only was it eventually adopted by other manufacturers, it also required new techniques for collision repairs. With frame construction, auto body repairs are typically made by using hydraulic jacks to push bent frame members back into position. With unibody construction, it became necessary to pull the parts back into position—resulting in the portable body puller tool. Milwaukee's Blackhawk Tools, which eventually became today's Actuant Corporation, developed tools that could be used to re-establish the shape of the auto body precisely. They also produced a stationary frame fixture with adjustable pull towers, for more severe damage to unibody vehicles.[189] These devices were adopted worldwide.

This 'X-ray' image of the Nash 600 shows the major Unibody features. The Nash 600 body was manufactured in Milwaukee by Seaman. Image from Wikipedia Commons

[tt] Lincoln's 1936 Zephyr, the Cord 810, and Chrysler's 1934 Airflow had adopted semi-unitized construction in which the body structure was welded to the frame. Nash was the first to introduce a "monocoque" vehicle structure in which the chassis is integral with the body.

Although the subject of some debate, unitized body construction also offers safety benefits to the occupants if properly designed. Unibody vehicles tend to score considerably better in front crash and rollover tests and score only slightly lower in side impact tests, than their body-on-frame competitors.[190]

The 600 designation for the Nash came from the manufacturer's claimed range on a 20-gallon tank of gasoline. While 30-miles per gallon may have been somewhat optimistic as an average rating, the 600 was more efficient than other automobiles of the era because of its lighter weight, use of overdrive gearing, and a relatively good aerodynamics. In the 1941 Gilmore Grand Canyon Run, an economy challenge for US automakers sponsored by the American Automobile Association, the Nash 600 took first place in the large car category—averaging 25.8 miles per gallon.

The vehicle provided to be the right car for its time because soon gas rationing became the norm during the years surrounding the Second World War. The Nash's 25 mpg figure compares favorably to the average US passenger vehicle in 1940 of only 13.8 mpg,[191] and yet the Nash 600 was a full-size automobile.

Although Nash soon started to turn its attention to defense orders from the US Government, it produced a sizable number of its economical, low-priced 600 models.[192] The unitized auto body design was adopted by all manufacturers for automobiles and eventually most sport utility vehicles.

This advertisement touts Nash's performance in the 1941 economy challenge. From author's collection.

NILS ERIK WAHLBERG was a pioneer in the automobile industry. He was born in Finland of Swedish parents in 1885, but fled to Switzerland in 1907 to avoid being drafted by the Russian Imperial army. At the time, he was pursuing technical studies in the Polytechnic School of Helsinki. He eventually earned an engineering degree from the University of Zurich in Switzerland.

Nils E. Wahlberg, from a 1941 Nash Press Kit

Upon completion of his degree, Wahlberg came to the United States and accepted a job with Maxwell-Briscoe. He also worked for a time at the Thomas Motor, Packard and Oakland motorcar companies. Oakland was a division of General Motors that was eventually absorbed into GM's Pontiac division.

When Charles W. Nash, then president of General Motors, resigned in 1916 to form the Nash Motors Company in Kenosha, Wisconsin, he took Wahlberg with him as vice president for engineering and research. Wahlberg worked in Kenosha but commuted from his home in Chicago, and continued to work in Kenosha even when Nash merged with Kelvinator in 1937 and it moved its headquarters to Detroit.

Nils E. Wahlberg is credited with leading the development of several significant automotive innovations that were incorporated into the auto bodies produced by Nash in Milwaukee. Most of these innovations were later adopted by the automotive industry:

- The modern fully-unitized/monocoque auto body construction, similar to that widely in use today in the industry.
- The Nash "Conditioned Air System," and eventually the "Weather-Eye" and the "Nash All-Weather Eye System," the first fully integrated automotive heating and ventilation systems.
- The first scientifically developed aerodynamic design for a mass-produced passenger sedan in the 1949 Nash Airflyte.
- The ergonomic driver-oriented instrument panel, the "Uniscope," which placed all instruments into a single unitized assembly that was mounted above the steering wheel.
- Fully reclining seats that converted the passenger compartment into a serviceable bed.
- The first to offer factory-installed safety belts and a safety-padded instrument panel in the 1949 Airflyte.
- The development of the first American compact car, the Nash Rambler, which was introduced in 1950.

Wahlberg retired from Nash-Kelvinator in 1952, two years before Nash and Hudson merged to form American Motors. He relocated to the Washington DC area following his retirement. Nils Wahlberg died in 1977.

CLEAVER-BROOKS COMPANY

John Cleaver founded the J.C. Cleaver Company in 1928 to manufacture a new packaged boiler that he designed. Cleaver was born in 1906 on a farm near Oregon Illinois. As a young lad on the family farm, one of his daily tasks was to chop the wood to fire-up the farm's large cast-iron steam boiler, which was used to heat their farmhouse. A chance remark by a physics teacher in high school about the expansive nature of converting water to steam led Cleaver to experiment with small steam boilers. His first few were failures, but he eventually designed a successful tank car heater.

His first boilers were fabricated by Milwaukee's Reliance Boiler Works and assembled in Illinois. Sales of the kerosene-fired packaged boilers were good, even during the early depression years, which caught the attention of Raymond E. Brooks. Brooks, who had left a sales position at the National Equipment Company, was looking for an opportunity. He bought out Cleaver's partners and moved the company to Milwaukee. The company introduced the first packaged boilers to the US Market in 1931. The company was reorganized as the Cleaver-Brooks Company in 1932.[193]

In response to an inquiry from the US Army Corps of Engineers, Cleaver-Brooks designed desalination equipment to produce potable water from seawater. Designed similar to stills, the units evaporated the salt water and then condensed the vapor—resulting in pure distilled water. During the Second World War, the company was asked to make thousands of units to produce fresh water for troops in the Pacific. Their early units produced only twenty-five gallons of water per gallon of fuel—design modifications increased the efficiency such that the water to fuel ratio improved to 100 and finally to 150 before the war was over.

Orders greatly diminished following the war, but the company continued to research the technology, eventually developing a flash distillation process where the salt water is evaporated under a vacuum. In 1955, the company provided the distillation units for the aircraft carrier *Independence*, with a 200,000 gallon per day capacity.

In 1961, the desalination equipment division was badged, "Aqua-Chem." The division produced the desalination equipment for the *Kittyhawk, Constitution, Enterprise* and *Savanah.* The latter two ships were nuclear-powered vessels and required extremely high purity water for their reactors. By the mid-1960s, Aqua-Chem had installed more than four thousand desalination plants around the world with a combined capacity of 60 million gallons per day.[194]

In 1967, Cleaver-Brooks and Aqua-Chem were merged under the Aqua-Chem name.

Long Tube Multi-Stage Flash Evaporators

In 1940, Aqua-Chem, as a Division of the Cleaver-Brooks Company, entered into a contract with the United States Corps of Engineers to develop a machine to purify water. The program was successful and more than 2,250 units were manufactured for use by the Armed Forces throughout the World during the Second World War. After the war, Aqua-Chem pursued its research in the conversion and purification of saline waters by conducting tests on its own and in collaboration with the US Government. Substantial technological advances were made which broadened the application of desalination equipment to include commercial applications. These advances were based on a progressive development program of corporate research and development, production and field trials of successively larger evaporators, and finally commercial production of a full range of desalting plants.

Large-scale prototype evaporators were built and operated to confirm theoretical assumptions and to obtain actual performance data in lieu of extrapolation.

One of the most significant achievements in this field by Aqua-Chem was the development of the long tube multi-stage flash evaporator. The multi-stage flash evaporator produces distilled water from feedwater by heating it until it is ready to flash. The flashed vapor is drawn to the cooler tube bundle surfaces where it is condensed and collected as distillate. Flashing occurs when heated brine is turbulated in a chamber that is maintained at a lower vapor pressure than that of the entering heated brine. Heat is given up by the brine and a portion converted into vapor until the temperature of the brine reaches the saturation temperature corresponding to the chamber pressure. In other words, the heated brine is flashed off by a pressure reduction. Entrained brine droplets are removed from the vapor by entrainment separators and the pure vapor condenses into distillate on the condenser tubes. The distillation process operates from a low vacuum in the first stage to a high vacuum in the last stage, with stage-to-stage pressure differential being the key to the repeated flashing.

The first long tube multi-stage flash evaporating plant was installed by Aqua-Chem in an Oxnard, California power plant in 1959. It was later moved to Santa Catalina Island. It was a one hundred thousand gallon per day 28-stage flash evaporator.

Prior to the advent of the flash evaporator, the only large plants for desalination were multiple effect submerged tube units. The flash evaporator enabled the development of higher economies, lower cost, and new methods of scale control, which made desalination possible in many parts of the world. The long tube principle eliminated the need for a water-box at each stage, thus lowering costs and pumping power requirements. The flash evaporator, as opposed to submerged tube units, greatly reduced the scaling problems associated with seawater evaporation and enabled the use of newer and more effective means of scale control.

Because of their developments, Aqua-Chem was established as a world leader in the design and manufacture of desalination units. The company eventually produced more than six thousand desalination units that provided a total fresh water capacity exceeding one hundred million gallons a day. Aqua-Chem desalination plants operated throughout the world—in Europe, the Middle East, South American, Africa, Australia and Antarctica, as well as on islands in the Indian, Pacific and Atlantic Oceans. They also built installations in Saudi Arabia, Gibraltar, Ecuador, Chile, Bermuda, Wake Island, and many other arid areas.

Long tube multi-stage flash evaporators, developed by Aqua-Chem, were the mainstay of the desalination industry. This 1966 installation for the Southern Peru Copper Company had a capacity of 720,000 gallons per day. It used turbine extraction steam from an adjacent power plant. Photo provided by Fred A. Lobel, Aqua-Chem.

CHAPTER 7: 30TH STREET INDUSTRIAL CORRIDOR

As noted before, when manufacturing companies needed to expand their capacity most fanned out to the peripheries of the City of Milwaukee. Several companies moved along what is now referred to as Milwaukee's 30th Street Corridor—a strip of land that extends along the tracks of the Chicago, Milwaukee and St. Paul Railway. While the corridor is currently named after 30th Street, the railroad line actually traverses a slight diagonal line as it moves north and goes from roughly 30th Street to 35th Street. When built in 1869, the railroad line was officially called the "Chestnut Street Line." Chestnut Street was the former name of Juneau Avenue and the Milwaukee Road once had a rail yard at that location.

This 6.2-mile line is essentially a connector track that linked the Milwaukee Road's main line in the Menomonee Valley with the railroad line between Milwaukee and La Crosse, referred to locally as the "Beer Line." The tracks met up in the area once called Schwartzburg before it was renamed the Village of North Milwaukee. The village was eventually absorbed by the City of Milwaukee.

This chapter covers the significant manufacturers that built their factories along this corridor.

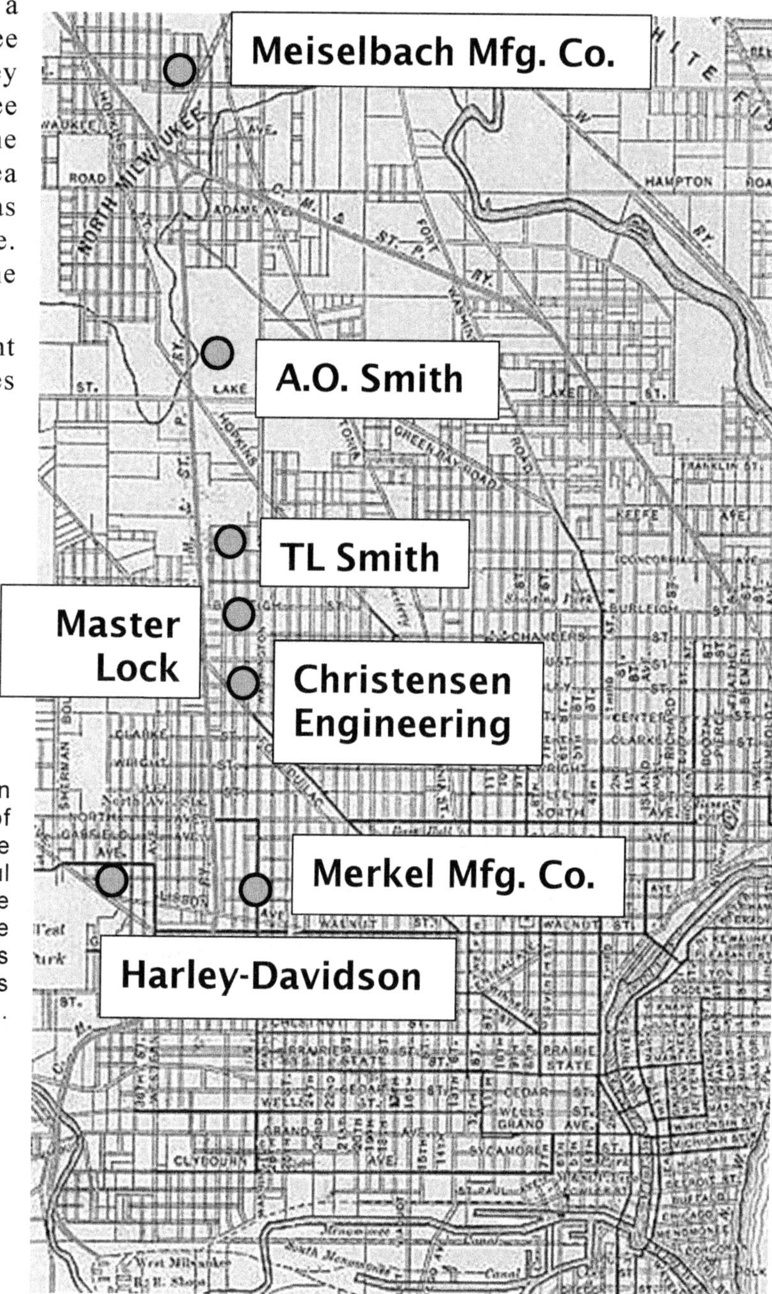

This map of Milwaukee's northwest side is an extract of a TMER&LCo streetcar map of 1898. It has been highlighted to show the lines of the Chicago, Milwaukee and St. Paul Railway that traversed the area. The manufacturing company locations are approximate. Note that several street names have changed. For example, Lake Street is now known as Capitol Drive.

MEISELBACH MANUFACTURING COMPANY

Augustus D. Meiselbach moved his manufacturing company into the Milwaukee-area hamlet of Schwartzburg on October 12, 1896. The community, named after an early German settler who purchased land in the area from Byron Kilbourn, was incorporated as the Village of North Milwaukee in January of the following year. The area was largely populated by German immigrants.

Meiselbach had briefly manufactured bicycles out of a small shop in the Menomonee Valley at 19th Street and St. Paul. Sales grew quickly and he needed to expand his operation. He purchased two buildings in what was to become North Milwaukee—buildings that formerly housed a failed furniture manufacturer owned by C.M. Hunt and B. A. Kipp.

William Harley worked for the company for a time, presumably while Meiselbach Manufacturing was still located at 19th and St. Paul because it was very close to Harley's home. Harley, of course, went on to partner with his friend, Arthur Davidson, to form the Harley-Davidson motorcycle company. Harley's time at Meiselbach would have been quite helpful in providing him with a working knowledge about bicycle frames, wheels and related components—knowledge that he later applied to the design of early motorbikes.

An article in the Milwaukee Journal in 1897 reports that Meiselbach's bicycle manufacturing plant had the capability of producing 750 complete bicycles every working day. The firm employed 375 workmen and anticipated more than doubling the number of employees. The company made three grades of bicycles—cheap, medium and high grade—and had models for men, women and juveniles. The company sold its bicycles throughout the United States and beyond.

In 1899, Meiselbach joined with other bicycle companies around the country, including CJ Smith of Milwaukee, to form a trust that became the American Bicycle Company. Meiselbach was named general manager and his factory became one of the manufacturing plants for the company.

The bicycle trust was an attempt by suppliers to limit competition and increase prices. As the so-called "safety bike" became popular, numerous manufacturers entered the field. By 1898, there were over three-hundred bicycle manufacturers in the United States and manufacturing capabilities exceeded the market demand, resulting in intense price competition. Albert Augustus Pope, the owner of Columbia Bicycle, urged the major companies to consolidate. Initially, forty-two manufacturers joined to form the American Bicycle Company—eventually, the company encompassed over seventy-five companies.[uu]

Augustus Meiselbach was soon named general superintendent of all of the American Bicycle company's plants. He anticipated producing 200,000 bicycles in his Milwaukee facilities to meet the demand for the American Bicycle Company. The effort failed. By 1901, the American Bicycle Company reported that sales were collapsing. Interest in the bicycle had peaked and sales were on the decline. The company declared bankruptcy the following year.[195]

The bicycle market at the start of the last century would today be viewed as a bubble, which burst in a predictable fashion. Many lost their jobs as a result, as employment in the large Milwaukee bicycle factories was reduced. The bicycle factories were quickly repurposed to serve the automobile industry.

This photograph shows a portion of the Meiselbach workforce. Notice how young the workers are in the foreground—some of which appear to be in their early teens, or younger. From their equipment, it appears the young lads worked on spoking bicycle wheels. Photograph courtesy of the Milwaukee County Historical Society.

[uu] One of these later companies was the Gormully & Jeffery Manufacturing Company which made Rambler brand bicycles. Thomas B. Jeffery of the concern sold the company to the Bicycle Trust in 1900, after holding out initially, to focus on producing the Rambler automobile.

THE SHOLES 'VISIBLE' TYPEWRITER

While the American bicycle industry was restructuring, August Meiselbach was pursuing other ventures. One was the typewriter industry. As noted in Chapter 1, Milwaukee's Christopher Latham Sholes had developed a "visible" typewriter before he died—one in which print was immediately visible as soon as typed on the page. A patent was obtained for the typewriter in 1891, a year after his death. In carrying out their dad's legacy, his sons, Louis C. and Frederick Sholes formed the C. Latham Typewriter Company in Kenosha and began to manufacture the Sholes Visible Typewriter. However, their venture was not a commercial success.

On St. Patrick's Day of the year 1900, August Meiselbach purchased the rights to the device and began producing it at the reconditioned Wagner Pulley Works factory the following year. The typewriter sold for $60. It is believed several thousand were produced. Its principal selling points were that it was designed by the inventor of the first successful typewriter, and it featured an open channel that kept the typebars in alignment as it struck the paper. Those remaining in existence are sought after by collectors.

By the early 1900s, however, the typewriter business had become very competitive. Most manufacturers were producing visible typewriters, so the advantage of the Sholes' designed typewriter was limited. Meiselbach's venture was not a commercial success and he closed the factory in 1903. The plant was eventually sold to Franz Johann Leopold Dorl, who formed the Visible Typewriter Manufacturing Company of Kenosha, Wisconsin. However, the typewriters the Dorl Company produced were not designed by Sholes.

Picture of a Sholes Visible typewriter manufactured by the A. D. Meiselbach Typewriter Company. Photograph courtesy of Martin Howard, who maintains the typewriter in his collection. Used with permission.

The A.D. Meiselbach Motor Wagon

As Meiselbach's bicycle and typewriter interests were ending, Augustus started another venture. He formed the A.D. Meiselbach Motor Vehicle Company in North Milwaukee. A 1904 article stated that the company planned for the erection of eight brick and frame buildings on five acres adjoining the right-of-way of the Milwaukee Road for the manufacture of all styles of motor vehicles.[196]

Meiselbach initially developed a line of heavy-duty trucks with friction drives. The friction drive limited the maximum speed to ten miles per hour, a feature that the company capitalized on by advertising that the vehicle "Can not be overspeeded." While today it is hard to imagine a truck limited to such a lumbering speed, it must be remembered that trucks were replacing horse-drawn vehicles with even slower speeds.

The Meiselbach Trucks' friction drives used beveled pinions covered with tarpaper. They were limited to a single speed forward and reverse. While crude, they were simple to operate. The trucks featured gasoline engines from Wisconsin Motor Manufacturing, which initially also operated out of North Milwaukee.

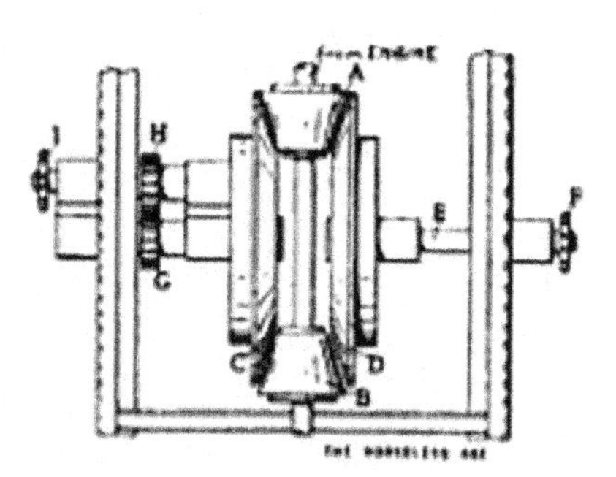

Illustration of the Meiselbach Friction Drive mechanism.

An early ad for the Meiselbach Truck

In 1910, the company changed its name to the Crown Commercial Car Company—likely because of anti-German sentiment. By this time, the company's offerings had become more sophisticated, featuring a water-cooled, four-cycle, four-cylinder engine built by Wisconsin Motor and Manufacturing Company, a conventional clutch and four-speed transmission, and a worm gear driven rear axle. The trucks were advertised in several sizes rated by hauling capacity—from one-half to eventually 5 tons. It was advertised as "The car that delivers the goods." Crown also advertised that it would mount bodies from horse-drawn vehicles to its chassis, or design car bodies to the buyer's specifications.

Crown employed engineer George Van Rottweiler, who formerly worked for Daimler Motorem Gesellschaft and Milnes—the manufacturer of the Mercedes vehicles in Germany.[197]

This postcard was used to advertise the Crown Commercial Car—essentially a small delivery truck.

MERKEL MANUFACTURING COMPANY

If you mention motorcycles and Milwaukee to almost anyone, the name Harley-Davidson comes up. Very few people have heard of Joseph Merkel and his "Flying Merkel." Yet, Merkel preceded Harley-Davidson in the market and developed one of the most innovative of the early US motorcycles.

Merkel Manufacturing Company was established in about 1902 and was located at 1095 26th Avenue in Milwaukee.

THE MERKEL MOTORCYCLE

Merkel started designing motorcycle engines in 1902 and built his own motorcycles beginning the following year. His engine used ball bearings rather than bronze bushings for its crankshaft, and a cam-actuated mechanism for its intake valves. Merkel also pioneered a throttle-controlled engine oiler.

Merkel is perhaps best known for his patented spring front fork that he called a "truss fork"—it was the forerunner of the modern telescopic front fork. His bikes also featured a "mono-shock" rear suspension. Eventually, these features were promoted with the company's slogan, "All roads are smooth to the Flying Merkel."

In 1905, Markel entered the racing scene and designed several racing motorcycles that were so successful that they set a number of performance standards.[198]

38 MOTORCYCLE ILLUSTRATED JULY 15, 1908.

RESULTS COUNT

Three Single Cylinder Merkels entered in the F. A. M. Endurance Run

TWO PERFECT SCORES and

GOLD MEDAL for the Highest Honors for Single Cylinder Machines ridden by a Private owner

NEVER LATE—Ascended all hills without assistance. Also WON many events and highest honors previously at Providence, Cleveland, Washington, etc., etc. SPRING FRAMES and BALL BEARING Motors helped do it. Power, Speed, Comfort, Reliability, all the time in a MERKEL Spring Frame Motorcycle with Ball Bearing Engine

MERKEL MOTOR CO., 26TH AVE., MILWAUKEE, WIS.

REPRESENTATIVES:

Pacific Coast: Oliler & Worthington, 1030 So. Main St., Los Angeles, Cal. Chicago and Cook County, Ill.: I. H. Whipple, 290 W. Jackson Blvd., Chicago. New York, Jersey City and Long Island: E. J. Willis Co., 8 Park Place, New York City. Brooklyn, N. Y.: John Pfleiger, 1604 Bushwick Ave., Brooklyn, N. Y. Cleveland and Northeastern Ohio: L. J. Mueller, 6417 Woodland Ave., Cleveland, O.

This ad for the Merkel motorcycle appeared in 1908—the final year the vehicle was produced in Milwaukee.

In 1908, Merkel merged his company with the Light Motor Company of Pottstown Pennsylvania, and all operations were moved there. Light Motor, in turn, was acquired in 1911 by the Miami Cycle Manufacturing Company and production moved to Middletown Ohio. The last Merkel motorcycle was manufactured in 1917.[199]

Schlitz Brewing Company advertised itself as the "Beer that Make Milwaukee Famous." Merkel adopted that slogan and applied it to his motorcycles. While most Milwaukeeans would expect Harley Davidson to have claimed that distinction, in the early 1900s the Merkel was better known.

Flying Merkels are reported to have been some of the most reliable motorcycles of the era. While they were more costly than many motorcycles in the early 1900s, they used German-made bearings and other high-quality materials, which enhanced their reliability.

Surviving Merkel motorcycles are prized by collectors.

In 1905, Merkel decided to build automobiles, renaming his company the Merkel Motor Company. He made three models, each with different horsepower and wheelbase sizes. They were offered at a price range of $1,500 to $3,500. Vehicles could be ordered with either air- or water-cooled engines. Merkel produced automobiles on a small scale until 1907, when he merged his company with the Light Motor Company of Pottstown Pennsylvania.

A 1906 Merkel Runabout Automobile, one of three models offered.

JOSEPH FREDERIC MERKEL was born in Manistee, Michigan, in 1872. He was employed for a time on a logging railroad when he was 14 years old. At age 15, he performed an apprenticeship in a machine shop, after which he enrolled at Michigan Agricultural College, which is now Michigan State University, and studied mechanical engineering.

Merkel moved to Milwaukee in 1897 to accept a draftsman position at E.P. Allis—while he apparently worked on small internal combustion engine design during his spare time. Soon he was producing small internal combustion motors for bicycles, which led to the formation of Merkel Manufacturing Company. In addition to designing and building numerous motorcycle designs, many of which were innovative for its time, Merkel produced a motor-powered tricycle in 1900 and, in 1906, his company built automobiles featuring a thirty horsepower engine. Approximately 150 automobiles were manufactured by the company.[200]

In 1908, Merkel merged his company with the Light Motor Company and the new Merkel-Light Motor Company moved activities to Pottstown, Pennsylvania.

Merkel moved to Rochester, New York in the 1920s to take over experimental design for the Cyclemotor Corporation. Merkel earned a lot of praise from the motorcycling industry in the early 1920s when he convinced the New York legislature to assess lower highway fees on motorcycles since they caused much less wear and tear to the road than automobiles.

Merkel also was a designer for the Ever Ready Corporation on Long Island for several years before joining the General Railway Signal Company's engineering staff, where he worked for thirty-five years.

He had numerous patents in his name—the early ones which were motorcycle innovations and the later for various railway relay and signaling apparatus.

Markel died in Rochester in 1958.

He was inducted into the AMA Motorcycle Hall of Fame in 1998.

HARLEY-DAVIDSON COMPANY

Of all the machines manufactured in Milwaukee, the Harley-Davidson motorcycle is perhaps the most recognizable and enduring. Initially sold in 1903, it competed against numerous domestic motorcycle manufacturers before eventually becoming the dominant manufacturer of motorbikes in the United States, as well as in the world.

Harley-Davidson got started in a very modest way. The Davidson and the Harley families were next-door neighbors until about 1896, both living in the 200 block of 9th Street in Milwaukee. Arthur Davidson and William Sylvester Harley, sons of the families, were a year apart in age and became boyhood friends—a friendship that turned into a major enterprise.

William, or Bill Harley as he was known, started working for the Meiselbach Manufacturing Company when he was fifteen. Harley's time at Meiselbach as a "cycle fitter" provided him with a working knowledge about bicycle frames, wheels and related components—knowledge that proved to be quite helpful later in his life as he began working on the design of early motorbikes. When Meiselbach ceased bicycle manufacturing, Harley took a job at Barth Manufacturing—a metal fabricating company that made floor jacks, patterns for castings, and other equipment. His friend Arthur Davidson joined him there.

While watching a performance at Milwaukee's Bijou Theater, Arthur and Bill happened to see a Parisian actress named Anna Held start the show by streaking across the stage on a French motor-bicycle. That motorized bicycle used a single-cylinder engine designed by Jules-Albert de Dion, considered the first lightweight, internal combustion engine.[vv] The Davidson and Harley lads mentioned this to the workers at Barth Manufacturing and expressed an interest in building one themselves. One of their fellow workers, a German-born mechanic named Emil Kroeger, said he was familiar with the engine and could help them.[201]

With some rudimentary knowledge of internal combustion engines, in the summer of 1901 Bill drew up the plans for a small gasoline engine designed to power a bicycle. He entitled the drawings, "Bicycle Motor." Bill and Arthur worked on manufacturing the various components, using a lathe owned by Johann Melk—father of Henry Melk who was another neighborhood friend. It was largely a trial-and-error process and they made limited initial headway.

Soon they both left Barth and took jobs with Pawling and Harnischfeger—Arthur Davidson as an apprentice patternmaker and Bill Harley as an apprentice draftsman. The two continued to spend a good deal of their own time tinkering in Arthur's family basement.

The following summer, Arthur Davidson visited his grandmother's farm in Cambridge Wisconsin and met Ole Evinrude, whose family also owned a farm near Cambridge. The two became good friends and fishing buddies and began scheming about designs for a small motor that would power a rowboat. Back in Milwaukee, Arthur Davidson and Ole Evinrude started their own pattern-making company in 1902. While the company was short-lived, it provided both men with valuable experience.

[vv] The de Dion-Bouton engine was licensed to more than 150 manufacturers and was a popular choice among assemblers of early motorized bicycles. The small lightweight four-cycle engine used a battery and coil ignition that was less problematic than others which used the hot tube ignition. The 2-inch bore and 2.8-inch stroke resulted in an output of just over one horsepower (1 kW). It was also widely copied by many makers including US Brands Indian and Harley-Davidson. Wikipedia, retrieved 2015.

This interesting photograph was taken at Evinrude and Davidson's pattern shop in 1902. Ole Evinrude is shown at the center of the top of a spoked wheel, Arthur H. Davidson at the bottom. The other two were employees of the company. The wheel was likely a wood pattern created by the company, to serve as the mold for a casting. The photo is included with the permission of Jean Davidson, author of "Growing Up Harley-Davidson."

Arthur and Bill continued to work on the engine on-and-off for two years, but progress continued to be slow. While they were motivated, they had full-time jobs and other interests. When they had problems getting the right fuel flow from their rudimentary carburetor, which used a tomato-soup can for the throttle body, Arthur's friend Ole Evinrude stepped in to help design the successful carburetor.[202]

In early 1903, Arthur Davidson's older brother, Walter, returned to Milwaukee, from where he was working as a machinist on the railroad in Parsons Kansas, to attend his oldest brother's wedding. He was intrigued by Arthur and Bill's small, yet unfinished engine, and stayed to help them complete it. With his help, the engine was finished later that year and the three men mounted it on a bicycle frame, driving the rear wheel with a leather belt. While it ran, it was a disappointment. The engine didn't provide enough power to the bicycle to climb hills without using the pedals. But Walter Davidson was hooked. He decided to abandon his job in Kansas and continue working on the motorbike.

By 1903, there were more powerful motorized bicycles appearing on Milwaukee's streets, including Milwaukee's Merkel motorcycle. Bill Harley and the Davidson brothers studied the Merkel, as well as other motorbikes, and realized that if they were to be successful they needed a more powerful engine. They also observed Merkel's loop-frame design, which more easily accommodated an engine than a typical diamond framed bicycle frame, and saw that a similar design would help them mount a larger motor.[203]

In early 1903, the team began work on an engine that would produce more power. To provide additional room for their work, they moved into a 10-ft by 15-ft shed in the Davidson family's backyard. Their older sister, Janet, hand-lettered the words Harley-Davidson Motor Company over the door. While the shed may have been used for assembly, it appears that many of the parts were machined at the shops of the Chicago, Milwaukee, and St. Paul Railway, where the Davidson's oldest brother worked. Their new engine had a displacement just under 25 cubic inches and drove a 9.75-inch flywheel. The trio fabricated a loop-frame for their new motorcycle to match the new motor and Bill Harley developed a simple belt-tensioning device that allowed a quick disconnect of power to the rear wheel. The larger motor mounted in an advanced frame took them into the motorcycle category for the first time. By the end of 1903, they had completed three motorcycles, which immediately sold.

The business continued to be a part-time "friends-and-family" operation. Bill Harley served as an engineer-draftsman, designing the parts. Walter Davidson was the machinist for the firm, Arthur Davidson worked as a salesman, and William Davidson contributed by offering advice and fabricating critical parts on the tools in the railroad shop where he worked.[204] The Davidson sisters Elizabeth kept the books and Janet provided lettering and artwork. An uncle loaned them money to get them through a rough time, which also provided them with the funds needed to expand their backyard shed.

This is the backyard shed behind the Davidson home, as it appeared in 1904. It was located on 38th Street and Highland Boulevard. This is where the first Harley-Davidson motorcycles were developed and assembled.

The lads all had full-time jobs, until Bill Harley took time off to attend engineering school at the University of Wisconsin in Madison. Still, they managed to produce another three motorcycles in 1904. They hired their first full-time employee in 1905 and built seven motorcycles that year, plus an undetermined number of bare-boned engines that were sold through an advertisement in the *Automobile and Cycle Trade Journal*. In that year, Arthur also located a dealer in Chicago who would sell their motorcycle—Carl H. Lang.

Meanwhile, Bill Harley pursued his mechanical engineering degree in Madison, working his way through school by taking a job in a Madison architect's office and waiting tables at a fraternity house. There was only one course on internal combustion engines offered at the time, but it helped provide Harley with some needed theoretical design skills. In 1906, he designed his first V-twin engine for Harley-Davidson. He graduated in June 1907 and returned to Milwaukee, where he took an engineering job at the Wisconsin Bridge and Iron Works, but continued to provide engineering services for Harley-Davidson.

In September of that year, the Harley-Davidson Motor Company finally incorporated, and in June 1908, Bill Harley quit his job at Wisconsin Bridge and began working full time as Harley-Davidson's chief engineer. One of his first projects was to redesign and improve the Harley-Davidson motorcycle. The new design was launched in 1909. Bill also began to refine his V-twin engine design. The new "F-head" engine was launched in 1911 and continued in production for the next eighteen years.[205]

This photograph is of the earliest surviving Harley-Davidson motorcycle, as it appears in the Harley-Davidson Museum in Milwaukee. From Author's Collection.

Early Racing and Endurance Runs

In September 1904, one of the first Harley-Davidson motorcycles competed in a race at Milwaukee's State Fair Park. It was stock-motorcycle driven by Edward Hildebrand and placed a respectable fourth.[206] It was referred to in the newspaper as a "Harley." The race was won by Frank Zirbes of Racine, who was riding on a Mitchell. Harley-Davidson does not appear to have actively participated in the race in any way. In fact, Arthur Davidson was opposed to promoting motorcycle racing—especially the early 'motordrome' races in which motorcycles raced other participants around steeply pitched wooden tracks, which he considered far too risky.

Walter Davidson, however, enjoyed participating in motorcycle races. After cutting his teeth in some local motorcycle race events, in 1908 Walter entered in a two-day, 365-mile endurance run from the Catskills to New York City, sponsored by the Federation of American Motorcyclists. Racing on a stock machine with no team to support him, he placed first of the sixty-one entrants in the field.

His next race was an economy run, in which each entrant was given a single gallon of fuel. Walter won the race with his Harley-Davidson, achieving a remarkable (for the time) one hundred and eighty-eight miles on a gallon of fuel. By participating in these events, Walter Davidson helped to spread the reputation of Harley-Davidson to other parts of the country.

While Harley-Davidson didn't sponsor factory riders in motorcycle races, soon Harley owners were going wheel-to-wheel against other motorcycle racing enthusiasts—and winning races. An advertisement by Harley-Davidson in *Motorcycle Illustrated* magazine stated, "No, we don't believe in racing and we don't make a practice of it, but when Harley-Davidson owners win races with their own stock machines, hundreds of miles from the factory, we can't help crowing about it."[207]

In reality, Harley-Davidson provides at least unofficial support for its dealers, employees and others in motorcycle racing. This policy changed as motorcycle road-race events became popular. Such races were generally run over public roadways and did not require specialized motorcycles designed purposefully for track racing. This form of racing appealed to Harley-Davidson, since it allowed it to demonstrate the reliability of its production motorcycles. In 1914 H-D initiated a full-scale racing campaign and hired engineer William "Bill" Ottaway to oversee the program.

The initial 1914 H-D racing motorbike was a modified production model which used an engine that the company sold for police use, Referred to as an 'A' motor, it could better handle the stress and heat of long races. Ottaway soon realized that other modifications were necessary, if the company was to compete against Indian, Excelsior and other early motorcycle companies successfully.

The result was the Harley-Davidson 11-K racer. The engines featured larger intake ports, manifold and carburetor, along with stiffer valve springs, a special camshaft, and steel flywheels. The oil pump was cast into the gear case cover and provided better lubrication.

On October 5, 1914, Red Parkhurst raced a Harley-Davidson 11-K in the Federation of American Motorcyclists One Hour Championship Race at the Alabama State Fairgrounds Raceway. Despite stiff competition from Gene Walker and Gail Joyce, who road Indians, and Joe Wolters riding an Excelsior, Parkhurst won. While there was a post-race challenge involving Parkhurst's acceptance of a handkerchief from a spectator to clean his goggles, Harley-Davidson prevailed.

Early H-D racing advertisement

Harley-Davidson quickly began advertising its first "Championship" win, which demonstrated that it could compete on equal footing with other motorcycle manufacturers. This was a big step forward for the Harley-Davidson racing program.

Red Parkhurst was the first rider hired by the Harley-Davidson factory in 1914 to ride the new racer developed by Bill Ottaway.

In 1915, Harley-Davidson offered the 11-K for sale, touting the features of the redesigned V-twin engine and new automatic mechanical oil pump, as shown in this advertisement.

In 1912, the company started construction on what eventually became a six-story factory on Juneau Avenue in Milwaukee. As Harley-Davidson's factory capacity enlarged and its reputation increased, sales boomed. In the following year, H-D's annual production run of seventeen thousand motorcycles was sold out in a little over twelve weeks—a fact the company proudly boasted about in its advertising.

Harley-Davidson introduced its sidecar option in 1914. Many, if not all, of its early sidecar bodies were built for the company by the W.S. Seaman Company of Milwaukee. In 1916, the frame on the J model was reinforced for sidecar use and the steering head bearings enlarged. The 'Standard Pleasure Sidecar added $80 to the price of the motorcycle. It proved to be an extremely popular option, providing economical transportation options. In 1919, Harley-Davidson manufactured 23,279 motorcycles, of which nearly seventy percent (16,095) included a sidecar.[208]

Delia Crewe and her dog Trouble traveled from Waco to Milwaukee to New York in 1915, with numerous side trips—5,378 miles in which the 1914 H-D motorcycle reportedly performed flawlessly. Photograph courtesy of Harley-Davidson Motor Company Archives. Used with permission of Harley-Davidson.

After war broke out in Europe, the British Ministry of Munitions issued an order that no further work on motorcycles or cars would be allowed after November 15, 1916, without a special permit, in order to devote factory space to munitions work and other wartime goods. Earlier that year, Bill Harley helped to introduce a military motorcycle equipped with a machine-gun sidecar for use in combat by General John 'Blackjack' Pershing against the revolutionary Poncho Villa. Harley-Davidson's timing was good because this vehicle soon helped supply the needs of the US military during the First World War.

In addition to the specialized machine-gun sidecar, Bill Harley designed a gun-carrier sidecar, an ammunition sidecar and a medic sidecar. It appears that most, if not all, of the sidecars were produced by Milwaukee's W.S. Seaman Company.[ww] As the United States entered the war in 1917, the military purchased approximately one-third of Harley-Davidson's annual production. In the following year, military sales rose to over half of production capacity. Harley-Davidson continued civilian production during the war, while the Indian motorcycle company devoted its full production to military sales.

Because of Bill Harley's work in designing motorcycles for military applications, he was selected as head of the Society of Automotive Engineers committee on standardized military motorcycles—which envisioned a motorcycle dubbed the 'Liberty Motorcycle.' The war ended before the full program could be carried out.

The day after the signing of the Armistice, Corporal Roy Holtz of Chippewa Falls, Wisconsin, was the first US serviceman to enter into Germany—riding on a Harley-Davidson motorcycle.

After the war ended, the US military decided to sell the motorcycles overseas, rather than ship them back to the States. It was reported by the Milwaukee Journal that "when the bidding started, a remarkable thing happened. It was the highest tribute ever paid to a manufacturer of motorcycles. It may well have been the highest tribute ever paid to any manufacturer. For, after carloads of crated Harley-Davidsons had been eagerly bought, and when other makes were being offered, the purchasing agents of the allied

[ww] Seaman produced the sidecars under subcontract from the Velie Motor Car Company, a long-time Seaman customer. Seaman also manufactured ammunition boxes, gun mounts and other war material under the contract.

governments announced that they would buy second-hand Harley-Davidsons before buying the new motorcycles of other manufacturers."[209]

By 1920, Harley-Davidson was the largest motorcycle manufacturer in the world, being sold through a network of over two thousand dealers in sixty-seven countries. In that year, it also established the 'HOG' association—named after the H-D racing team's mascot (a pig).

The Depression years were not kind to motorcycle manufacturers and by 1931, the only remaining American manufacturers were Harley-Davidson and the Indian Motocycle [sic] Manufacturing Company.

When the United States again prepared to go to war, Harley-Davidson was called to produce motorcycles for the military. In 1941, almost all motorcycles produced were for military applications. It is estimated that close to ninety-thousand Harley-Davidson WLA motorcycles were manufactured for the armed services during the war years. The WLA model had a high compression, 45-cubic inch, flathead V-Twin engine. The US Army used the motorcycle for police and escort work, courier duties, and some scouting, as well as limited use to transport radio and radio suppression equipment. During the Second World War, motorcycles were almost never used as combat vehicles or for troop mobility, and so were rarely equipped with sidecars—as was common for German motorcycles. Nevertheless, the WLA acquired the nickname "Liberator," since soldiers liberating occupied Europe were seen riding Harley-Davidson motorcycles.[210]

This photograph of military police on their Harley-Davidson WLA motorcycles was taken in Australia in January 1944. From author's collection. (The author's father is shown at left.)

Harley-Davidson Developments in Motorcycle Design

Harley-Davidson has a well-deserved reputation for its conservative motorcycle design. The company typically wasn't the first to introduce new features, but when advances were adopted they were robust and functioned well.

H-D's engineering department, under the supervision of Bill Harley, continued to introduce a steady davistream of features. The earliest advancements included such features as the chain-drive, the three-speed transmission, a kick-starter, and a spring fork—all of which were introduced prior to the First World War.

In the 1920s, several new machines were introduced, including the smooth-running Sport Model. A new V-twin engine was designed—the 74 cubic inch side-valve VL model. Additionally, better electrics were employed.

A number of innovations were designed by Harley and his assistants that never made it to production. For example, the engineering team designed a four-cylinder engine, as well as an all-aluminum racing motor and industrial engines. The motorcycle market wasn't considered strong enough to risk introducing these engines and we are left to wonder what impact they may have made. However, it is hard to criticize the company for its restraint during this period, given that most other motorcycle manufacturers failed, while Harley-Davidson endured.

It is reported that Bill Harley had eighty-six known patents issued in his name and assigned to Harley-Davidson Motor Company, and another four issued in collaboration with others.[211] It is almost certain that other company engineers collaborated on many of the innovations represented by these patent filings. Harley-Davidson was typical of most companies during this period, in which almost all patents were filed under the chief engineer's name. There is no question, however, that Bill Harley was an extremely innovative and productive engineer.

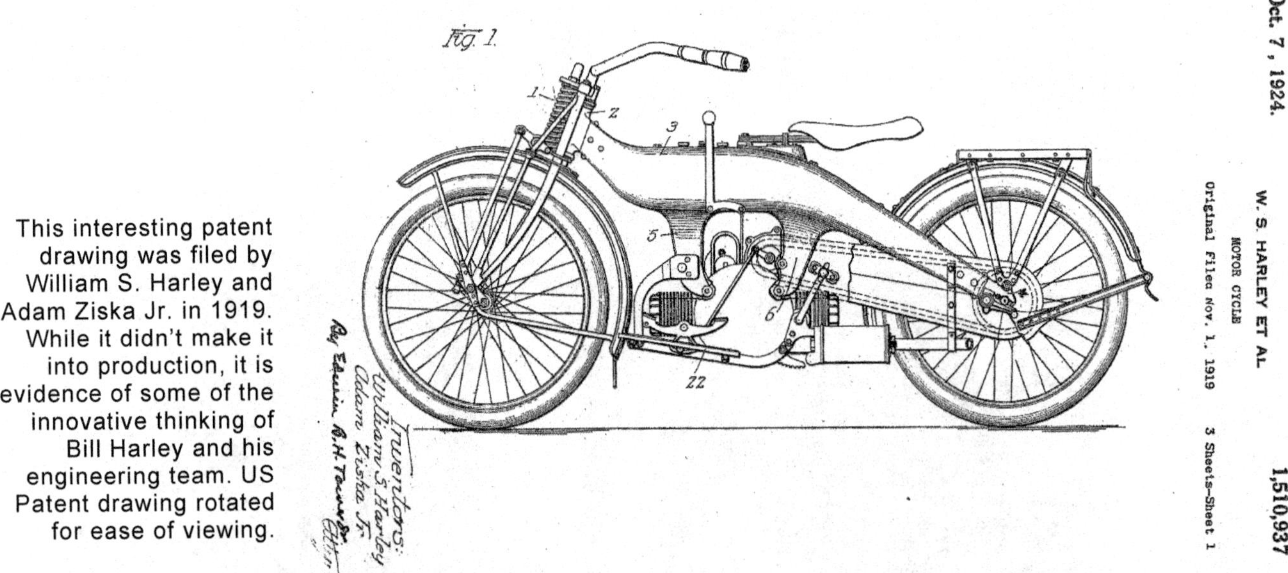

This interesting patent drawing was filed by William S. Harley and Adam Ziska Jr. in 1919. While it didn't make it into production, it is evidence of some of the innovative thinking of Bill Harley and his engineering team. US Patent drawing rotated for ease of viewing.

In his *Biography of William Sylvester Harley*, author Herbert Wagner provided some insights into how Bill Harley worked with his engineering team. He noted that "the engineering department at Harley-Davidson was separated from other parts of the factory by a locked door. Even if you got through that door you still had to pass under the watchful gaze of Bill Harley's personal secretary, Joe Geiger, who was in charge of all drawings and records and who scrutinized anyone requesting entry into this inner sanctum of American motorcycle design. The drafting room was connected to Bill Harley's office and he could observe it through a large plate glass window. His office, filled with experimental parts and drawings, was likened to "a mother's bedroom overlooking the nursery of her children." This was where Bill Harley felt most content and at home in the plant. This is where he communicated with his team of engineers and draftsmen and where he appointed tasks, suggested changes, approved work, and put his own hand to the drawing board."[212]

The Knucklehead

Referring to someone as a 'Knucklehead' is an insult meant to infer that the recipient isn't particularly bright. When referring to a motorcycle engine, however, the term has evolved into a much different meaning. It refers to what might be the crowning achievement of Bill Harley and his engineering team—a motorcycle engine that has come to define what a Harley-Davidson motorcycle looks, and perhaps more importantly what it sounds, like.

In 1930, Bill Harley designed the 74-cubic inch, side-valve, VL model. While an improvement over the previous V-twin engines, it tended to overheat in demanding conditions, such as when the rider traveled long distances on the improved roads of the US highway system. Harley recognized that a better lubrication system was essential. He also saw opportunities for other engine improvements. In order to meet these challenges, in 1936 the company introduced the "61 OHV" engine. It has been referred to as Bill Harley's finest achievement.[213] Not only did it have a much-improved lubrication system, it had a modern overhead valve design—hence the OHV term. The valve covers that resulted from the design were eventually referred to by H-D's riders as 'Knuckleheads.' The name soon caught on and the engine today is commonly known as the Knucklehead.

Developed with foreman Ed Kiechbusch and a team of talented individuals in Harley-Davidson's engineering and experimental departments, Bill Harley oversaw and directed each step of the engine's design. A patent was issued in Harley's name for the new recirculating lubrication system (US Patent 2,111,242).

The combination of a well-lubricated, air-cooled engine with easy breathing overhead valves, and a distinctive appearance, was well received by the motorcycle buying public—and helped Harley-Davidson survive the Depression. To demonstrate the engine's performance, motorcycle race rider Joe Petrali established the world motorcycle speed record in Daytona Beach using the engine in 1937 at 136.183 miles per hour. In the same year, Fred "Iron Man" Ham broke the 24-hour endurance record with the engine, traveling 1,825 miles on the Muroc Dry Lake, which is now part of Edwards Air Force Base. Temperatures that day ranged from 30 degrees Fahrenheit at night to the 90s during the day.[214]

The Harley-Davidson Model EL motorcycle, which incorporated the Sixty-One Knucklehead engine, was the last completely new motorcycle designed by Bill Harley and built while the original four founders oversaw the company. Bill Harley's engineering team introduced a bored and stroked engine based on the Sixty-One in 1941, called the 74 OHV. Harley died in 1943 before he could continue his legacy.

Subsequent Harley-Davidson engines, nicknamed the 'Panhead' (1948-65), the 'Shovelhead' (1966-83), the 'Evolution' (1984-1998) and the Twin Cam (1999-present) are reportedly based on the same engine design that Bill Harley and his team introduced in 1936.[215] These engines all were derived from the 1936 'Knucklehead,' with its air-cooled V-twin configuration, using overhead valves activated by pushrods.

A Harley 1936 Model EL or "61ci OHV," more commonly called a 'Knucklehead.' Photograph was taken by the author at the Harley-Davidson Museum, Milwaukee, in 2015.

WILLIAM SYLVESTER HARLEY was born in Milwaukee, Wisconsin in 1880, the son of William Harley, Sr. and Mary Smith from Littleport, Cambridgeshire, England. His parents had immigrated to the United States in 1860.

When 21, William Harley—known as Bill—drew up plans for a small gasoline engine meant to propel an ordinary pedal bicycle. He and his friends, Arthur and Walter Harley, worked for two years to construct that single-cylinder engine and mount it on a motorcycle frame. The trio soon started work on a larger and more powerful engine that Bill designed. It was mounted on a loop-frame and formed the basis for the first motorcycle of Harley-Davidson Motor Company. They completed three motorcycles in 1903.

Bill Harley received a degree in mechanical engineering from the University of Wisconsin–Madison in 1907, and soon started full-time employment with the company as its chief engineer—a position he served until his death in 1943. Harley is credited with numerous early motorcycle innovations, including the company's V-twin engines, transmissions, various frame and seat designs, lubrication systems and related components. Harley had at least eighty-six patents issued in his name and assigned to Harley-Davidson Motor Company, and another four issued in collaboration with others.

During the First World War, Harley was selected as head of the Society of Automotive Engineers committee on standardized military motorcycles. He designed a number of innovations in the use of motorcycles for military use.

During the 1920s, Bill Harley and his assistants designed the Harley-Davidson Sport Model, various racing motorcycles and hill-climb bikes.

Harley married Anna Jachthuber in 1910 and had three children with her: William J. in 1912, Ann in 1913, and John E. in 1915.

William Harley died on September 18, 1943, and is buried at the Holy Cross Cemetery and Mausoleum in Milwaukee. He was inducted into the Motorcycle Hall of Fame in 1998.

In 2003, a Harley-Davidson statue was unveiled in Littleport, Cambridgeshire, to commemorate the centenary of the famous motorcycle company. Bill, along with Arthur, William and Walter Davidson, were inducted into the Hall of Honor by the United States Department of Labor in 2004. The honor is given to individuals "who have all improved working conditions, wages, and overall quality of life for American workers."

A.O. SMITH CORPORATION

As noted in Chapter 2, A.O. Smith's early success in building automotive frames led to the company becoming the largest auto frame manufacture in the United States. Business was so good that they were forced to turn down orders. The company decided to build a new factory—selecting a 135-acre site on Milwaukee's north side adjacent to the tracks of the Chicago, Milwaukee and St. Paul Railway and just south of Capitol Drive. In 1913, Arthur O. Smith died and his son, Lloyd Raymond Smith (known as L.R. or Ray) assumed control.

The production of automobile frames was a labor-intensive process and the company had a difficult time meeting the high demands of the automotive industry. In 1915, as demand for automobile frames continued to grow, Ray Smith called his engineers together and discussed preliminary plans for a plant that would greatly increase the company's capacity—and would largely automate the process. One of these engineers was R. Stanley Smith, the son of Ray's uncle Alonzo. Stanley quickly rose to the position of chief engineer for the growing concern.

This photograph shows a portion of A.O. Smith's operations, looking southwest from Capitol Drive. The railway tracks intersected the site and provided transportation for materials and finished frames. The mechanized automotive frame assembly facility is shown at the top (south) of the photograph. Evinrude and Cutler Hammer were also relocated in this same industrial neighborhood—just north of Capitol Drive.

Led by Stanley Smith, A.O. Smith engineers got busy with the task of designing the automated frame factory. Then on May 7, 1915, the Germans sank the Lusitania.

Work on the automatic frame design was interrupted for a time to allow A.O. Smith to turn its resources to producing war materials for the United States Government—caisson wheel hub flanges, frames for army trucks, metal frames for saddles for the cavalry and casings for bombs. At first the company had trouble with these casings but, through research, they found a better way. They introduced a greatly improved method of electric arc welding, in which the weld metal had the same ductility or "stretch" as the steel. In two years' time, A.O. Smith became the largest manufacturer of casings for aerial bombs and shells. This was noteworthy, of course, but it was A.O. Smith's experiments in arc welding that turned out to be one of the company's greatest contributions to industry.

As the war ended, work resumed on the automated frame factory. An overall design was established and a patent was filed on January 21, 1918. It would later become known as the *Mechanical Marvel.* It could produce ten thousand automotive frames daily.

A.O. Smith's innovations didn't stop with the automated frame factory. In 1930, the company employed over 400 engineers—they were put to work on numerous other developments, including:

- The first arc-welded, high-pressure vessel used to refine oil. A.O. Smith pressure vessels were used in a wide variety of chemical processing, refinery, and related applications (1925).
- A method of economically forming and welding large-diameter steel line pipe. This new mass production technique was instrumental in launching the natural gas industry and transcontinental oil pipeline business (1927).
- A process of fusing glass to steel. In perfecting this process, the company produced the first large, single-piece glass-lined brewery tank (1933). Over the next 32 years, A.O. Smith made more than 11,000 glass-lined tanks.
- The glass-lined water heater, which became the standard of the industry and made hot water an affordable convenience for homeowners (1936).
- Bomb casings, aircraft propellers and landing gear, torpedo air flasks, and other material to support the Allies during the Second World War. By 1945, the company had built 4.5 million bombs, 16,750 sets of landing gear, and 46,700 propeller blades, as well as nose frames for the B-25 bomber, water heaters, jeep frames, and components for the atomic bomb project (1942-45).
- The Harvestore® structure, a glass-fused-to-steel silo for dairy farms and livestock operations (1949). Over the next fifty years, A.O. Smith installed more than 70,000 such storage structures on farms throughout North America.

During its history, A.O. Smith has morphed from a bicycle company, to an automotive frame manufacturer, to a producer of large weldments, to a manufacturer of water heaters. It has been observed that this history of moving from product-line to product-line would be a textbook axiom for failure—a company that was a "Jack-of-all-trades and master of none." Francis Walton, author of *Miracle of World War II, How American Industry Made Victory Possible,* explains A.O. Smith's success as due to its ability to "master its basic trade" which was "the working of metal"—and that "the shape or use of the ultimate product hardly mattered."[216] This remarkable ability to continuously reinvent itself appears to have been due, in no small part, to A.O. Smith's large engineering staff and its commitment to research and development.[xx]

[xx] In 1930, A.O. Smith employed more than 400 engineers and constructed an Art Deco-inspired research and engineering building in Milwaukee, one of the first dedicated R&D operations in the United States. The company advertised that it had a 50 to 1 ratio of engineers to salesmen.

A.O. Smith Automatic Frame Plant—the 'Mechanical Marvel'

Today, it's fairly normal to see robotic machines on automotive assembly lines. However, this was unheard of in the 1920s—the subject of science fiction. But at the end of 1916, L.R. Smith challenged his team of engineers to design a factory that would produce millions of frames annually—and to do so in an almost completely automated facility. This was a "game changing, bet the farm" directive. This team, which was led by Stanley Smith and included Richard Stresau, James Adams, Jr., Henry Miller, Edward W. Burgess, Birger T. Andren, Thorvald Hansen and Frederick Orton, designed an assemblage of mechanisms that were specifically designed to produce automotive frames.

L.R. Smith approved the plan for what was to become the South Frame Plant in late 1919. It was placed into operation in May 1921—and operated for an incredible thirty-seven years.

Each automobile frame required five hundred and fifty-two separate operations, which were accomplished by the frame assembly line, all automatically within a rigid framework of a ten-second cycle. The line was capable of manufacturing ten thousand frames each day and averaged well over eight thousand—or over an incredible four million operations daily.

The level of automation was such that the assembly line required only one hundred and eighty operators. While that sounds like a high figure, before the automated assembly line went into production the company required two thousand operators to produce three thousand frames daily. Putting that comparison another way, each operator on the automated frame line could produce fifty-five frames a day—whereas before automation each operator only produced 1.5 frames daily. That's a staggering thirty-seven hundred percent increase in productivity!

The mechanized assembly line allowed A.O. Smith to reduce the cost of automobile and truck frames. General Motors, which was planning to build its own frames, dropped its plans and entered into a five-year contract with A.O. Smith.[217]

Numerous patents were filed for the various mechanisms, as well as for the entire assembly. The patents were issued to Reuben Stanley Smith and assigned to the A.O. Smith Corporation. The patent for the assembly was filed in 1918 and issued on November 15, 1921, as Patent Number 1,397,020. It was entitled "Method of and Apparatus for Forming and Assembling Metal Elements." It would eventually be referred to as the *Mechanical Marvel.*

In 1979, the Mechanical Marvel was designated a Historic Landmark by the American Society of Mechanical Engineers.

A portion of the automatic automobile frame plant of A.O. Smith. In the foreground, automatic riveters are connecting frame members. Photograph courtesy of Walter Smith of Smith Precision Pumps.

In order to accomplish this feat, Reuben Stanley Smith and team needed to design numerous mechanisms, including automated riveters as well as a mechanism to feed the rivets to the machine. The patent drawings are detailed with eighty-one figures that describe the various mechanisms, all designed specifically for the automated frame assembly line.[218]

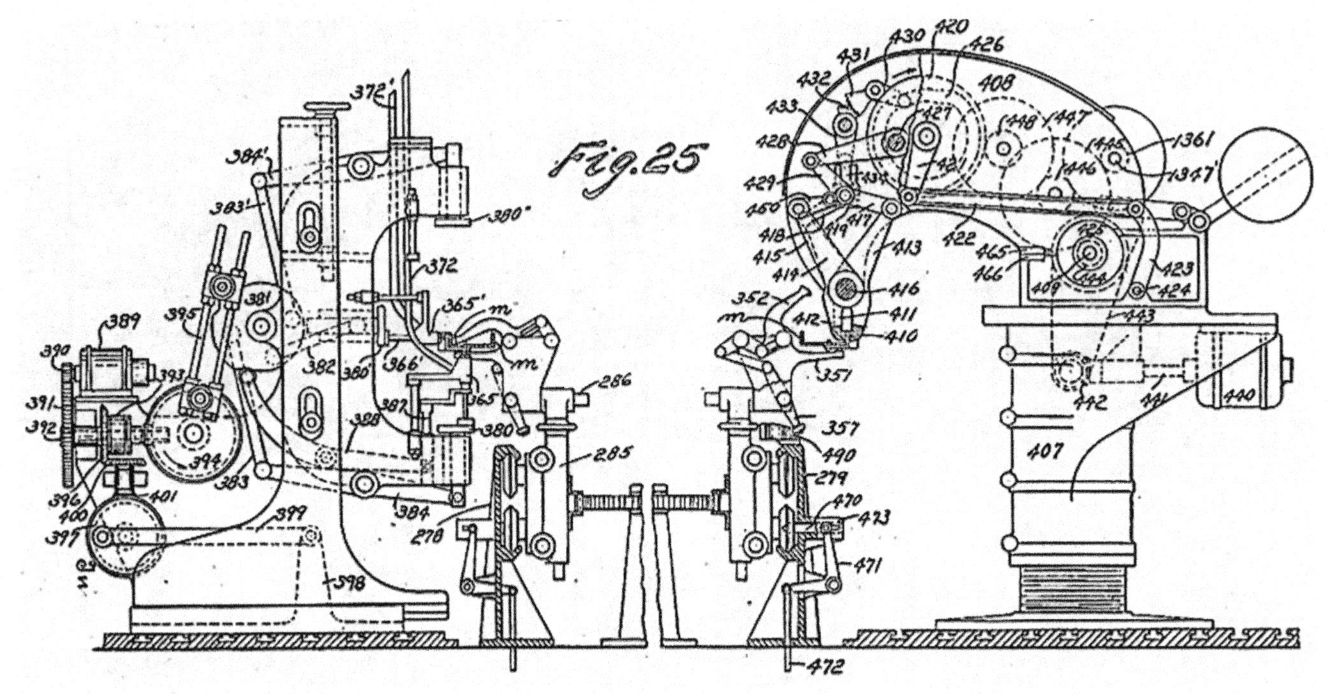

Figure 25 of the Patent filing is a cross-sectional view showing a rivet inserting machine and one of the rivet heading machines, with parts partially broken away to show how the mechanism functioned.

Fortune magazine called the plant "the most advanced single exhibit of automatic function in the world." Its successful operation, perhaps more than any other outstanding accomplishment in A.O. Smith history, solidly established the company as a world leader in automobile and truck frame development.

The company produced over one-hundred million automotive car frames and over fifty million truck frames in Milwaukee, before exiting the business in 1997 with the sale of its automotive products company to Tower Automotive.

Revolving Carousel Assembly Device

Among the numerous mechanisms that were part of the automatic frame assembly facility was a device that Reuben Stanley Smith called a "turret," but which would now be more commonly referred to as a "carousel."

Most of the machines that operated on the automotive frame assembly line operated on a synchronized cycle, and proceeded in a linear fashion down the assembly line. However, there were certain smaller components of the automotive frame that required numerous machining processes. The engineers at A.O. Smith accommodated this by designing various revolving carousel assembly devices that could perform the necessary machining operations and then feed the part to the frame assembly line.

In describing one such device in the 1921 patent application, which was designed to machine the cross bars for the automotive frames, it was noted that "The cross bar blanks being comparatively short in length are well adapted to easy manipulation by the revolving turrets. The turrets are provided with means for clamping the cross bar blanks delivered to them in succession from the line conveyor, so that such blanks are properly held for positioning in correct relation to the machines and tools arranged about the revolving turret. The movement of the turret is intermitting, so that the blanks supported in the clamping devices thereon are presented to the successive machines associated with the turret.[219]

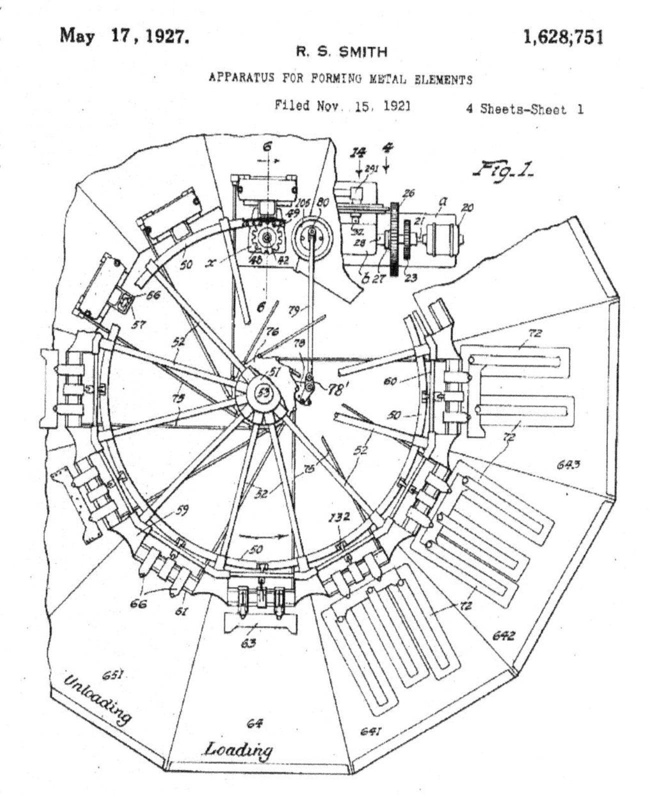

1921 patent drawing for Reuben Smith's turret assembly device

It was noted that the mechanism made economic use of the floor space adjacent to the automated frame assembly line, and feed parts to the line as required to meet the synchronized time—yet the turret operations did not need to be in exact synchronism with the other machines on the general assembly line, which afforded flexibility.

The automatic frame assembly line had eighteen turrets for performing the necessary machining for the crossbars, as well as turrets for other operations such as the finishing of spring hangers.

It is not known whether this was the world's first application of a rotating carousel assembly mechanism. However, it was certainly an early example of an advanced assembly device that has become commonplace in manufacturing plants—a significant mechanical engineering innovation.

Revolving Turret Spring Hanger Finishing Machine, located on A.O. Smith's automatic automotive frame assembly plant. Picture Courtesy Walter Smith of Smith Precision Pumps.

Development of Heavy-Coated Welding Rods

As the United States entered into the First World War, A.O. Smith received orders for various war materials including caisson wheel hub flanges, frames for Army trucks, and casings for bombs.

The company ran into some challenges in producing casings for the one hundred pound bombs—the largest bombs being produced for aircraft at the time. Welding the casings with oxyacetylene gas flame produced a fairly crude weld, and there was an acute shortage of oxyacetylene during the war. A.O. Smith was familiar with arc welding using a coated weld rod imported from England. Arthur Percy Strohmenger of England received a patent for "electric deposition of metals" in 1909, using a heavily coated electrode. The coated welding rod, called the *Quasi-Arc* weld rod, was coated with asbestos to resist the heat of the arc, allowing an ionized atmosphere to be produced at the welding tip capable of producing welds of reasonable metal properties.[220] However, German submarine activity precluded shipping the weld rods to the United States.

A.O. Smith turned to its engineers to find a suitable substitute. After a month of experimenting, a team led by Reuben Stanley Smith which included Orrin Andrus and Dick Stresau, developed a welding rod with a coating of paper pump and sodium silicate, both of which were inexpensive and readily available. The resulting weld rod was cleaner, simpler, swifter and more accurate in temperature control than gas-flame welding.[221] It also avoided the use of asbestos. While the patent was filed under Stanley Smith's name (filed in 1918 and granted in 1919), it appears that the key insights were those of 18-year-old Orrin Andrus, a lab assistant.[222]

Andrus was familiar with carbon arc street-lighting, where an arc forms between two carbon rods that burns at a very high temperature and causes the rods to incandescence. He and Stanley Smith experimented with a rod of carbon lampblack, coated with sodium silicate, but the resulting welds were brittle and cracked. Andrus then came up with the insight of wrapping the rods with newsprint, soaked with sodium silicate. The coated rods were then baked in an oven to dry the coating. The resulting welds were of better quality than produced by the Quasi-Arc rods. They discovered that the paper in the coating created an ionized gas at the tip that eliminated virtually all of the oxygen, resulting in welds of unprecedented ductility.

The Smith weld rod launched A.O. Smith into the business of producing weld rods and the company became recognized as a pioneer in electric arc welding, as well as other welding techniques.

In addition to the patent for the coated electrode, Reuben Stanley Smith obtained a patent for an arc welding apparatus that could weld longitudinally and circumferentially. The apparatus was ingeniously designed so the device being welded could be moved in different directions automatically for a variety of weld configurations. He also filed a patent for a "Process and Apparatus for Producing Electrically Welded Joints." The process used water for cooling and allowed for continuous seam welding of pipes that produced a strong and ductile weld.[223]

With these advancements in hand, A.O. Smith began exploring other applications. It began using its arc welding techniques in the manufacture of rear axle housings for automobiles, as well as for couplings for oil well casings. The company also used the welding technology to build high-pressure vessels used to refine oil, producing some of the largest cracking towers for the industry. In 1925, it fabricated a single-piece heavy-walled pressure vessel entirely by welding and publically tested it before placing it in service at an oil refinery. The company produced pressure vessels for the industry capable of withstanding pressures as high as sixty-five hundred pounds per square inch.

This high-pressure vessel was produced by A.O. Smith for cracking oil into its components in a refinery. The company provided the vessels for what was the world's largest oil cracking unit at the time.

A.O. Smith also began using its welding techniques for the production of pipe for the petroleum industry. Prior to this innovation, piping was manufactured from solid cast billets that had to be reamed, drawn and straightened into the desired dimensions—a costly and time-consuming process.[224] A.O. Smith was able to roll sheets of steel into pipe with welded longitudinal joints. After testing, the company was able to demonstrate the integrity and quality of the welds. Soon A.O. Smith was producing up to thirty-five miles of oil pipe per day, supplying most of the oil pipeline industry's requirements.

The process for wrapping the coated weld rods was labor intensive, requiring that several layers of paper be spirally wrapped around each rod before sodium silicate was applied. New coatings were formulated and in the 1920s, A.O. Smith engineers Clifford B. Langstroth and George G. Wunder developed the first extruded cellulosic electrodes, which significantly reduced their cost.

In 1927, the company introduced resistance flash welding; a type of resistance welding that does not require a filler metal. The welding process involves setting the pieces of metal to be welded at a predetermined distance. Current is then applied and an arc is produced across the gap. Once the pieces of metal reach the proper temperature, they are joined, effectively forging the pieces together. A.O. Smith applied the technology to produce longitudinal pipe joints. After the joint was formed and any surplus metal removed, an internal mandrel was run through the pipe, which rounded the pipe to its final dimension, and tested the pipe at the same time.[225] The highly productive process was capable of manufacturing longitudinal seams of 40-foot lengths of large-diameter pipe in thirty seconds.[226]

This photograph, taken about 1928, shows a record shipment of A.O. Smith pipe—requiring three separate trains.

When the United States entered into the Second World War, A.O. Smith was again called upon to produce bomb casings—this time in the "blockbuster" class—ten times the size of the bombs it produced for the prior war. The nose and base of the bomb were manufactured using forging hammers of the company's own design. The sides were produced using ordinary steel tubing—all assembled with arc welding. The resulting design reduced cost and allowed the company to produce a thousand of the biggest bombs in a single day.[227]

During the war, A.O. Smith and other manufacturers produced pipe for the "Big Inch" and "Little Big Inch" petroleum pipelines; considered one of the greatest engineering feats of the war. Before the war, petroleum products were transported from the oil fields of Texas to refineries in the northeastern United States via oil tankers. When the country entered the war, German U-boat submarines threatened this link. To protect this vital resource during the war, the "Inch" projects were constructed as emergency war measures. Constructed between 1942 and 1944, the Big and Little Big Inch pipelines were 1,254 miles and 1,475 miles long respectively, with thirty-five pumping stations along their routes. The project required 16,000 people and 725,000 tons of materials—most of which was large diameter steel piping.

Because of its welding expertise, A.O. Smith was consulted by William (Bill) Knudsen during the war about using welding techniques, rather than riveting, to construct the Liberty ships. Knudsen, who oversaw the nation's manufacturing companies in providing material for the war effort, approached the company. He was told that it would be feasible to weld the thick steel plate for the ships, and doing so would greatly increase productivity. Knudsen directed that the Liberty ships to use new welding techniques, based in part on this advice.[228] Rudolph Furrer of A.O. Smith was assigned by the company to the US War Metallurgy Committee to advise the army on ordinance research and development, where he contributed to the design of the Liberty ships and tanks, and for the casings for the atomic bombs. Furrer was posthumously inducted into the Army Ordnance Hall of Fame in 1993, because of his contributions to the war effort.

When early Liberty ships suffered hull and deck cracks, and a few were lost to structural failures, it appeared that the new welding techniques used to produce the ships were the cause of the failures. However, tests by the British Ministry of War Transport demonstrated that the failures were caused by the grade of steel plates used, which suffered from embrittlement at the temperatures being experienced in the North Atlantic, and not by failures of the welds. Various reinforcements were made to the Liberty ships to minimize the crack failures, and the successor Victory ship was designed to better deal with fatigue.[229]

The decision to produce the Liberty, and later the Victory ships, using modern welding techniques allowed the United States to build over 3,200 of these ships during the course of the war. The first ships required about 230 days to build, but the average eventually dropped to an amazing forty-two days, using welding and mass production techniques.

Electric welders at work on the Liberty ship SS Frederick Douglass in 1943 at the Bethlehem-Fairfield shipyards in Baltimore, Maryland. Source, US Library of Congress's Prints and Photographs division.

First Glass-Lined Water Heater

It was perhaps inevitable that after developing techniques for large welded tanks in a city known for its breweries A.O. Smith would eventually be called upon to provide tanks for the brewing industry. Brewery tanks were traditionally fabricated of wood or copper. However the price of copper was high, especially in the years leading up to the Second World War.

The temptation to produce tanks of steel must have been strong. But steel corroded, especially when used in brewing service. The answer, of course, was to line the tanks. Given that A.O. Smith had developed processes for coating weld rods with various ceramic materials, they had experience with the processes required. The company's engineers experimented with various vitreous enamel linings—when fired it provided a glass-like surface. Fusing the material to the interior of steel tank walls provided an ideal solution. It was impervious to the brewing products and wouldn't affect flavor.

The company introduced its first large, single-piece glass-lined brewery tank in 1933. Over the next thirty-two years, A.O. Smith made more than 11,000 glass-lined brewery tanks.

Here, workers apply a coating to the interior of a large tank for the brewing industry. The vessel will later be fired to harden the coating and fuse it to the surface walls

Having perfected the technology for fusing glass to steel, A.O. Smith looked to other applications. In 1936, the company patented the process of glass lining a steel water heater tank, in which a vitreous enamel coating is fused to the steel tank at 1,600° F. The company called the coating, *Permaglas*, which inhibited rust and corrosion, yet was flexible enough to avoid cracking. It began mass-production of residential water heaters in 1939. Soon the use of glass-lined tanks for hot water heaters became the standard for the industry.

Reuben Stanley Smith was chief engineer at A.O. Smith at a time in which the company was incredibly inventive. He was normally referred to as "Stanley" or "Stan." Forty-six patents were filed in his name during the years that he worked at the corporation (between 1914 and 1923). His list of patents includes one for an assemblage of machines to produce automotive frames automatically—later recognized as the "Mechanical Marvel" because of its unheard-of level of automation and complexity. The automated assembly line was capable of producing ten thousand automobile frames a day. In 1979, the American Society of Mechanical Engineers designated the assembly line a Historic Mechanical Engineering Landmark.

Stanley also was the inventor of numerous other designs involving welding, various internal combustion engine innovations associated with the Smith Motor Wheel for bicycles,[yy] one of the first metallic automobile wheels, and various manufacturing apparatus.

He worked for the company under the direction of L.R. Smith, president of A.O. Smith. Ray was Stanley's first cousin and son of Arthur O. Smith, the founder of the corporation.

Stanley led a team of dozens, if not hundreds of engineers, in developing the various innovations. While his name is listed as the filer for the numerous patents, it is evident that many other engineers and technicians were also involved in the developments. It was often the practice for only the engineering leadership to be listed on corporate patent filings. While some of the team members that worked on these innovations are known, other individuals who may have contributed significantly are not. Noting this isn't meant to detract from Stanley's incredible ingenuity, as well as his resourcefulness and leadership.

In the early 1920s, Stanley left Milwaukee and moved to the Los Angeles area of California, where most of his family lived. In 1931, he started his first company, the Smith Meter Company. Smith Meter manufactured a flow meter for petroleum products. Stan developed a four-piston meter for gasoline transfer. He sold the patent rights to Charles Wright, president of the Badger Meter Company in Milwaukee who was married to one of Stanley's first cousins. Stanley subsequently developed what eventually became known as the "Smith Meter" for measuring the flow of petroleum gases. His cousin, Ray Smith of the A.O. Smith Company, loaned him the money to develop the meter. A.O. Smith eventually acquired the company.

In 1938, Stanley founded the Smith Precision Products Company to manufacture pumps. He died in 1948.

Photograph of Reuben Stanley Smith, courtesy of Walter Smith of Smith Precision Pumps

[yy] The rights to manufacture and sell the Smith Motor Wheel was eventually sold to Briggs & Stratton and is discussed in Chapter 4 of this book.

T.L. SMITH COMPANY

As noted in Chapter 3, the T.L. Smith Company built a new factory in about 1916 at 2835 North 32nd Street in order to increase manufacturing capacity. It was at this location that the company developed its first high-discharge transit mixer.

The company also built a series of progressively larger fixed mixers for large construction projects. As the maximum size of its fixed mixers increased, the company periodically announced the design and manufacture of the "World's Largest Heavy-Duty Cement Mixer."

By the 1940s, the combination of Milwaukee's T.L. Smith Company, Koehring Machine Company, and the Chain Belt Company, along with Kwik-Mix of Port Washington, Gilson Manufacturing Company of Fredonia and Leach Company of Oshkosh, produced seventy percent of the concrete mixers in the United States.[230]

In 1949, TL Smith announced a sales, operating and financial alliance with Essick Manufacturing Company of California, but TL Smith's Milwaukee factory continued to operate until the mid-1970s. The factory was eventually sold to Koehring.

This TL Smith Transit Concrete Mixer is mounted on a 1937-1940 International-Harvester D Series truck. Notice the Smith logo prominently displayed on the water tank mounted in front of the mixer. The company continued to manufacture its line of transit mixers in Milwaukee until the mid-1970s. Photograph from the Thomas S. Pierce collection, used with permission.

World's Largest Concrete Mixers

T.L. Smith Company manufactured a number of large concrete mixers that it claimed were the world's largest. While it is difficult to substantiate these claims so many decades after they were made, it is clear that these were indeed massive machines.

In one such pronouncement, an article in the Milwaukee Journal stated, "To Milwaukee's cap as the greatest manufacturing center of heavy machinery in the world, another feather was added [ ] when the T.L. Smith Co. shipped the world's largest concrete mixer—twice as large as any mixer previously built." The mixer, produced in late 1927, weighted fifty-five thousand pounds and stood 15½ feet tall. It was shipped to the Knoxville Power Company, a subsidiary of the Aluminum Company of America, to supply concrete for the construction of a hydroelectric dam.[zz] During that decade, T.L. Smith mixers were also used on the Wilson dam at Muscle Shoals, Bartlett's Ferry dam at Columbus Georgia, the Martin dam in Alabama, the Exchequer Dam across the Merced River in California, and the Isle Maligne dam in Canada.[231] Often several mixers were employed at concrete dam construction sites, to allow large continuous concrete pours for the massive structures.

In 1949, a similar claim of the world's largest was made in both the Milwaukee and the Escanaba Michigan newspapers. It was reported that T.L. Smith is the designer of a new six-yard tilting concrete mixer, which the company claims, "is by all odds the largest heavy-duty concrete mixer in the history of the world." It was designed by the chief engineer of the company at that time, George A. Rockburg. It was noted that the machine weighed 41,800 pounds, complete with feed chute, main supporting frame and a one hundred horsepower electric motor. It was built for a lime company in Camden, New Jersey.[232] It appears this machine was actually smaller than the record mixer the company built in 1927, although its capacity may have been greater.

This advertisement from T.L. Smith from 1949 provided an image of this six-yard tilting concrete mixer—the World's Largest at the time, as noted in the ad copy.

[zz] This mixer was likely used on the Calderwood Hydroelectric Dam, built in the Little Tennessee River Valley. It was one of the last dams to be completed in the Tennessee River watershed before TVA took control of the watershed in 1933.

CHRISTENSEN ENGINEERING

Niels Christensen didn't allow a lawsuit with Westinghouse stop him from manufacturing and selling his streetcar braking system. He simply outsourced production for a time to his former employer, Allis-Chalmers, until he could re-establish his own manufacturing facilities. In 1907, he built a new plant near 30th Street and North Avenue, along what is now referred to as Milwaukee's 30th Street Industrial Corridor.

Christensen's new company and the National Brake & Electric Company engaged in a pricing war for streetcar braking systems that proved disadvantages to both.[233] As a result, Christensen diversified operations into other products, including compressors, internal combustion engines (marketed primarily for agricultural needs ranging from 2.5 to 75 horsepower), pumps, and hoists. The company also began manufacturing Christensen's compressed air starter for internal combustion engines and, by the 1920s, it focused its operations exclusively on air starter production. Christensen Engineering appears to have been eventually acquired by Allis-Chalmers and folded into its operations.

In about 1926, Niels Christensen accepted a position at Midland Steel Products of Cleveland to work on the design of truck and automotive air brakes. He relocated to the Cleveland area with his family, where he continued to innovate. His story is continued in Chapter 9: The Ones that Got Away.

THE CHRISTENSEN SELF-STARTER

In the early part of the 1900s, Christensen set out to design a device to solve significant problem—the need to hand crank internal combustion engines to get them started. He developed an innovative starter for internal combustion engines that used compressed air.

Christensen initially developed the device to be used for automotive engines, including the popular Ford Model T, although it was more commonly used on more expense vehicles, which typically had larger engines and required more cranking force. Using the advertising slogan, "Start Your Engine from Your Seat," the self-starter eliminated the necessity, and the risk, of getting out of the vehicle and hand cranking it.

The starter introduced a rich mixture of gasoline into the cylinders by means of a compact air compressor and a separate starting carburetor. The system had a compressed air storage tank, which could start the engine multiple times.

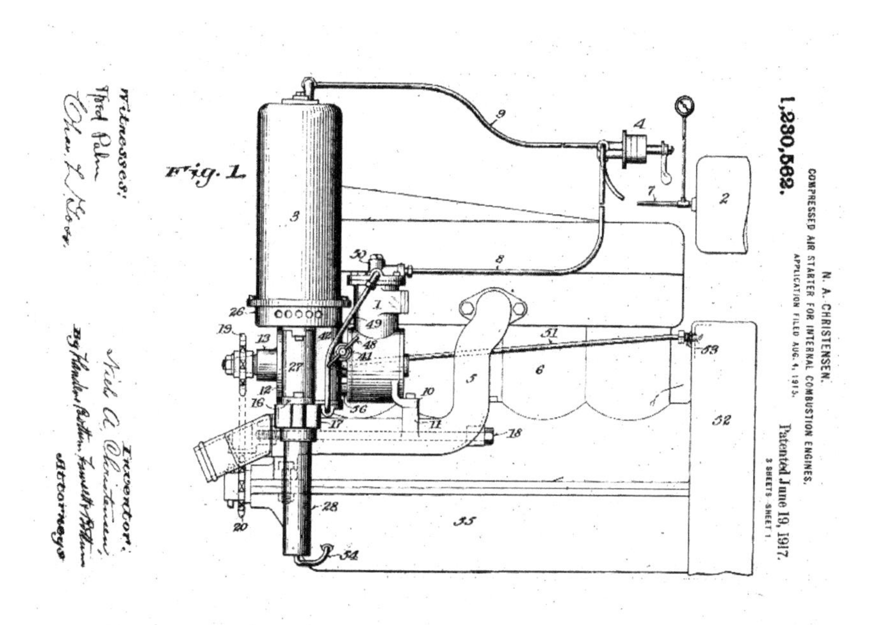

The patent drawing at right shows the Christensen compressed air starter mounted on a Ford Model T engine.

However, Christensen's device had a competitor—Charles Kettering developed and patented an electric self-starter in 1911, which was adopted by Cadillac in 1912. His Delco starter quickly became the preferred way of starting automotive engines.

It appears that Christensen's self-starter found its niche, at least for a time, for starting large truck motors and aircraft engines. Kettering's electric self-starters had not evolved to the point where they were available for larger engines. As a result, thousands of Christensen self-starters were installed in large vehicles. The starters were reportedly installed as standard equipment in fire trucks in twenty-six cities in the United States, including all fire department vehicles in the City of Milwaukee and many in Chicago.

The Christensen self-starter system was also popular for starting aircraft engines. In addition to its use by western aircraft companies, the company reported that system was used by the British Admiralty, the Japanese government, the French Aviation Service, and the Danish Aviation Service.

670 *AERIAL AGE WEEKLY, February 19, 1917*

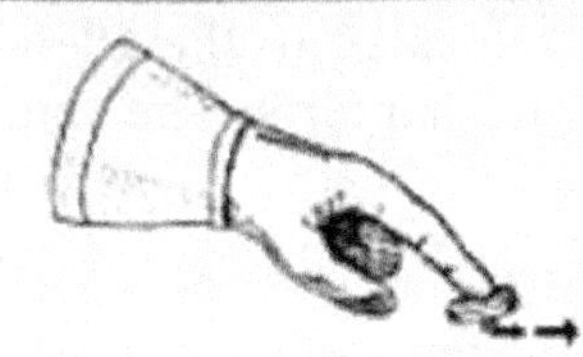

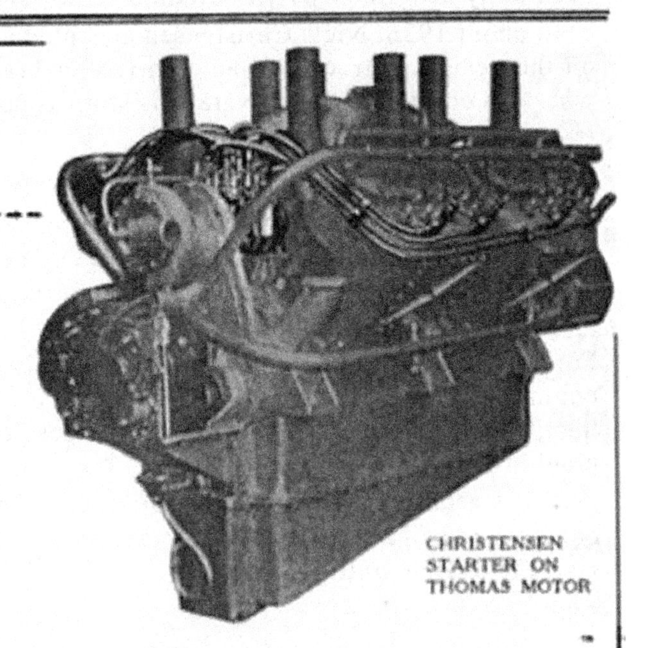

The Touch of a Button

AND

She Goes:—*Instantly!*

A Self Starter that combines unfailing action with simplicity and light weight; a Starter that cannot be put out of commission by excessive vibration; a Starter that is absolutely dependable, regardless of weather conditions or altitude.

Christensen Automatic Motor Starter

THE marked efficiency of the Christensen Self Starter is perhaps best suggested by the results of a test on a Hall-Scott Motor made at the plant of the Standard Aero Mfg. Co., at Plainfield, N. J. There on a tank of air of 250 pounds pressure, 28 complete starts were made without recharging the tank, and the last start was made on the low pressure of 100 pounds.

Send for Catalogue

CHRISTENSEN ENGINEERING COMPANY, Milwaukee, Wis., U.S.A.

An advertisement for Christensen's Automatic Motor Starter, as appeared in 1917 in the *Aerial Age Weekly magazine.*

BADGER METER MANUFACTURING COMPANY

Badger Meter Manufacturing Company was established in 1905 by four Milwaukee businessmen to fabricate frost-proof water meters for measuring water consumption in Midwestern homes. Co-founder John Leach eventually became president of the company. Badger's innovation was a meter with a soft, replaceable cast-iron bottom plate that ruptured when the water in the meter froze, thus relieving pressure on the meter and safeguarding its mechanical parts. Since frozen water pipes were an all too common occurrence in Wisconsin's bitter winters, Badger Meter found a ready market. By 1910, it was selling close to four thousand of its eight-dollar meters a year under the motto "Accuracy Durability Simplicity Capacity."

Badger Meter operated initially out of a two-story machine shop in Milwaukee's downtown at what is now 929 North 3rd Street. By 1918, it was producing ten thousand meters annually and its line of meters included bronze as well as cast iron offerings. The company also added disc, turbine, and compound water meters to its line of products. It had only twelve employees and when large orders came in, everybody worked extensive overtime—reportedly stopping only "for a lunchtime 'pail' of beer and a free dinner" on the company's tab at the end of the day.

This is one of the first Badger Meters produced - Serial No. 7. It was originally installed on Nov. 2, 1905 at 2526 N. Palmer Street in Milwaukee. It was in service for more than 50 years and is the oldest Badger Meter water meter known to be in existence. It is on display in the basement of "Granny's House" at the MPM Streets of Old Milwaukee exhibit. Notice the replaceable bottom plate. Photograph courtesy of Badger Meter.

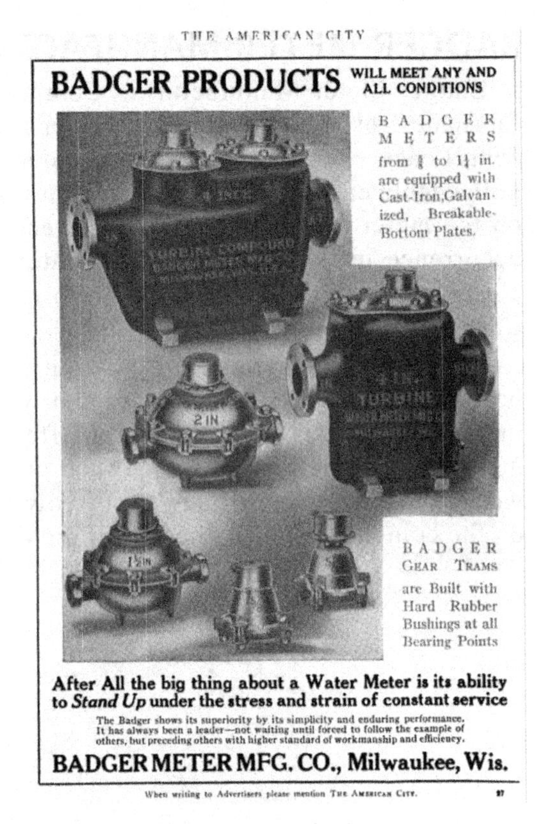

This Badger Meter advertisement from 1916 magazine entitled, The American City shows a number of different sized meters. Notice that the smaller meters were advertised as "equipped with cast-iron, galvanized, breakable bottom plates." The plates were designed to rupture if the fluid within became frozen, thus protecting the rest of the meter. The bottom plates could then be replaced without removing the meter.

In 1919, Badger Meter finally expanded. It moved to a new facility at 30th Street and North Avenue, alongside the railroad tracks in what is now known as Milwaukee's 30th Street Industrial Corridor.

The new factory included the company's first foundry, which allowed it to cast meter housings and parts for its own metal components. In addition, the foundry acted as a job shop for other Milwaukee manufacturers, including the production of bronze castings for A.O. Smith Corporation and auto hubs and fingers for Milwaukee Automotive Supply.

In the 1920s sales increased greatly as many cities expanded their water departments and service areas. Badger Meter received an order from the City of Chicago for four hundred meters a day. Sales and profits remained good until the US Depression reduced sales to a minimum. The company was forced to reduce the number of employees and drop wages in half for those that remained. It wasn't until Franklin Roosevelt's "New Deal" provided municipalities with funds for their water facilities in 1937 that Badger Meter was back on a solid financial footing.

In 1939, Badger Meter diversified into the production of meters to measure the amount of motor oil and lubrication used in service stations. As the Second World War started, however, most activities were directed into military applications. The equipment and skills that Badger Meter had developed turned out to be perfect matches for manufacturing bomb fuzes. In early 1942, Badger Meter received its first order for 200,000 fuzes. Badger produced nose and tail fuzes, which generally carry a small propeller. As the bomb descends, the propeller begins to turn and in the process arms the firing pin. When the bomb strikes its target, the firing pin strikes a cap, which sets off the primer charge and initiates the explosion.

In order to meet the production targets established by the War Department, fuse production accounted for ninety-five percent of Badger's total output. By the end of 1942, the company's employment doubled to five hundred and fifty—many of which were women.

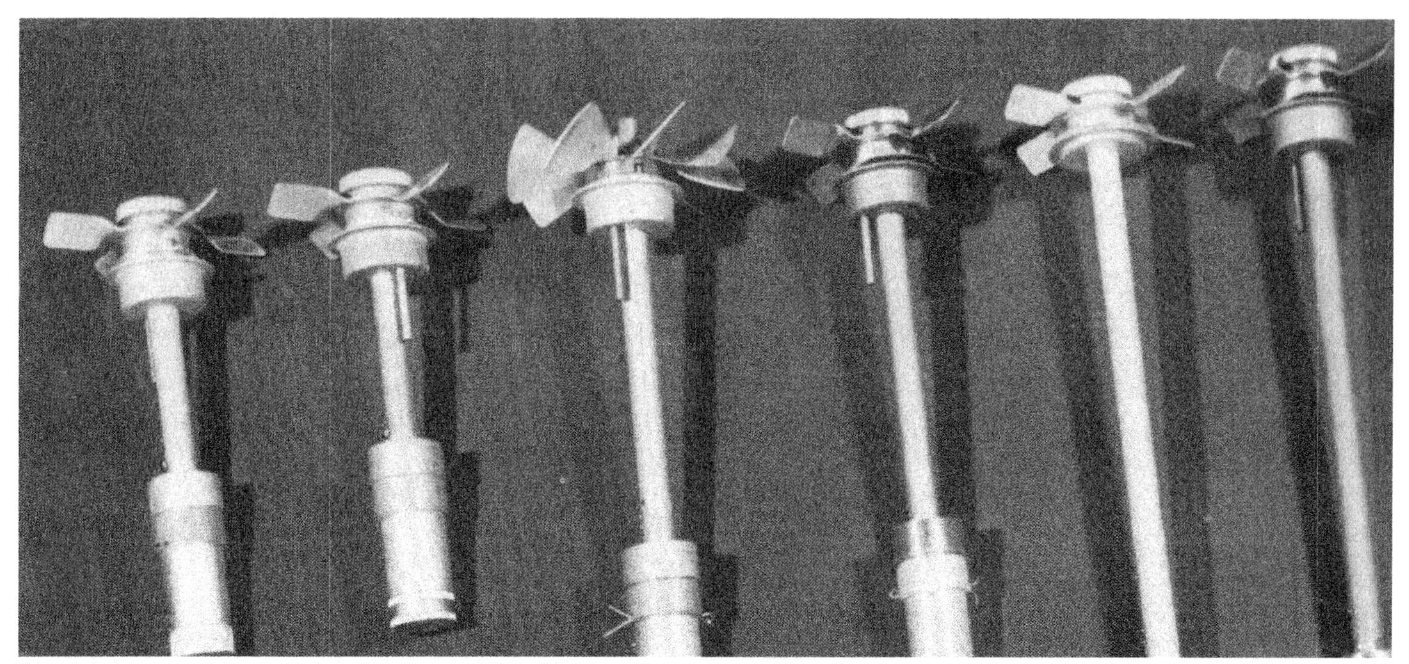

Bomb fuses produced during the Second World War—in various sizes, from a Badger Meter publication.

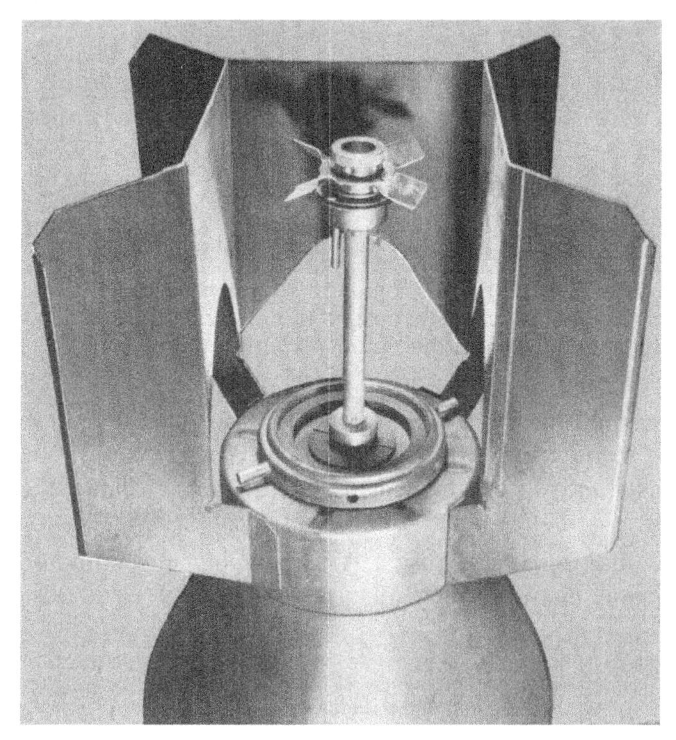

A tail-fuse, mounted on a bomb, partially cut-away to show fuse details

During the Second World War, Badger Meter employed a substantial number of women to supplement its workforce. The blue, all-cotton dresses they wore to prevent the buildup of static around the electrically sensitive fuses earned them the nickname Badger's *blue belles*. Photograph used with the permission of Badger Meter.

Following the war, the housing boom that was unleashed by the returning soldiers resulted in a high demand for Badger's water meters and sales continued to be strong throughout the next decade. In 1946 the company introduced its "GMOP" meter—a grease and an oil-measuring meter that became the standard for domestic makers of lubricating equipment.

During the Korean War, Badger Meter returned to making bomb fuses. After making an estimated twenty-one million fuses, Badger abandoned the munitions business in 1960.

MASTER LOCK COMPANY

Master Lock was established in 1921 by Milwaukee-area locksmith, Harry E. Soref. Harry, who was born in Russia in 1887, joined his parents in Milwaukee as a young lad. His parents decided that he was too old to go to school, so he was put to work in his father's furniture store. When the furniture business didn't hold Harry's interest, he left the city. He spent several years in a series of jobs, including work for a traveling circus and managing a pool hall in Indiana. Eventually, Harry returned to Milwaukee and worked in a cousin's hardware store. While employed there he began inventing various tools. Locks fascinated Soref and he began working as a locksmith and producing master keys, which he sold by mail order.

During the First World War, Soref worked as a security consultant for the military and designed a padlock to protect military equipment. After the war in 1919, he invented a padlock constructed with multiple layers of steel, which were stacked together and riveted to form the casing and hold the internal lock mechanism. The individual layers, or laminations as Soref called them, were fabricated with a punch press equipped with dies to form the desired shapes.

Soref applied for a patent for his laminated lock casing in 1921, which was granted in 1924. His lock was far superior to inexpensive locks, which used stamped metal casings, but far cheaper to produce than locks fabricated from solid metal casings.

Harry Soref attempted to interest lock makers in his invention, but found little interest because of what was perceived as a complicated manufacturing process. Soref eventually interested two investors into providing the funds to manufacture the locks himself, forming the Master Lock Company in 1924. The company operated initially out of rented space from Pabst Brewery at 9th and Juneau, which was available because Prohibition restricted brewing operations.

Soref's lock was recognized for its superior strength and resistance to breakage and tampering. Prohibition soon played another role for the young company. Padlocks were needed to lock down pubs and other establishments that continued to sell alcoholic beverages. In a single shipment in 1928, Master Lock supplied almost 150 thousand padlocks to the City of New York for that purpose.[234]

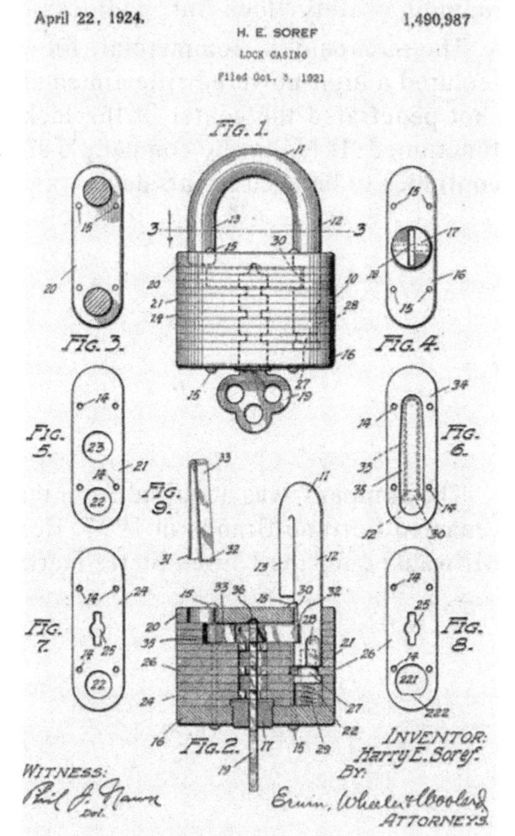

Illustration from Henry E. Soref's patent filing for a lock casing, showing the assemblage of individual laminations that formed the casing and held the mechanism.

By 1939, Master Lock had outgrown its rented space and constructed a factory at 32nd and Center Streets, along what is now known as Milwaukee's 30th Street Industrial Corridor.

Master Lock's new factory was located at 32nd and Center Streets. It featured a 32-foot high, three-dimensional padlock over the north end of the plant.

Master Lock has grown to become the largest global manufacturer and marketer of padlocks,[235] In addition to its high-security pin tumbler padlocks, the company produces a wide array of products for use in schools, hospitals, and offices, and on vending machines. It also makes various products for industry, including safety "lock-out" padlocks.

The company's commercial for the 1974 Superbowl game featured a high-powered rifle aimed at a Master Lock padlock. The shot penetrated the center of the lock, but the lock still held—and functioned. It led to the company's slogan "Tough Under Fire" that continues to be used to this day.

The company was acquired from the Soref family by American Brands in 1970. American Brands was renamed Fortune Brands in 1997. However, Master Lock continues to design and manufacture locks in Milwaukee for the United States market.

HARRY E. SOREF was a locksmith, padlock inventor, and founder of Master Lock Company. Harry was born in Bilozerika, Galena, Russia to Isaac and Ella Soref in 1887. His father immigrated to the United States, via Canada, in 1893 and sent for his wife and son when he was established.[236]

Harry worked in his father's furniture business for a time, but became bored and left to find something more challenging. He traveled to California and eventually hooked up with a traveling circus, but left to manage a pool hall in Gary, Indiana. He returned to Milwaukee after a year or two and established a store named Global Bedding, before going to work for his cousin who sold hardware. While at the hardware store, Harry began inventing various tools. The hardware enterprise apparently also resulted in an interest in locks, because shortly Harry Soref began working as a locksmith and producing master keys, which he sold by mail order.

Soref recognized that the inexpensive padlocks on the market were easily broken and offered little security. Their stamped metal casings were easily damaged. His daughter reported that he got an insight into a better lock design when viewing a deck of cards. In 1919 he designed a padlock with a casing fabricated with a dozen or more layers of stamped metal sheets which were riveted together to hold the internal lock mechanism. He filed for a patent in 1921 and in that year formed, with two partners, the Master Lock Company.

Operating from rented space in the Pabst Brewery, the company advertised its "laminated" padlock as being similar to bank vault doors and battleships. The superior strength of the padlock, and its relatively low-cost, resulted in a robust market for the laminated padlock.

Harry Soref became a recognized authority on locks. He worked as a security consultant for the United States military during the First and Second World Wars. It is reported that escape artist Harry Houdini consulted with Soref in Milwaukee about techniques to escape from chains secured by padlocks. In 1913, the American Association of Master Locksmiths presented Soref with a gold medal for "making the greatest contribution to the development of locks in more than 50 years." A company exhibit at the 1933 Chicago World's Fair touted the features of laminated steel locks.

Soref held over eighty patents for padlocks and other security devices, as well as for lock-making machinery.

Harry Soref died in 1957. He is buried in Spring Hill Cemetery in Milwaukee.

Photograph of Harry Soref, courtesy of his family.

CHAPTER 8: WEST, WEST AND WAUKESHA

As Milwaukee companies sought additional space to grow their operations, some moved west—to West Milwaukee and to a community that became known as West Allis.

The most dramatic of these moves was by the E.P. Allis Company. Its merger in 1901 with Fraser & Chalmers, the Gates Iron Works and the industrial business line of the Dickson Manufacturing Company, presented the company with an immediate need for manufacturing space. That same year it announced plans to build expanded facilities in North Greenfield—a location that was four miles directly west of the company's Reliance Works. The news brought with it a movement to change the name of the community to "West Allis," which occurred in 1902.

The former North Greenfield community had been named after a railway station built in 1880 by the Chicago Northwestern Railway. The community was largely defined by the Chicago, Milwaukee and St. Paul Railway on its north and the Chicago and Northwestern Railway on its south. Railway access, as well as readily available land, attracted industry to the area. In 1891, the Wisconsin State Agriculture Society selected the community as the location for its State Fair, which resulted in an increase in the public transportation facilities from Milwaukee to the area.

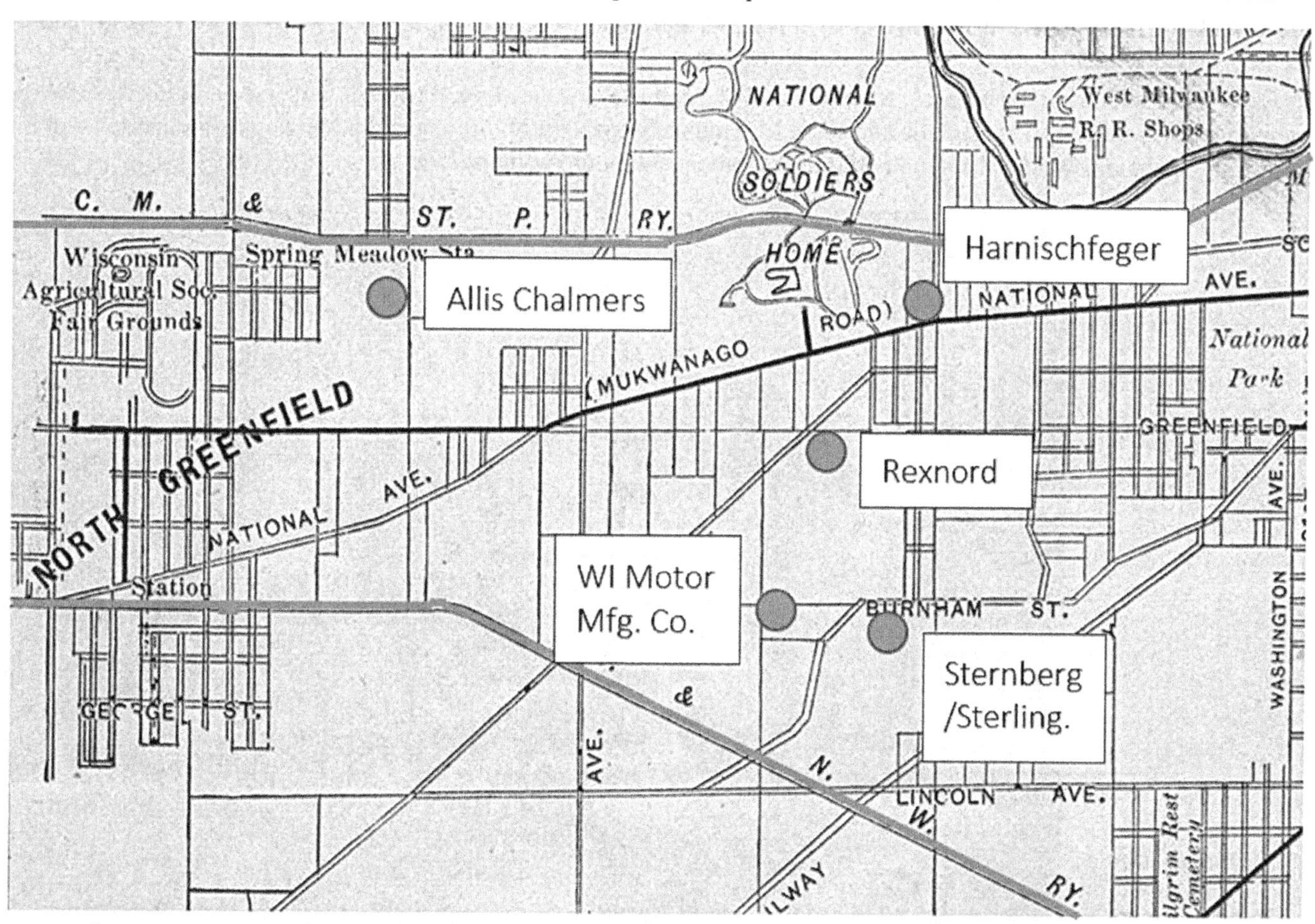

This map of Milwaukee's west side is an extract of a TMER&LCo streetcar map of 1898. It has been highlighted to show the lines of the Chicago, Milwaukee and St. Paul Railway, as well as the Chicago and Northwestern Railway that traversed the area. Notice the Wisconsin Agriculture Society's Fair Grounds on the far left of the map. The Town of North Greenfield eventually became the City of West Allis. Company locations are approximate

In 1902, E.P. Allis started building its new factory. Other industrial facilities were also drawn to the area, including Fred Prescott Company and the Rosenthal Corn Husker Company. The new factories attracted workers to the area, giving rise to a housing boom. The new residents also entices merchants to the area and soon Greenfield Avenue became a shopping district. In early 1906, the Wisconsin governor declared the community a "city of the fourth class." It had a population of 2,306.

Soon, industry spilled over to a community just east of West Allis—an area once referred to as "west" of the City of Milwaukee. The village of West Milwaukee was incorporated in 1906, shortly after Pauling and Harnischfeger located in the area. Soon new roads and streetcar lines were built to service the area and the population grew to over nine hundred.

Located further west along the Chicago and Northwestern Railway is the City of Waukesha. The city was originally incorporated in 1846, well before the railway, as the village of Prairieville. In 1847 the community was renamed Waukesha and by 1896 it had achieved status as a city.

While predominately a farming area, a Waukesha limestone quarry provided stone for many public and private buildings beyond the area's borders.

In the 1870s, the *springs* era began when Colonel Richard Dunbar visited his sister-in-law and used the water from a spring on her Waukesha farm. He believed the water helped to cure his "diabetes mellitus" and in 1869 returned to the community to purchase the land and establish the Bethesda Mineral Spring. He promoted the healing properties widely and the area became known as the "Saratoga of the West." It became a nationally recognized location for those seeking relaxation and restoration of the mind and body. The area flourished socially, with grand resorts, theaters, casinos and, of course, plants to bottle the water.

Early manufacturing businesses supplied useful products for farming, railroad, and other industries. While industrial development in West Allis and West Milwaukee were largely influenced by Milwaukee industries seeking room for expansion, Waukesha's industrial development was more homegrown.

Image from a postcard of Colonel Dunbar's Bethesda Mineral Springs.

ALLIS-CHALMERS CORPORATION

The year 1901 was a year of significant change for the E.P. Allis Company. Not only did it decide to move to what is now the City of West Allis in that year, it also announced a major merger. As noted, that merger resulted in the consolidation of E.P. Allis with the Fraser & Chalmers Company (mining and ore milling equipment), Gates Iron Works (rock and cement milling equipment), and the industrial business line of Dickson Manufacturing Company (engines and compressors). The newly consolidated company took the name *Allis-Chalmers*.

Edwin Reynolds had a major role in the design of the new West Allis factory. He and his team laid out a modern plant with material and work designed to flow from one end of the massive facility to the other.

An Allis Chalmers promotional view of the West Allis plant after the expansion of 1906. The building at the lower left, originally designed for pattern storage, was later remodeled for office space.

At a time when most large factories had grown up in a haphazard manner, the West Allis Works of Allis-Chalmers was planned by Edwin Reynolds that all manufacturing work, from blueprint to finished product, moved in one direction—eastward toward shipping. This plant, which was designed to employ ten thousand people, was a model of twentieth-century efficiency. Edwin Reynolds retired from the company in 1901, but continued in a consulting role to oversee the factory's construction. He worked as a consultant for Allis-Chalmers until his death in 1909.

By 1905 this new West Allis plant provided over 1.4 million square feet of floor space—or just about the same as the 1.45 million square feet of the four 1901 merger members together. It covered one hundred and thirteen acres and employed more than three thousand workers. It is doubtful whether any other plant in the US at the time was better designed, tooled, and equipped for building heavy machinery.

The West Allis plant, which was also the main office for the company, was like a small city. Utility services required for such a manufacturing complex were equal to those needed to supply a community of 50,000 people. There were twenty-one miles of railroad tracks and five miles of roadways in the plant's 160 acres. The facility also included hospital units equipped to give first aid, and a main hospital that offered complete physical examinations supervised by full-time physicians. In addition, the company maintained a fire department operated by trained members of the plant's protection force.

The merger strengthened and expanded the company's product offerings of steam engines, pumps, compressors, blowers, mine hoists and ore processing machinery lines. In 1904, Allis-Chalmers acquired the Bullock Electrical Manufacturing Company of Cincinnati, Ohio. With this acquisition, the company was able to furnish electric generators for its steam engines, hydroelectric turbines, gas engines and steam turbines.

The company struggled financially and entered receivership in 1912, emerging in early 1913 as the Allis-Chalmers Manufacturing Company with General Otto H. Falk as its president. Falk, who retired from the Wisconsin National Guard in 1911 with the rank of Brigadier General, was related to the Falk Company of Milwaukee—which his brother Herman chaired. Falk immediately set out to consolidate manufacturing operations, closing four factories including the former Fraser & Chalmers plant in Chicago, as well as the Reliance Works in Milwaukee's Walker's Point neighborhood.[237]

By 1914, Allis-Chalmers entered the agricultural equipment market, with the introduction of is first farm tractor. The company also added construction machinery, lift trucks, asphalt plants, self-propelled combines, and diesel engines. It also produced transformers, circuit breakers and metalclad switchgear. By the First World War, Allis-Chalmers received a subcontract to build shell casings. However, the German-American Alliance was critical of this activity because of Falk's German heritage.

Under General Falk's leadership, the newly restricted Allis-Chalmers continued to grow. Numerous acquisitions occurred during the 1920s and 1930s and Allis-Chalmers became one of the largest manufacturing companies in the United States.

The onset of the Second World War caused the company to expand into the production of war material, using Allis-Chalmers tools and expertise. Significant during this era was the company's contributions to the "Manhattan District Project"—the country's secret project to design and build an atomic bomb. Following the war, the company continued to grow in both product range and engineering experience.

Allis-Chalmers was incredibly innovative throughout the first half of the 1900s. This section highlights just some of their most notable engineering accomplishments.

This Allis-Chalmers steam-pumping engine is maintained in almost museum-like condition by the City of Jacksonville's (Florida) Water Department. It was designed by Irving Reynolds and installed in 1917. Photograph was taken by the author, 2015.

AUTOGENOUS MILLS

The Involvement of Allis-Chalmers Corp. in the mining industry goes back to the late 1800s when the E.P. Allis Co. was building stamping mills, pumps, and hoists for the industry. In the 1901 merger of Allis with three other companies, the Gates Iron Works of Chicago added, as part of its product lines, both pebble mills and ball mills for grinding the mined products to manageable size.

Autogenous grinding, which can be also called rock-grinding, uses ore or rock from the same deposit as the mill feed as grinding media. This method contrasts with using steel rods or iron balls as grinding media, or natural or manufactured iron, free pebbles, or other shapes and materials foreign to the mill feed.

Autogenous mills can be used at various stages in the grinding process, whether primary, secondary, or intermediate, for rock grinding. Historically, secondary autogenous grinding, also called pebble milling, dates back to the 1900s with installations in South Africa and Central America. A bulletin published by Allis-Chalmers in about 1917 discussing pebble mills refers to the possibility of using ore instead of flint pebbles as grinding media. Since the early 1950s, there have been a number of installations made in Canada using secondary autogenous grinding.

Development work on large-diameter grinding mills capable of grinding either primary crusher discharge or run of mine ore has established the applicability of autogenous grinding to the primary stage. An operator's dream come true: Big mills do in a single stage the work of two or three stages of crushing, rod milling, and sometimes ball milling. This is called primary autogenous grinding. Applications have been on Canadian iron ores and asbestos rock, for example, along with installations for grinding iron ore in the United States as well as copper both in the US and abroad.

The advantages of autogenous grinding mills are a reduction in grinding costs, a simplification of process flows, and the reduction in the production of extreme fines on the reduced product.

Allis-Chalmers built and installed six of the world's largest autogenous mills for use at a Minnesota taconite mining operation.

This huge autogenous grinding mill towers over an inspector at Allis-Chalmers as he checks the fully assembled unit in preparation for testing at the Company's West Allis plant.

Innovations in Hydroelectric Generation

Allis-Chalmers announced its entry into the hydraulic turbine business in 1904, hiring Clemens Herschel to support the effort. Herschel was regarded as one of the world's finest hydraulic engineers. The company built its first turbine a year later—a Francis type wheel rated at 1,250 horsepower with an 88-foot head for the Rochester Railway & Light Company.[aaa]

In 1915, the company installed a test facility to support continued development of hydraulic equipment, followed by more extensive test equipment in succeeding years.

In 1919, Allis-Chalmers was awarded an order to provide a new unit for the Niagara Falls hydroelectric power plant. This was not the first hydroelectric unit installed at Niagara Falls—units manufactured by Leffell and Westinghouse had already been installed. However, Allis-Chalmers's 1919 unit was so successful that an order for a larger unit was soon received. In 1923, Allis-Chalmers installed what was then the largest hydroelectric unit in the world. Designed by hydraulic engineer, William Monroe White, it had a rating of 70,000 horsepower—later uprated to 75,000 hp.

The first Allis-Chalmers unit installed at the Niagara Falls station was rated at 37,500 horsepower. Soon the company received an order for an even larger unit.

This iconic photograph provides a perspective of the massive size of the spiral casing for the second Allis-Chalmers unit for Niagara Falls.

The casing was preassembled on the shop floor in West Allis. Notice the vehicle parked adjacent to the casing at the lower right of the photograph.

The spiral casing directs water through wicket gates, where it drives the Francis-style water wheel. The photograph, from late 1922 or early 1923, is courtesy of Allis-Chalmers.

[aaa] This unit operated successfully until 1952.

In the early 1930s, Allis-Chalmers was selected to design and manufacture hydroelectric units for the Hoover Dam project (initially referred to as the Boulder Dam). These units eclipsed the size of the massive Niagara hydroelectric units. Built for the Public Works Administration during the depths of the Depression, Allis-Chalmers provided the first four units—each rated at 115,000 horsepower. The first units went into operation in 1936. At full load, each of the Francis-type hydro wheels discharges three-thousand cubic feet of water each second.

By 1940, Allis-Chalmers had constructed seven of the eleven units at Hoover Dam. From 1939 until 1949, the Hoover Power Plant was the largest hydroelectric power plant in the world. The seventeenth unit at the dam was installed in 1961. The hydroelectric units provide a total capacity of 1,850,000 horsepower, or about 1,380 megawatts.

In September 1940, tourists gather around one of the generators in the Nevada wing of the powerhouse of Boulder Dam (later Hoover Dam) to hear its operation explained. The generators extend above the powerhouse floor. The spiral casings are located below the floor, encased in concrete. The Francis-type water wheel assemblies are located within the spiral casings. Photograph taken by a US Bureau of Reclamation employee.

Allis-Chalmers was one of the few companies that had the expertise to produce hydroelectric turbines with the capability of handling the full spectrum of hydrostatic heads. Their Kaplan, Francis and Pelton styles turbines allowed them to match hydro equipment to meet the needs of specific dam installations in an optimum manner.

Allis-Chalmers engineers were well regarded for their hydraulic design expertise. In addition to Clemens Herschel and William Monroe White, the company also employed Arnold Pfau and Forrest Nagler. Pfau came to Allis-Chalmers from Escher-Wyss in Switzerland and was a "recognized expert on butterfly valves and governor design." Forrest Nagler joined Allis-Chalmers in 1906 and made significant contributions to fixed blade propeller runners of the Kaplan runner style.[238]

Nikola Tesla's Disk Turbine

Allis-Chalmers introduced numerous successful innovations into the marketplace, many of which are discussed in this book. The disk turbine, however, was not one of them. Yet its development is interesting, largely because its inventor was Nikola Tesla.

Tesla had gained incredible fame for his work in electricity, before his association with Allis-Chalmers. He is credited with the development of the alternating current system, which made electrification practical, since it made it economic to transmit electricity long distances. Almost all of Tesla's early inventions centered on electrical transmission, electric motors and the Tesla coil. He is credited with hundreds of patents covering a wide-range of innovation.

He also had one colossal failure—Tesla's vision of a global wireless transmission tower proved to be his undoing and led to his bankruptcy.

Later on in Tesla's career, he began exploring other mechanisms, one of which was the Tesla disk turbine. A Tesla turbine consists of a set of smooth disks located within a tight casing, with nozzles applying a moving fluid at the edge of the disk. The fluid drags on the disk by means of viscosity and the adhesion of the surface layer of the fluid. As the fluid slows and adds energy to the disks, it spirals into the center exhaust.

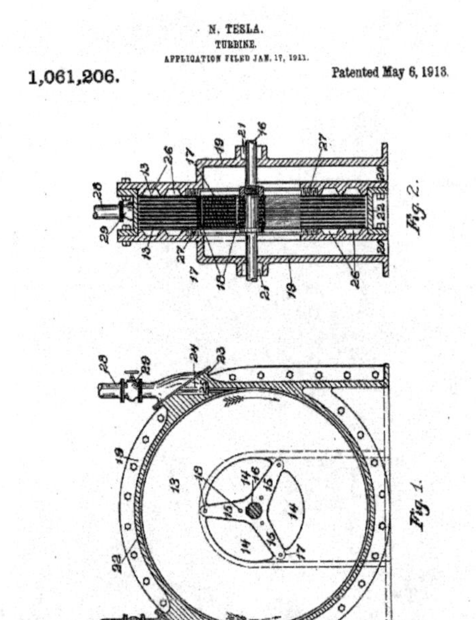

Tesla announced his turbine in 1902, describing it as a small, simple and powerful device. He completed his first model in 1906 and filed for a patent in 1911. He continued to refine his design and sought resources to build larger models. After approaching a number of companies, he entered into an agreement with Allis-Chalmers to develop the turbine.

Drawing from Tesla's 1913 Patent Filing for his Turbine (US1061206 A). Two inlets were shown, which would allow the turbine to be rotated in either direction.

Between 1919 and 1922, Tesla was commissioned to work with engineers at Allis-Chalmers to build and test prototypes. Tesla, however, had a different idea. While a report was being prepared by company engineers, he went directly to the president and to the company's board of directors and "sold" his ideas before the company's engineers could issue their report. Tesla received authorization to proceed.[239]

Three turbines were built. Two had twenty disks that were eighteen inches in diameter. A larger turbine was also built, consisting of fifteen disks that were sixty inches in diameter.

Hans Dahlstrand, a consulting engineer for the Allis-Chalmers steam turbine department, reported on the effort to build the larger turbine:

> *We also built a 500 kW steam turbine to operate at 3,600 revolutions. The turbine rotor consisted of fifteen disks 60 inches in diameter and one-eighth inch thick. The disks were placed approximately one-eighth inch apart. The unit was tested by connecting to a generator. The maximum mechanical efficiency obtained on this unit was approximately 38 percent when operating at steam pressure of approximately 80 pounds absolute and a back pressure of approximately 3 pounds absolute and 100 degrees F superheat at the inlet. When the steam pressure was increased above that given the mechanical efficiency dropped, consequently the design of these turbines was of such a nature that in order to obtain maximum efficiency at high pressure, it would have been necessary to have more than one turbine in series.*[240]

When the unit was dismantled for inspection, it was discovered that the disks had distorted significantly. The engineering staff concluded that the discs would ultimately have failed due to creep rupture—essentially the disks would have continued to expand due to high rotating speed and temperature, until a failure occurred. The disks also tended to warp, and the engineering staff noted vibration problems.

Whether it was the comparatively low efficiency, or the concerns for disc rupture and vibration, Allis-Chalmers engineers were apparently critical of the concept. However, it does not appear that they were reluctant to work out the issues. Higher strength steels, for example, would likely have mitigated the disk creep issue.

Tesla continued to advocate his turbine design and ultimately proposed a gas-turbine version. However, he never provided sufficient information to Allis-Chalmers about the concept to allow them to attempt construction.

Tesla apparently terminated the relationship with the company in 1922—essentially walking away from the joint effort.

It is unknown what happened to the three prototype Tesla steam turbines, nor could an image be found of any of them. Tesla continued to advocate for his gas-turbine version, however, it is not known if one was ever constructed.

The Tesla disk design has been used for pumping viscous and highly abrasive liquids. The pumps, known as "boundary layer disk pumps," continue to be manufactured and sold by a company known as DISCFLO.

MILWAUKEE RIVERSIDE PUMPING STATION

Milwaukee's Riverside Pumping Station was placed in service in 1924 to alleviate a severe water shortage that occurred shortly after the end of First World War. Prior to the installation of new pumps at the Riverside station, the Milwaukee Water Works had only one major pumping station—the North Point Station described in Chapter 2, which used equipment designed by the E.P. Allis Company—the predecessor of the Allis Chalmers Corporation.

Like the North Point Station before it, Milwaukee's Riverside Pumping Station also used triple-expansion steam engines to drive its water pumps. The engines were called "triple expansion" because each of the engines had three cylinders. Steam at high pressure was expanded in the smallest cylinder, which pushed down a piston to drive the water pump. The exhaust steam from the first cylinder was then used to power the second and third cylinders, respectively.

These were massive pumping engines—five stories tall and located in a 59,000-square-foot building. A viewing gallery was provided for visitors. These ornate engines were an impressive sight. Those who saw the triple-expansion engines marveled at seeing one-hundred tons of machinery at work while hearing only the clinking of the valves.

The three pumping engines had a capacity of sixty-nine million gallons per day.

When pumping engine number one was tested in 1924 during its acceptance, it broke the world's record for efficiency for a vertical triple-expansion Corliss engine. The previous record had been established by a similar engine in Cleveland.

Water for the pumps at the Milwaukee Riverside Pumping Station was initially pumped directly from Lake Michigan through a reinforced concrete tunnel. After Milwaukee's Linnwood Water Treatment Plant was completed in 1938, the water supply was fully treated and filtered before being conveyed through the tunnel to the Riverside Station.

By the 1920s, centrifugal water pumps driven by electric motors were beginning to replace steam engine-driven water pumps. The pumping engines for Milwaukee's Riverside Pumping Station were among the last such pumping steam engines constructed. Author Elmer W. Becker wrote in his book, *A Century of Milwaukee Water,* that city engineers disagreed over the type of pumps for the new plant. While many favored centrifugal pumps, they were relatively new for city water service and Milwaukee had considerable favorable experience with their previous steam engine-driven water pumps supplied by Allis. It didn't hurt that Allis-Chalmers was a major Milwaukee-area manufacturing concern.[241]

At the same time, steam turbines were replacing steam engines for electric generation. The last steam engine constructed by Allis-Chalmers was completed in 1930.[242]

The City of Milwaukee eventually replaced its steam-powered water pumps with centrifugal pumps powered by electric motors. The steam engines at the Riverside Pumping Station were taken out of service in 1968 and replaced with nine centrifugal pumps with a combined capacity of 240 million gallons a day. They are now remotely controlled from the Linnwood Water Treatment Plant, as well as from the Howard Avenue Water Treatment Plant.[243]

A photograph of one of the three E.P. Allis pumping engines at Milwaukee's Riverside Pumping Station. When pumping engine number one was tested in 1924 during its acceptance test, it broke the world's record for efficiency for a vertical triple-expansion Corliss engine. Photograph from the author's collection.

WORLD'S LARGEST GYRATORY CRUSHERS

The Gates ore crushers, brought to Allis-Chalmers in a merger in 1901, had long constituted the standard crusher line of the company. By the 1920s, however, this design was outdated by newer crushers also developed by Allis-Chalmers, which allowed the company to remain the principal producer of large ore crushers. Because of its extensive engineering experience, the company could handle specialized orders, including those for foreign installation.

The shipment of two 500-ton Superior-McCully gyratory crushers to Chile in 1926 is an example. Designed and built for the Chile Exploration Company, a subsidiary of Anaconda Copper Mining Company, these machines had to be specially constructed. On November 8, the Company's Executive Committee adjourned to inspect the crusher that had been erected in the shop, "it being the largest and most powerful gyratory crusher ever built." The crushers had receiver openings of 60 by 150 inches, and practically the entire machine was constructed of cast steel for excessively hard ore. The steel castings and forgings for the crushers were among the largest ever made in the United States.

To meet the special transportation problems, the machines were built in sections. The lower main frame of each crusher was made in two parts, with a combined weight of 88,000 pounds. The top shell was in two horizontal sections, with the lower ring weighing 114,900 pounds and the upper part halved in two sections, each weighing 92,000 pounds. All of these sections had to be handled with a special crane on the steamer *Chilcop*, special arrangements were made to receive them at the port of Mejillones, and special cars were built in Chile to transport the parts by a narrow gauge mountain railroad to an altitude of 9,500 feet. This was a specialized job from beginning to end.

The 1926 Chilean crushers were no more than preparation for the installation of four sixty-inch all-steel gyratory crushers in the same country three years later. Each of these crushers weighed one million pounds. To be transported on the narrow gauge railroads, no single piece could weigh more than sixty tons. When installed, each crusher could handle 2,500 tons of copper ore per hour. The imposing sight of twenty-five railroad cars transporting this shipment from West Allis to New York merely indicated the size of order and degree of specialization that the Allis-Chalmers Manufacturing Company was prepared to undertake in the twenties.

By 1952, however, the Company broke its own record with another giant crusher that took two years to build at the West Allis plant. It was built for processing of taconite ore on the Mesabi Range in eastern Minnesota. This Superior gyratory crusher had a capacity of crushing taconite ore at the rate of 3,500 tons an hour from pieces as large as five feet in one dimension to ten inches in final size.

A world's record was achieved in 1926 when Allis-Chalmers built two 60-inch all-steel Superior gyratory crushers for crushing copper ore in the Andes Mountains of Chile. Sectionalized for easier transportation, each complete crusher weighed one million pounds.

Allis-Chalmers Multiple V-Belt Drives

During the early 1920s, an age-old problem of power transmission continued to trouble textile manufacturers because the flat belts slipped and chains jerked, tearing delicate threads of the cloth and tying up production, as well as dripping oil onto the thread, spools, and cloth. These problems were brought to the attention of Walter Geist, the engineer in charge of flourmill power transmission systems at Allis-Chalmers. After working unsuccessfully for over two years with rope drives, Geist satisfactorily solved the problem of power transmission. He discovered that the difficulty came from ropes or belts that were designed to ride the bottom of the grooves. His solution to the problem lay in the principle of wedge contact. His experiments in developing a satisfactory textile mill drive centered on the concept of a V-shaped belt and sheaves. Geist patented in 1928 a design for using them in multiples, similar to the multi-grooved pulleys of the English rope drive system.

Multiple V-Belt Drives, pioneered by Allis-Chalmers, eliminated the problems involved in the use of conventional long belt drives attached to overhead line shafts. Motors could be connected to individual machines by short center drives and the rubber V-Belt acted as a shock-absorbing medium between the motor and the machine. Photograph courtesy of Allis-Chalmers Corporation.

The potential advantages of the multiple V-Belt Drive were of great significance to manufacturers. In factory layouts it meant a valuable saving of space and the elimination of line-shafting with all its handicaps, such as shutting out light, endangering operators, and taking up excessive space. The location of a machine could now be determined by the pattern of manufacturing rather than by the source of power. The potential efficiency of transmission was very high, running between 96 and 99 percent on well-engineered drives. Furthermore, belt slippage was prevented because each one of the multiple belts, either when starting or when under increased load, could momentarily sink deeper into its groove as well as extend longitudinally. The reduced tension made possible by this drive allowed for longer life in the motor bearings, too.

The Multiple V-Belt Drive, or "Texrope" drive as Allis-Chalmers called it, was also much safer for employees, since it was compact and could be easily and inexpensively placed behind a guard. Moreover, the drive was silent. Just when the effects of noise on employee efficiency were beginning to be appreciated, real strides towards noise elimination were made possible by the multiple V-Belt Drive of Allis-Chalmers.

In the beginning, the drive was planned for a maximum capacity of fifteen horsepower, but soon the company ventured into larger capacities, and by 1934 had built a drive with a capacity of 2,000 horsepower. Very simply, the introduction of the multiple, short-center V-Belt Drive revolutionized mechanical power transmission practice. It is was generally recognized as one of the major steps forward in the history of power transmission, eliminating from the factory the forest of single flat-belt drives.

First Rubber-Tired Tractor

In the early decades of the 20th century, the tractor had not been universally adopted for farm use and was not in a position to replace the horse on the farm as the automobile and the truck had replaced him on the highway. This was due, in part, to the limitations and inefficiency of the type of wheel equipment used. The lugs on the steel wheels damaged meadows, orchards and barnyards. Signs stating "Tractors with Lugs Prohibited" were appearing on most well surfaced roads.

The sheer inefficiency of the lug-type tractor wheel is indicated in that the tractive efficiency—the ratio between the power delivered at the drawbar and the power produced by the engine under field conditions—varied from a low of 40 percent to a high of about 65 percent. Very simply, it took power to push the lugs in, and power to pull them out. The result was that, even on level ground, the tractor was compelled to constantly climb a rather steep grade, so to speak. As the speed of the tractor was increased, more of the total horsepower was required merely to move the tractor. At higher speeds, the tractor with steel lugs tended to utilize almost the total output of the engine, leaving little power for useful work. The consequence was that conventional tractor work had to be done slowly, inefficiently, and with a high rate of fuel consumption.

Engineers had flirted with the idea of putting rubber tires on tractors. Experiments were conducted with both hard rubber tires and high-pressure pneumatic tires, similar to those used on trucks. But when attempts were made to plow with this equipment, it was found that the tractor could perform only under the most favorable ground conditions and it was essentially useless on wet ground. However, the tractor engineering staff at Allis-Chalmers, with the assistance of Firestone Tires, finally arrived at the solution to the problem. They conceived of the idea of a low-pressure tire with flexible casings that would allow the tread to spread out and distribute the load, thus giving the needed traction. The development of this low-pressure air tire was a significant breakthrough for the entire industry. As The Farm Implement News put it on October 13, 1932, "Just about the time this industry seems to have dropped into a rut and reached a static joint with no outstanding developments in sight, something arises to change its course. Rubber may be the pivot of the next turn."

A testing program was established. Allis-Chalmers tractors equipped with air tires were put on selected farms so that the tests could be conducted under a wide variety of work operations and conditions. The reports were uniformly enthusiastic. Rubber actually seemed tougher than steel in this sort of use. The tractors rode more comfortably and the air tires were easier on the tools used. They provided greater fuel economy, presented greater tractive surface, and most importantly, permitted greater speed of travel in the fields. The farm tractor, equipped with pneumatic tires was no longer limited in use, but had become a general utility machine to be employed wherever power was required.

Tests by the University of Nebraska Tractor Test Laboratory found that rubber wheels resulted in a 25 percent improvement in fuel economy, and allowed for smoother, faster driving with less wear on parts—and on the driver.

While the field tests were very successful, sales were lackluster. The country was in the midst of the Great Depression and farms just didn't have the funds to purchase new equipment. In part to publicize the rubber-tired tractor, Allis Chalmers participated in tractor tests. Rubber-tired tractors started winning plowing competitions—sometimes by a wide margin. The success with these tests caused the company's advertising department to enter tractors into high-speed races. At a time when most tractors were limited to about ten miles per hour, this seemed nearly comical to man farmers. But the idea caught on.

The demonstration racing tractors had modified drive ratios. Prior to this time tractor transmissions were not designed to accommodate high speeds.

In the first demonstration race, a driver took an Allis-Chalmers tractor around the track at 35 miles per hour—apparently stunning the crowd. They were so impressed with the gleaming A-C tractors with its big Firestone tires that the company decided to hire Barney Oldfield. Oldfield achieved 64 miles per hour at a country fair, resulting in great publicity.

Allis-Chalmers hired additional drivers to barnstorm the Midwest, participating at State Fairs to demonstrate the speed capabilities of rubber-tired tractors.

The other tractor manufacturers and rubber companies quickly joined A-C in producing rubber-tired tractors.

In the promotional photograph above, Ab Jenkins is congratulated by Harvey Firestone. As the sign states, Jenkins set a new world's record of 65.45 mph. Jenkins ultimately achieved an average speed of 68 mph on an Allis-Chalmers tractor at the Bonneville Salt Flats in Utah over a measured course.

Allis Chalmers Contributions to the Manhattan District Project

Like most Milwaukee-area manufacturing companies, Allis Chalmers turned its attention to military applications during the Second World War. Given its size and manufacturing capabilities, the company played a major role in US wartime production. In the late 1930s through mid-1940s, Allis-Chalmers manufactured Liberty ship steam engines, steam turbines, generators, and electric motors for naval ships. It produced artillery tractors for army use, as well as tractors for the Corp of Engineers for building airfields, building roads, constructing fortifications and for other applications. In addition, the company produced an extensive assortment of electrical switches and controls for various applications.

Allis-Chalmers was also one of the companies contracted to build equipment for the Manhattan District Project—the top secret US project to produce the atomic bomb. The company's experience in mining and the manufacture of large machinery made it a logical choice for uranium processing equipment. The company was awarded over sixty separate orders for equipment associated with the effort, including heavy operating machinery for the diffusion and enrichment of uranium.[244]

Forest Nagler served as chief engineer for Allis-Chalmers on the project. Nagler interfaced with engineers from the Stone & Webster consulting firm, which was selected as the main subcontractor to separate large quantities of fissionable uranium-235. Nagler described the contributions of Allis-Chalmers in this way:

> *We built most, if not all, of the heavy production equipment used in the magnetic separation method evolving from the experimental application of the cyclotron, and later built a large number of pumps which formed the heavy operating machinery in the diffusion process which was so outstandingly successful.*

The massive undertaking actually employed three different processes to separate uranium-235 from natural uranium. U-235 is an isotope that makes up less than one percent (0.7%) of uranium. Electromagnetic separation, gaseous diffusion, and thermal diffusion were all successfully employed and used in sequence to obtain the required level of uranium enrichment.[bbb]

It is reported that Allis-Chalmers produced more equipment by weight for the Manhattan District Project than any other company. Part of the company's Halley plant was constructed specifically for the effort.[ccc]

In order to help ensure secrecy, the workers and supervisors were not told the nature of what they were working on. Movement was restricted and workers were limited to access to their own workspace. Certain areas of the plant were walled off, so that others could not observe the assembled equipment.

General Leslie Groves, the director of the Manhattan Project, visited the plant frequently to oversee operations.

[bbb] Material from the thermal diffusion process was fed into the gaseous diffusion process in the K-25 plant, which produced a product enriched to about 23%. This was, in turn, fed into Y-12 electromagnetic separation plant, which boosted it to about 89%; sufficient for nuclear weapons.

[ccc] The Hawley plant was an extension of the West Allis Works of Allis-Chalmers that expanded its footprint east to North Hawley Road.

Electromagnetic Separation Process Equipment

Allis-Chalmers made produced and wound magnetic coils for the "calutrons"[ddd] used in what was known as the Y-12 plant at Oak Ridge, Tennessee. The company also provided compressors designed to handle the uranium hexafluoride gas used in the process.

It was calculated that five thousand tons of copper would be required to create the huge magnetic field to deflect charged particles. Because of its use during the war, copper was in extremely short supply. Rather than contribute to the existing material shortage, a decision was made to use silver. Fourteen thousand, seven hundred tons of coinage silver was loaned from the US Treasury and transferred to the Allis-Chalmers factory, where the magnetic coils were fabricated. The silver was formed into forty-foot-long, half-inch-thick, strips, which were wound into magnetic coils.[eee]

The calutrons, which were assembled at the Y-12 plant, required an exorbitant amount of energy—Oak Ridge used one-seventh of all the electricity produced in the United States at the time.[245]

Some of the magnetic coils forming the calutrons failed during initial tests in late October 1943. Later that year, General Groves ordered some partially used coils be returned to Allis-Chalmers for cleaning and inspection.

Silver dust on equipment and the factory floor was swept and collected frequently. Following the war, all machinery was painstakingly disassembled and cleaned and the wooden flooring was removed and burned to recover silver remnants and return all of the silver to the Treasury.

Gaseous Diffusion Process Equipment

The gaseous diffusion process required pumps that could meet stringent requirements, including resistance from corrosion from uranium hexafluoride, a gas twelve times as dense as air. Several pump manufacturers, including Ingersoll Rand, Clark Compressor and Worthington Pump, were approached but all declined to participate, stating that it couldn't be done. Allis-Chalmers, however, agreed to build a new factory to produce the pumps, even though the final design was uncertain. Eventually, Judson Swearingen at the Elliott Company came up with a revolutionary design, with seals that would contain the gas without leakage. This design was released for construction by Allis-Chalmers.[246]

Employees at many defense plants in the United States were forbidden to talk about their work. This was particularly true at the Allis-Chalmers facilities where many of its 25,000 workers were told they could not discuss their work activities with anyone. Posters provided reminders in the factory lunchrooms and billboards were installed at factory gates.

It is unlikely that any of the workers knew they were working on the atomic bombs program, even though they played a critical role in ensuring its success.

[ddd] A "calutron" is device that used large electromagnets to separate uranium isotopes from uranium ore. The term was derived from CALifornia University cycloTRON. The device was principally designed by Ernest O. Lawrence, working at the University of California-Berkeley. Based on prior work with cyclotrons, he determined that when an electrically-charged atom was placed in a magnetic field it would trace a circular path with a radius determined by the atom's mass. Since U-235 was lighter than U-238, he theorized that it could be isolated by placing a collecting pocket in its path.

[eee] At today's prices, the silver bullion used in the effort is worth an estimated $7 Billion. It was loaned to the project with the expectation that it would all be returned to the US Treasury. Special accountants were assigned to the project to track the silver and ensure its recovery. Almost all the silver was recovered.

Photograph of the interior of the Allis-Chalmers erection floor, showing winding of magnetic coils for the magnetic separation process. Notice the temporary interior wall, designed to limit observation. Photograph courtesy of the Milwaukee County Historical Society.

A billboard outside the Hawley plant in 1944 was designed to emphasize the importance of security. Workers were forbidden to discuss their work, even though most had no idea of what they were working on. Photograph from Department of Energy by Ed Westcott.

Kiln Developments for the Cement Industry

Allis-Chalmers has been making equipment for the cement manufacturing industry since the early 1900s. Cement making equipment became part of Allis-Chalmers product lines with the merger of three companies with the Edward P. Allis Company in 1901 to form the Allis-Chalmers Company; in particular the Gates Iron Works of Chicago, which brought to the merger a line crushers, mining machinery, and cement making machinery.

By the 1920s, the cement equipment business became increasingly important at the company. During the period of 1920 to 1929, Allis-Chalmers probably did as much as 75 percent of the domestic business and was the prominent American firm in the foreign cement equipment business as well.

In the 1920s, there was an increased use of cement in building construction. This occurred at the same time that the United States began to pull its every-growing number of automobiles out of the mud by placing them on paved roads. Then, too, American builders were finally dissuaded of the idea that German cement was superior to American Portland cement. Suspicious of the alleged superiority of European cement, Ray C. Newhouse bought a complete set of German testing equipment and installed it in the Allis-Chalmers laboratory next to American testing machinery. Samples of cement, both domestic and foreign, were tested by the two methods. It was found that the technique used by the German testing apparatus, and not the cement itself, was responsible for the apparently higher strength. In fact, Newhouse proved that the strength of American cement was equal to or even greater than the German cement, which had previously been imported into the United States.

In 1928, six new Portland cement plants were built in the United States, five of them completely equipped with Allis-Chalmers products. Allis-Chalmers also exported a great deal of cement making equipment to Japan, the Far East, and South Africa.

In 1949, Dr. O.G. Lellep, a German inventor who had developed a new process involving the use of otherwise wasted hot exhaust kiln gasses to preheat the incoming raw cement mix, joined Allis-Chalmers as a consultant and arranged a licensing agreement for his process. This innovation substantially reduced fuel requirements of the kiln itself. Later Dr. Lellep developed a second pass of gasses through the bed of incoming material, virtually eliminating dust in the exhaust gasses. By 1958, a scale model pilot plant proved the applicability of this system to iron ore pelletizing for taconite ore, and it is now used by iron mining companies worldwide.

Allis-Chalmers kilns were also used for a direct reduction system for producing sponge iron and, experimentally, for coal gasification to produce a synfuel for generating electric power by utilities.

This photograph from 1964 shows large Allis-Chalmers kiln segments being loaded aboard the lake steamer *Maitland* at Jones Island. The kiln was made for the Fox River Cement Company in Festus, Missouri. Fourteen sections, each weighing 23 to 83 tons, made up the kiln, with a total weight of 960 tons. The steamer took the kiln sections to the Chicago area for transfer to shallow draft river barges, which floated the equipment down the Illinois Waterway to the Mississippi. Photograph courtesy of Allis-Chalmers Corporation.

Hiwassee Dam Unit 2 Reversible Pump-Turbine

The integration of pump and turbine for the Hiwassee Dam was the first of many reversible pump-turbines to be installed in power plant systems in the United States. When installed it was the largest and most powerful reversible pump-turbine in the world.

During periods of low electrical demand, the unit powers a pump to add water to the reservoir. When electrical demand is high, the unit operates as a conventional electric-generating hydro turbine. Called a "pump storage" hydro unit, this innovation to the Tennessee Valley Authority's system offered significant economies in the generation of electrical energy.

The pump-turbine unit was designed by engineers of the Tennessee Valley Authority and the Allis-Chalmers Company. When built by Allis-Chalmers Company, it was the first of its kind to be installed in the United States.

Hiwassee Dam is a concrete gravity overflow dam that is over three-hundred feet high and 1,376 feet long. It has a generating capacity of 185,000 kilowatts. The Unit 2 reversible pump-turbine was installed in 1956. It is located in Cherokee County, North Carolina.

The pump-turbine has a generator rating of 59.5 megawatts and a capacity of 3,900 cubic feet per second at a 205-foot head. During commissioning, extensive performance tests were performed for both pumping and generating modes of operation. These tests revealed a turbine efficiency of 88.4 percent and a pump efficiency of 90.0 percent—meeting or exceeding the guaranteed performance levels.[247]

This Allis-Chalmers photograph, taken on their assembly floor, was obviously meant to convey the immense size of the Hiwassee unit. The three men are standing inside the generator stator. The large device located behind and to the left of the stator is the stacked wicket gate. A-C had the capability of assembling large hydraulic turbines and generators prior to shipment, to ensure that assembly would go smoothly in the field.

Pump storage hydroelectric plants had been in use for electrical energy storage in Europe for some years prior to operation of Hiwassee Unit 2. However, these plants used either completely separate motor-driven pumps and turbine-generators, or a separate pump, turbine, and a generator/motor all on a single shaft. They did not use reversible pump-turbines, which are now standard for pumped storage plants.

Pump storage hydroelectric power plants have become increasingly significant due to the increasing amount of solar and wind power being generated in the United States, as well as the rest of the world. Reversible pump-turbines such as the Hiwassee unit use renewable energy from solar and wind, as well as other generating resources, to pump water into the reservoir during times when electricity is plentiful. The stored water is then used to generate electricity when electric demand is high.

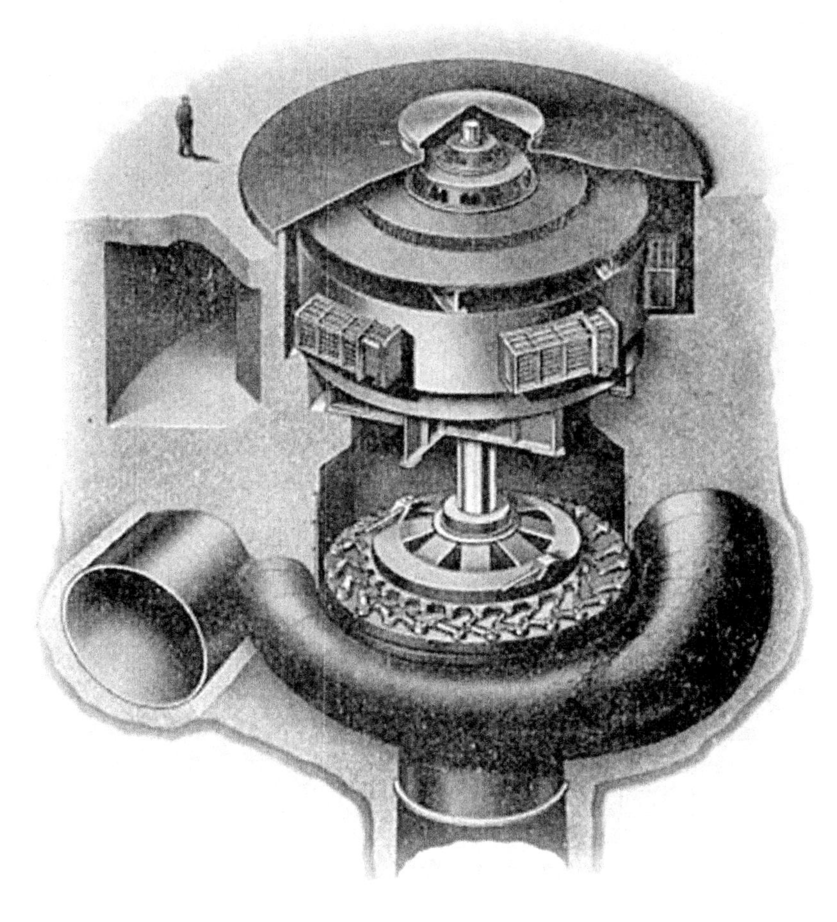

This is an artist's drawing of the Hiwassee reversible pump-turbine and motor/generator. It was designated a landmark by the American Society of Mechanical Engineers in 1981.

Fuel Cell Tractor

A new source of electrical power—fuel cells—came out of the laboratory to power a vehicle for the first time on October 15, 1959, when Allis-Chalmers demonstrated its fuel cell-powered tractor in Milwaukee. The research vehicle developed about 3,000 pounds of drawbar pull, enough to pull a multiple-bottom plow.

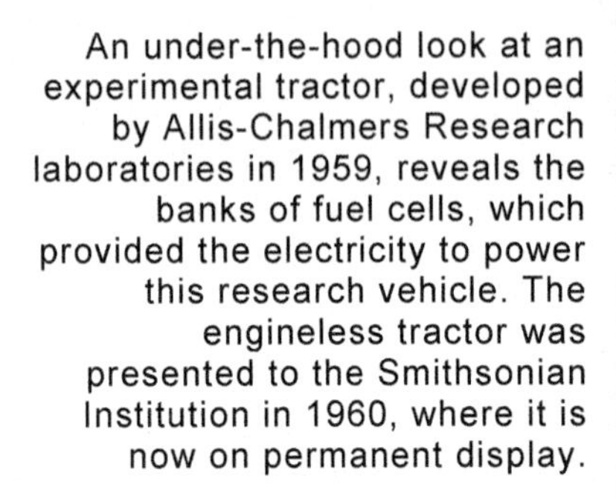

An under-the-hood look at an experimental tractor, developed by Allis-Chalmers Research laboratories in 1959, reveals the banks of fuel cells, which provided the electricity to power this research vehicle. The engineless tractor was presented to the Smithsonian Institution in 1960, where it is now on permanent display.

The electricity that powered the tractor came from 1,008 individual fuel cells. These were joined in 112 units of nine cells each. The 112 units were arranged in four banks and electricity could be taken from any combination of the banks. A mixture of gasses—largely propane—fueled the cells. The gasses were fed into the cells through a system of tubing; and, once in the cells, the gasses, including oxygen, reacted in an electrolyte. A catalyst coating the electrodes of each cell aided the reaction.

The chemical reactions within the cells caused a direct current of electricity to flow through an external circuit, which was connected by bus bar to a standard controller that allowed both forward and reverse travel. A compact controller, measuring eight by eleven by twenty-one inches, regulated the electricity supplied to a standard twenty-horsepower D.C. motor that was also supplied by Allis-Chalmers.

Allis-Chalmers Research Division developed the prototype of the fuel cells used in the tractor. The company announced its fuel cell a little more than a year before this demonstration of the first self-contained system. Since then, researchers developed larger, more efficient fuel cells. Indeed, fuel cells were used to supply electric power for the Apollo space vehicle in its flights to the moon. Due to project cutbacks in which Allis-Chalmers had been involved, the Company discontinued its work in fuel cells in 1970. The tractor, however, was put on display at the Smithsonian in Washington, D.C.

World's Largest Steam Turbine-Generator

Allis-Chalmers designed and built the world's first one-million-kilowatt steam turbine-generator unit for Consolidated Edison Company of New York City in the early 1960s.[fff] This massive power package went into service during 1965 at Consolidated Edison's Ravenswood Station, five years before one of equal capacity was built by any either manufacturer. The biggest such unit that Allis-Chalmers had built previously was rated at 400,000 kilowatts, several of which had been installed. Other manufacturers had built units rated at 620,000 Kilowatts at the time. Obviously the one million kilowatt unit was a tremendous step forward, involving a large increase in size that required numerous innovative design details. For example, it embodied six rows of 40-inch long exhaust blades which were of the freestanding, unlashed type—the first such advance in the U.S. by any manufacturer. Two European manufacturers had just introduced exhaust blades of this type, but of a shorter length. Many of the tips of these thousands of turbine blades spin faster than the speed of sound.

Because of the physical size of the parts in this unit, many of the previous design concepts common to steam turbine generators had to undergo a drastic change. In addition, inasmuch as the giant machine was of the cross-compound type and received steam from two large boilers (each four million pounds of steam per hour), an elaborate control system as specified by Con Ed had to be developed.

These are only some of the many design innovations that went into the building of this giant machine. Ellis P. Hansen was the manager of engineering in the Thermal Power Department in West Allis and overall responsibilities for the effort. However, the design of this unit was the combined result of the capabilities of a large number of talented engineers working together to make this possible. More than 3.5 million hours of labor by specially qualified workers were required to place this unit online. The unit was big enough to serve the residential electrical needs of 3,000,000 people.

The turbine-generator became known as "Big Allis," because of its size.

During a 1965 blackout in New York, the pumps providing lubricating oil to the turbine and generator bearings shut down due to loss of power, resulting in bearing failure and contributing to a lengthy outage. Power shortages during the outage led to significant notoriety for the unit.

[fff] One-million kilowatts is approximately 1.34 million horsepower.

The world's first one-million-kilowatt steam turbine-generator was designed and built by the Allis-Chalmers Corp. for installation in New York City. It was nicknamed, "Big Allis."

HARNISCHFEGER CORPORATION

Pawling and Harnischfeger's early history is discussed in Chapter 2. When fire destroyed their foundry building on April 15, 1903, the company quickly acquired a twenty-six-acre site located "out in the country" at 4400 West National Avenue in what is now the City of West Milwaukee. While the fire created a hardship, the company was growing and needed room to meet an expanding product line. Relocating to West Milwaukee provided that manufacturing space. The state-of-the-art facility also allowed the company to consolidate operations and to begin making parts that it previously had purchased from suppliers.

Henry Harnischfeger administrated the business affairs of the company and Alonzo Pawling oversaw engineering. But in 1911, Pawling's health began to fail and he asked his partner to buy out his interest in the enterprise. Upon Pawling's death three years later, the company was renamed simply the Harnischfeger Corporation. It retained the P&H logo as a trademark, however, because of customer recognition and out of respect for the company's co-founder.

The company struggled for a time financially. It was saved by a large order from Racine's J.I. Case Company for gasoline tractor engines.

Harnischfeger recognized the need to diversify its product line and introduced a ladder-type trenching machine, which was mounted on a four-wheel wagon and operated by a steam engine. While the machine was relatively unsophisticated, it was novel and automated the digging of trenches—far outpacing the manual use of picks and shovels. Soon, Harnischfeger introduced other trenching machines.

This drawing from a Harnischfeger brochure shows the P&H ladder-type trench excavator, which was one of Harnischfeger's first products for the construction and mining industry.

In 1910, company engineers designed a gasoline-powered dragline. It was a truck-mounted machine that could lift, pile-drive, clam and drag.

Harnischfeger also continued to introduce innovations in the overhead crane market, including a patented crane for hoisting and conveying lumber and similar products. The timing for new crane apparatus was good—with the advent of the First World War, the demand for new factory cranes increased dramatically. As demand for P&H products increased, Harnischfeger expanded the West Milwaukee factory. At the end of the 1920s, employment grew to fifteen-hundred.

In 1911, Pawling and Harnischfeger obtained a patent for a crane to be used for hoisting and conveying lumber and similar materials. Many cranes of the time were controlled from a crane-mounted operator cab, as shown here. Photograph from a Harnischfeger booklet to celebrate its 40th anniversary.

P&H also designed a backhoe and a shovel-type excavator that was mounted on tank treads. The products were popular for both construction excavation and mining applications. Soon the P&H brand was recognized as the pioneer in the excavator field.

The Great Depression hit the company hard. The market for cranes and excavation equipment dried up and the company lost money every year from 1931 to 1939. Harnischfeger was forced to deal with used equipment, as many of its cranes and excavators were returned for non-payment.

While the 1930s were difficult financially, the company continued to innovate. It introduced welded construction for its cranes and excavators—replacing riveted assemblies. Welded components resulted in stronger and lighter machines that also cost less to build. In order to increase efficiency, Harnischfeger patented and introduced an automatic welding system in 1936. It was described as "the first successful automatic welding machine using standard coated rods." The machines robotic arm grabbed weld rods from a magazine, moved it into position, and immediately initiated welds. It permitted higher welding currents and welding heats than manual welding.[248] Based on its favorable experience, Harnischfeger introduced its own line of welding machines and welding electrodes during the 1930s. For a time, it also manufactured diesel engines and prefabricated houses.

One of its most significant innovations was the introduction of an electromagnetic brake and control system, which it called Magnetorque®. It was designed by Harnischfeger engineers in 1946.

During the Second World War, the Harnischfeger plant operated at full capacity. Many of the workers who had left to join the service were replaced by women, who worked as welders and other jobs to build P&H cranes for lifting tanks and heavy artillery in defense plants, and to build cranes to position planes on aircraft carriers. The company spent millions on plant additions in order to meet the high demand.

In 1994, Harnischfeger purchased Joy Mining Machinery Company, which specializes in equipment for the underground mining industry. P&H Mining Equipment and Joy Mining Machinery become today's Joy Global in 2012. The surface equipment is branded with the P&H name, and the underground equipment is branded Joy. In mid-2016, Joy Global announced that it was being acquired by Tokyo-based Komatsu due to "challenging market conditions the company believes are likely to persist."

P&H Model 206 Excavator

In the early 1920s, Harnischfeger introduced its first excavator crane, the P&H Model 206 that was powered by a gasoline engine. It proved to be a highly popular digging machine. Weighing twenty-three tons and standing two stories high, it could move a ton of material with each pass of its three-quarters of a cubic yard dipper. It was offered with a Waukesha gasoline engine, a Buda diesel engine, or a P&H electric motor—depending upon customer requirements and preferences.

This drawing illustrated the 1922 patent for an excavator crane, powered by a gasoline engine (US Patent No. 1,570,108), filed by Lewis Wehner, chief engineer for Harnischfeger

Starting with this basic shovel design, Harnischfeger produced thousands of shovels for construction and mining operations since the early 1920s. With each evolution, P&H shovels became larger, more powerful and sophisticated.

The electric mining shovel pictured here, which is a descendent of the original P&H 206 excavator, has become a prominent Milwaukee-area landmark. It is a P&H 2100-series shovel manufactured by Harnischfeger in Milwaukee. The 2100 series line was first introduced in 1968 to meet the requirements of the coal, copper and iron ore industries. Approximately 380 have been built. This particular machine is a P&H 2100BL excavator rated at eighteen cubic yards dipper capacity. It was manufactured in 1979 for the Falcon coalmine in Jackson Kentucky. Three years later, it was purchased by Arch Minerals for another coalmine. In 1994, the shovel was acquired by Cleveland-Cliffs and refurbished for use in one of its iron ore mines in the Upper Peninsula. After nearly thirty years of service, the shovel was retired and placed on display at Joy Global's surface mining headquarters.

This P&H 2100BL electric mining shovel occupies a prominent place at the corner of National Avenue and Miller Parkway in West Milwaukee, home to Joy Global and the P&H brand mining equipment.

LEWIS WEHNER was chief engineer at Harnischfeger when many of the company's early innovations in excavating machinery were developed. He had at least four patents to his name for various excavating equipment.

Wehner was born in Salzungen, Saxe-Meiningen, Germany in 1876. He graduated from the Massachusetts Institute of Technology in 1903 where he specialized in Naval Architecture. He worked initially for Albro Clem Elevator Company in Philadelphia, before taking a job with Warren Webster Company where he worked on the design of feedwater heaters and other steam specializes for power plants. In 1901, he worked for the New York Ship Building Company, detailing ship hulls.

In 1903, Wehner accepted a job as chief draftsman for the Bucyrus Company. After seven years at Bucyrus, he was hired by one of its competitors, the Vulcan Steam Shovel Company, to serve as their chief engineer. Vulcan was acquired by the Bucyrus Company in 1911. Apparently, that acquisition caused Wehner to accept a job at P&H Harnischfeger.

In addition to the patent for the excavator that was to become the P&H Model 206 shovel, Wehner is also credited with the patent for one of its early trench-digging machines, a mobile machine for tamping dirt, concrete or like materials used in filling trenches, and building making roads, and a system for equalizing the load of steam shovels and like equipment.

He was elected as a junior member of ASME in 1901.

Wehner was married to Ottilie L. Wehner. They had two daughters, Ann Dorothy Wehner and Ruth Elizabeth Wehner.

A photograph of Lewis Wehner could not be located, although he may be pictured in the front row of the following photograph of Harnischfeger's engineering department.

Photograph of the early Harnischfeger engineering department.

WISCONSIN MOTOR MANUFACTURING COMPANY

Charles H. John and Arthur Frederick Milbrath began designing their first internal combustion engine in the first decade of the 20th Century. Recognizing the limitations of single cylinder engines, they designed a multiple cylinder engine—an in-line, four-cylinder engine. The engine worked well and impressed investors. In early 1909, they incorporated the Wisconsin Motor Manufacturing Company to manufacture the engine. Their first factory was located in North Milwaukee, but they soon outgrew their small shop and began building a new plant in West Allis the following year.

Wisconsin Motor's first production engine was their Type 'A' four-cylinder 'L' head engine. An enormous number of automobile companies were springing up across the United States, but most didn't have the resources to develop their own engines. As a result, Wisconsin Motor found a ready market for their engine from small automobile manufacturers. Wisconsin Motor Model 'A' engines were used in Stutz and Kissel automobiles, Clintonville Four Wheel Drive trucks, Meiselbach trucks and many other vehicles. Wisconsin Motor also produced a line of engines for the Case Company of Racine for its automobiles and eventually its tractors.

The company also began building engines for automobile racing, including an overhead 16-valve engine that was used by the Stutz Company racing team, the "White Squadron." In 1915 at the New York Long Island Raceway, the White Squadron finished first and second and helped to demonstrate Wisconsin Motor's engine design skills.

Wisconsin Motor introduced other engine models, both four- and six-cylinder, 4-cycle water-cooled engines with power ratings from twenty to 200 horsepower.

In addition to automotive applications, Wisconsin Motor's engines were also used in large cranes and power shovels of the Bucyrus-Erie Company of South Milwaukee. Seventy-seven Wisconsin Motor engines were used on the Panama Canal, the most prominent construction project of the era and one that provided both companies with significant publicity. Additionally, Wisconsin's engines were used by the Marion Power Shovel Company, which also supplied power shovels to the canal effort. As a result, engines from Wisconsin Motor were ubiquitous on the Panama project.

As many automotive companies disappeared, or consolidated with other companies, the market for independent automotive engine makers declined. For example, in 1917 the Stutz Motor Car Company started building its own engines. Wisconsin Motor looked to other markets for its engines, introducing a line of marine engines it called the "Wisconsin Whitecaps." The United States Coast Guard purchased a large number of these engines to patrol the coastline during the Prohibition era. An international market also developed for these engines.

Wisconsin Motor's contract with the Four Wheel Drive Automobile Company (which became known simply as FWD) was a particularly significant, long-term agreement. When the United States entered into the First World War, the US Army selected the FWD Model B truck as a standard vehicle, leading to large orders for both FWD and Wisconsin Motors. The FWD truck was popular following the war as well and Wisconsin engines were used by FWD exclusively throughout the 1920s.

Starting in 1929, Wisconsin Motor began developing air-cooled engines for various applications and were so popular that by 1939 the company devoted its activities solely to air-cooled engines, producing as many as twenty-two thousand engines per month. In the 1940s, they introduced a line of four-cylinder, air-cooled engines in a 'V' configuration. The engines were very popular with agricultural companies and Wisconsin Motor had sales contracts with J.I. Case for combines and balers, Massey-Harris for combines, Gehl Brothers on forage choppers and related equipment, and the Rosenthal Corn Husking Company.

During the Second World War, Wisconsin Motor was asked to provide air-cooled engines for numerous military applications, including powering mobile electric generators and construction equipment. Following the war, this led to a significant demand for the company's engines for construction equipment, and a continued strong demand for agricultural applications.

In 1937, Wisconsin Motor was purchased by the Continental Engine Company of Muskegon, Michigan, but continued to produce its engines under its own name. In 1966, Ryan Aeronautics purchased Continental Engine, including its Wisconsin Motor Division. Ryan Aeronautics, in turn, was acquired by Memphis-headquartered Teledyne Corporation in 1969.

A strike at the Wisconsin Motors manufacturing facility on Burnham Street in Milwaukee in the early 1990s led Teledyne to close the factory and move all engine operations to Tennessee.

An ad for Wisconsin Motor's line of air-cooled engines, as appeared in the Milwaukee City directory.

Development of Gasoline Engines for the Stutz Motor Car Racing Team

One of the first of Wisconsin Motor's internal combustion engines was delivered to Henry Stutz in 1911. It was a Wisconsin Type 'A' four-cylinder engine, rated at sixty horsepower. Stutz had just established the Ideal Motor Car Company in Indianapolis. He built his first prototype vehicle in just five weeks, incorporating the Wisconsin Motor engine. As soon as the vehicle was complete, he entered it into the very first Indianapolis 500-mile race. The vehicle averaged 68.25 mph, coming in 11th—a remarkable achievement for a stock engine running in an unproven vehicle.

Advertising his vehicle as "the car that made good in a day," Stutz introduced the vehicle for sale with the Wisconsin Type 'A' engine.

The following year, Stutz introduced its famous Bearcat model vehicle. According to factory literature from 1913, the Bearcat "was designed to meet the needs of the customer desiring a car built along the lines of a racing car with a slightly higher gear ratio." Owning a Stutz Bearcat became a status symbol for the wealthy of the era. In 1914, it was priced at $2,000, which was almost four times that of the basic American-made Ford Model T.

The Stutz 'Bearcat' used a Wisconsin Motor internal combustion engine rated at 60 horsepower. Pictured vehicle is a 1914 Bearcat. Photograph from Wikipedia Commons.

Charles John described the features of Wisconsin Motor's Stutz Racing Engine at a meeting of the Society of Automotive Engineers in Chicago in 1916. He noted the design philosophy was to build a lightweight, relatively small, yet powerful, internal combustion engine. To do so, the company reduced the weight of the reciprocating parts as much as practical by using heat-treated chrome vanadium connecting rods, aluminum alloy pistons, and alloy steels for the crankshaft, bolts, gears and other parts. The engine also had a forced feed lubrication system. All of this allowed the engine to run at higher speeds (revolutions per minute) than the competition. The engine was designed by Arthur Milbrath, chief engineer of the company.

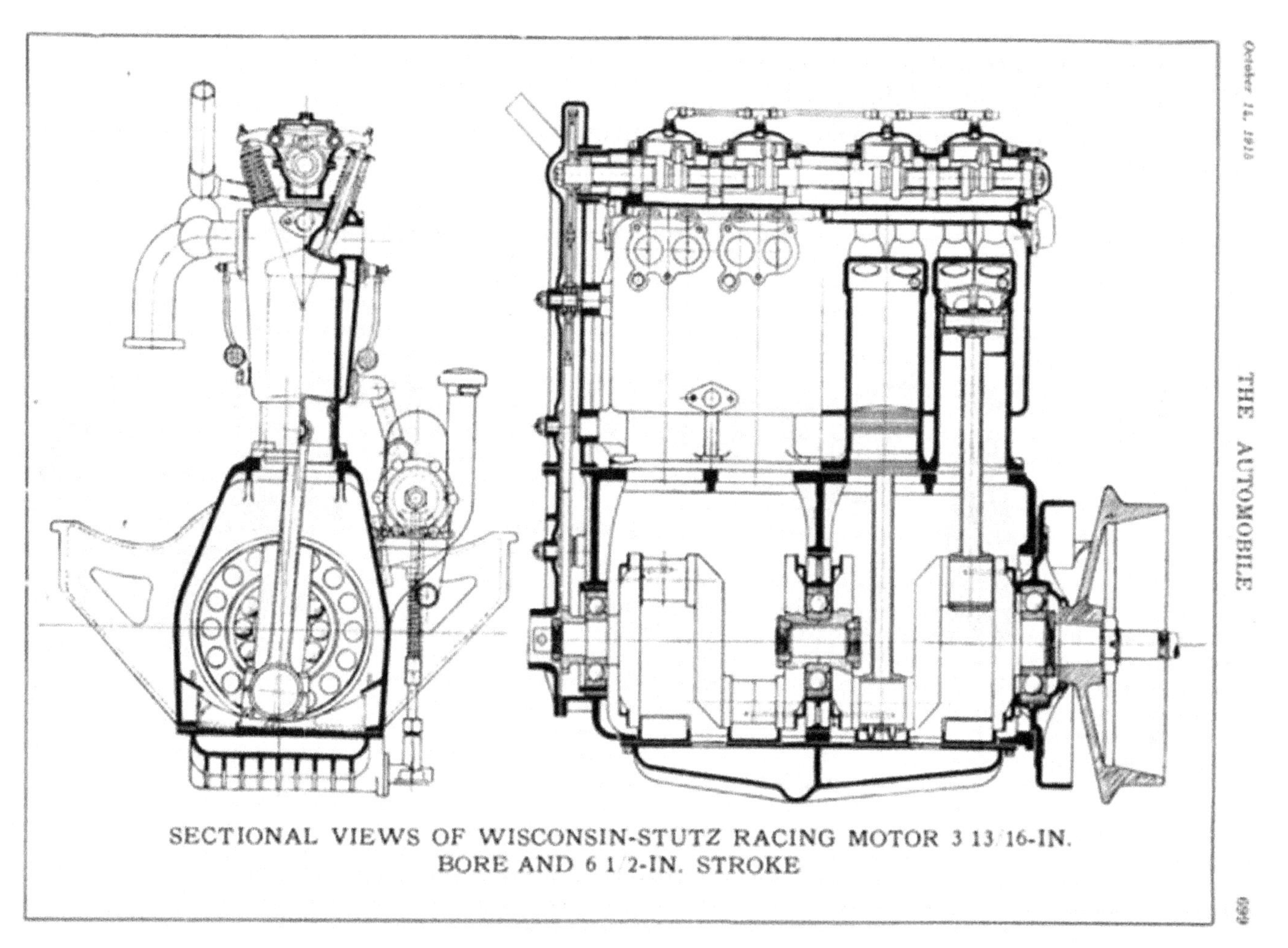

Sectional views of the racing motor designed and built by Wisconsin Motors for the Stutz Ideal Motor Car Company.

Engine for the Wisconsin Special—World's Land Speed Record at Daytona Beach, 1922

In the early 1920s, 'Sig' Haugdahl from Albert Lea, Minnesota, began participating in automobile racing events at fairgrounds and dirt tracks in the Midwest. Known as the "Norwegian Speed Demon," he was also a mechanic and machinist and operated a repair shop, which he dubbed the Auto Hospital in Albert Lea. It was there that he built a special vehicle in an attempt to beat the world's land-speed record.

The vehicle had a 250 horsepower, six-cylinder, all-aluminum, aircraft engine built by Wisconsin Motor Manufacturing Company of West Allis. Because of the vehicle's Wisconsin power plant, he called it the *Wisconsin Special.*

On April 7, 1922, Haugdahl drove his vehicle over a measured mile on Daytona Beach at a speed of 180.3 miles an hour—establishing a new land-speed record.

After that feat, he toured the country with the vehicle, driving demonstration laps around various dirt tracks.

Sig Haugdahl is shown being congratulated by the Daytona Beach mayor, after his record run.

Haugdahl's run was not recognized as a world record by the American Automobile Association, or other auto-racing groups. Two reasons were given. Haugdahl apparently had allowed his membership in the Association to lapse, although he claimed to have mailed in his dues prior to the race. More importantly, the record was timed on a one-way course. The Association requires that official land-speed trials be conducted over two runs in opposite direction, in order to offset any effect of the wind.

The lack of official recognition didn't deter Haugdahl from claiming that he had set the "three-mile-a-minute" speed record during his publicity tour. It was another five years before an official land-speed record was established above 180 mph.

Sig Haugdahl is shown in his *Wisconsin Special'* at one of the many dirt tracks that the vehicle was exhibited on during his publicity tour. The "3 Mile-a-Minute" claim is displayed under his name on the side of the vehicle.

ARTHUR FREDERICK MILBRATH was born in 1879 in Milwaukee. He received his technical training at "a private engineering school" in Milwaukee.[249] Milbrath was hired by C.J. Smith & Sons in 1895. He rose to the position of chief draftsman and held this position until 1907, at which time he left to join McLaughlin Carriage Company of Oshawa Ontario, in charge of its motorcar department. He stayed there only six months, returning to Milwaukee to form the Wisconsin Motor Manufacturing Company with Charles H. John.

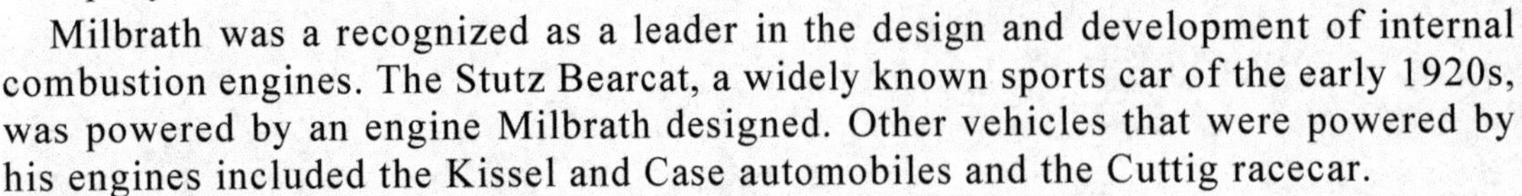

Milbrath was a recognized as a leader in the design and development of internal combustion engines. The Stutz Bearcat, a widely known sports car of the early 1920s, was powered by an engine Milbrath designed. Other vehicles that were powered by his engines included the Kissel and Case automobiles and the Cuttig racecar.

He designed and built engines for record-breaking racing cars driven by drivers Tom Rooney, Ralph De Palma, Gil Anderson, Bob Burman, Earl Cooper, Bill Endicott, Joe Jaegersberger, 'Sig' Haugdahl and 'Cannon Ball' Baker.[250] A six-cylinder, 250 horsepower engine designed by Milbrath was in the vehicle in which Haugdahl broke the world's land speed record in 1921 by going 180 miles an hour at Daytona Beach, FL.

During the First World War, Milbrath's engines were used in Four Wheel Drive Auto Company (FWD) trucks and Schacht trucks, and were standard equipment of the Coast Guard for many years.

After the war, Milbrath designed a complete line of four- and six-cylinder 'L' head and overhead valve engines for heavy-duty service, replacing steam engines in heavy excavating machinery. In 1930, he designed his first air-cooled engine and in 1940 produced a V-type, four-cylinder, heavy-duty air-cooled engine which became the power plant for such agricultural machinery as combines and hay balers and was also used for pumping water in irrigation systems.

Milbrath was granted numerous patents for his innovations in engine design. He was a past president of the Society of Automotive Engineers and had membership in the American Society of Mechanical Engineers and various local social clubs.

He died on February 15, 1955.

STERNBERG MANUFACTURING/STERLING MOTOR TRUCK COMPANY[251]

In 1904, William Sternberg traveled from Davenport, Iowa to Milwaukee to look for suitable manufacturing space. Sternberg had established a machine shop in Iowa, along with a bronze foundry and forge shop. His company manufactured cigar molds and other laborsaving devices. Sternberg also had devised a steam traction engine for agricultural purposes. But in 1903, a fire had destroyed his factory and he was unable to find a suitable replacement facility in Davenport.

In Milwaukee he located suitable space in the second and third floors of a factory owned by the Brodesser Elevator Company at Weil and Burleigh Streets on Milwaukee's northeast side. Signing a five-year lease, he opened business as a general machine shop. At least two of his adult sons, Ernst and William Jr., moved to Milwaukee with him. In 1905, Ernst formed a separate company with Robert G. Hayssen to manufacture automobiles. They named their company the Hay-Berg Motor Car Company and produced three vehicles. One of their vehicles, the Hay-Berg Roadster, was entered in a race at the Wisconsin State Fair Track in West Allis. The vehicle was designed with unsuitable gearing for racing, which limited its maximum speed. As a result, the vehicle trailed badly and dropped out of the race. The winner of the race was Henry Ford. The race results didn't help sales and soon Hay-Berg closed its operations.

Meanwhile, William Jr. Sternberg accepted a job at the A.D. Meiselbach Motor Wagon Company in North Milwaukee (see Chapter 7, for information about Meiselbach). He reportedly became intrigued by truck manufacturing and discussed it with his father and brother. The family decided to enter the truck market, producing its first vehicle in 1907.[ggg] Similar to the Meiselbach truck, it employed a friction drive, with final drive provided to the rear wheels through chains and sprockets. It used gasoline engines from Waukesha Motors. The company incorporated in late 1909 and built a small factory in West Allis the following spring.

It appears that at least three companies were producing similar vehicles in Milwaukee at this time—Meiselbach of North Milwaukee, Sternberg, and Brodesser. Sternberg entered into various road competitions in order to distinguish itself. For example, in late 1908, a one-ton capacity Sternberg truck traveled from Milwaukee to Chicago through difficult snow conditions on bad roads, without a mishap. It must have been a long and difficult trip. The driver was exposed to the elements during the winter and could only travel at about thirteen miles per hour on a vehicle with solid rubber tires.

The following year the company entered an endurance race against fifty-seven trucks from Chicago to Milwaukee and back in the Commercial Vehicle Reliability Run. There were four trucks from local manufacturers—one by the Stephenson Motor Car Company, a three-ton Kissel of Hartford Wisconsin, a one-ton Brodesser truck, and a Sternberg half-ton truck. While Buick won the race, the Milwaukee-area entrants completed the race reliably.

Sternberg 1-ton truck on route to Chicago, January 1909

[ggg] Brodesser also manufactured a truck in 1910. It is not known whether this was a collaborative effort with Sternberg, or an independently designed vehicle.

In 1915, Sternberg decided to change the name of the company to Sterling Motor Truck Company—presumably because of the increasing anti-German bias that was occurring at the onset of the First World War. By the following year, the company was building heavy-duty, worm-gear drive trucks, although chain-drives were still used for their largest vehicles. All their trucks used four-cylinder Waukesha Motor engines and four-speed transmissions.[hhh] The vehicles were equipped with cabs, windshields, headlights and oil sidelights. The frames were of bolted construction and were lined with solid white oak, which the company advertised as a feature that increased frame rigidity, and dampened road shocks. By 1917, the company was concentrating on 2½ to 7-ton heavy-duty trucks.

During the First World War, Sterling made almost five hundred Liberty Model B trucks for the United States Army. The Liberty trucks were built to a standard design. Sterling was one of fifteen manufacturers of the standard Army vehicles.

Factory floor of the Sterling Motor Truck Company, West Allis. Photograph from the Historic Photo Collection of the Milwaukee Public Library.

[hhh] Waukesha Motors ultimately produced 19,000 internal combustion engines for Sternberg/Sterling.

Forward Tilting Cab-Over-Engine Truck

Sterling began making cab-over-engine trucks in 1933. Cab-overs were popular for a while because regulations limited the overall length of trucks. Cab-overs have a shorter wheelbase, allowing longer trailers to be used for tractor-trailer trucks, and longer cargo areas for conventional trucks. The cabs were designed to tilt to allow access to the engine.

Initially, the Sterling cabs were designed to tilt backward, resulting in their designation as "Camel Backs." The front of the cab including the windshield and dash, seats, and fenders were stationary, while doors and the remainder of the cab tilted backward on hinges. While that design worked for tractor-trailer trucks, it wasn't a good solution for cargo-bed on-frame designs. By 1935, Sterling had introduced a forward tilting cab-over-engine vehicle. While at least one of its competitors had introduced a forward tilting cab prior to that, the Sterling design was the first to tilt the entire cab assembly, including the radiator and other vehicles parts, which provided full access to the engine for service.

Tilting was a two-person manual operation, using the leverage of two eight-foot rods that were inserted in the front of the cab. Later models used pneumatics to tilt the cabs. These cab-over-engine trucks by Sterling were the forerunners of all forward-tiling cabs produced in the United States.[252]

Sterling forward tilting cab-over-engine truck, shown outside of the Sterling company factory in West Allis. Note to accessibility the design provides to the engine for maintenance. Photograph from the Ernest Sternberg Collection at the American Truck Historical Society.

Large Military Trucks

During the Second World War, Sterling produced numerous trucks for various specialty heavy-duty applications such as the construction of air force bases, aircraft crash recovery service, airport fire crash containment equipment, and torpedo crane carriers. In addition, Sterling produced massive 12-ton, 8x8 trucks and tractors with front and rear dual-chain drive bogie units and hydraulic steering. The 8x8s (eight wheels both front and back) were powered by American La France V-12 engines, which developed 275 horsepower, and were equipped with chain drives and aggressive tires. These trucks weighed 50,000 pounds each.

A number of even larger special-duty 8x8 trucks were built for the US Army in 1945, shortly after the war ended. Most of these trucks were equipped with Ford 500 horsepower V-8 engines normally used in certain Sherman tanks. They had a top speed of 41 miles per hour. Some were intended for tank recovery operations and were the largest trucks built anywhere during the 1940s.

Sterling special-duty 8x8 military truck built for the US Army—shown undergoing field tests. Photograph from the Ernest Sternberg Collection courtesy of the American Truck Historical Society

KEARNEY & TRECKER

As noted in Chapter 2, Kearney & Trecker started operations in the Walker's Point area of Milwaukee's Menomonee Valley in 1898. It followed Allis Chalmers to West Allis in 1901 by building a one-hundred square foot factory at what is now 68th and National Avenue.

Kearney & Trecker's first milling machine was also produced in 1901. Kearney and Trecker were familiar with such machines from their experience at Kempsmith, and set about devising an improved design. Their first milling machine was equipped with a quick-change gear feed, box section, solid top knee and other important features—many of which were designed to hold the metal being machined solidly in position to keep the sharp milling tool from being moved out of alignment. [253]

In 1905, they developed the first milling machine with automatic lubrication. So successful was the innovation that it was extended to all gears and bearings on their milling machines. Edward Kearney and Theodore Trecker shared a patent for this innovation, issued in 1906.

The 1906 automatic lubrication patent was one of seven patents the two men shared for various improvements to milling machines. Edward Kearney had twenty-three additional patents in his name, some of which were shared with other innovators.

The partners designed and manufactured the famous *Milwaukee*-series knee-type milling machines. Since 1912, the "double overarm" arbor support design was the signature feature of their horizontal milling machines, along with the *Milwaukee* brand name.

By 1920, the K&T plant occupied 117,000 square feet and grew to 300,000 square feet by 1940.

KEARNEY & TRECKER AUTOMATIC TOOL CHANGING MACHINING CENTER

The Milwaukee-Matic II, introduced by the Kearney & Trecker Corporation in 1959, was the first machine offered commercially with automatic tool changing.

The machine included a tool storage magazine that carried a number of tools for insertion into a rotary spindle. All movements of the machine were under automatic numerical (computer) control.

A tool storage magazine rotated the selected tool into a tool change station. A double-grip tool change arm then rotated into engagement with the selected tool at the tool change station and with the previously used tool in the spindle. The tool change arm then moved axially to extract the two tools from the magazine and spindle, respectively, rotated 180 degrees to interchange the positions of the two tools, and then moved axially to insert the previously used tool into the magazine and the new tool into the spindle. The new tool was then automatically locked in the spindle for the performance of the next machining operation and the tool change arm moved into a parked position.

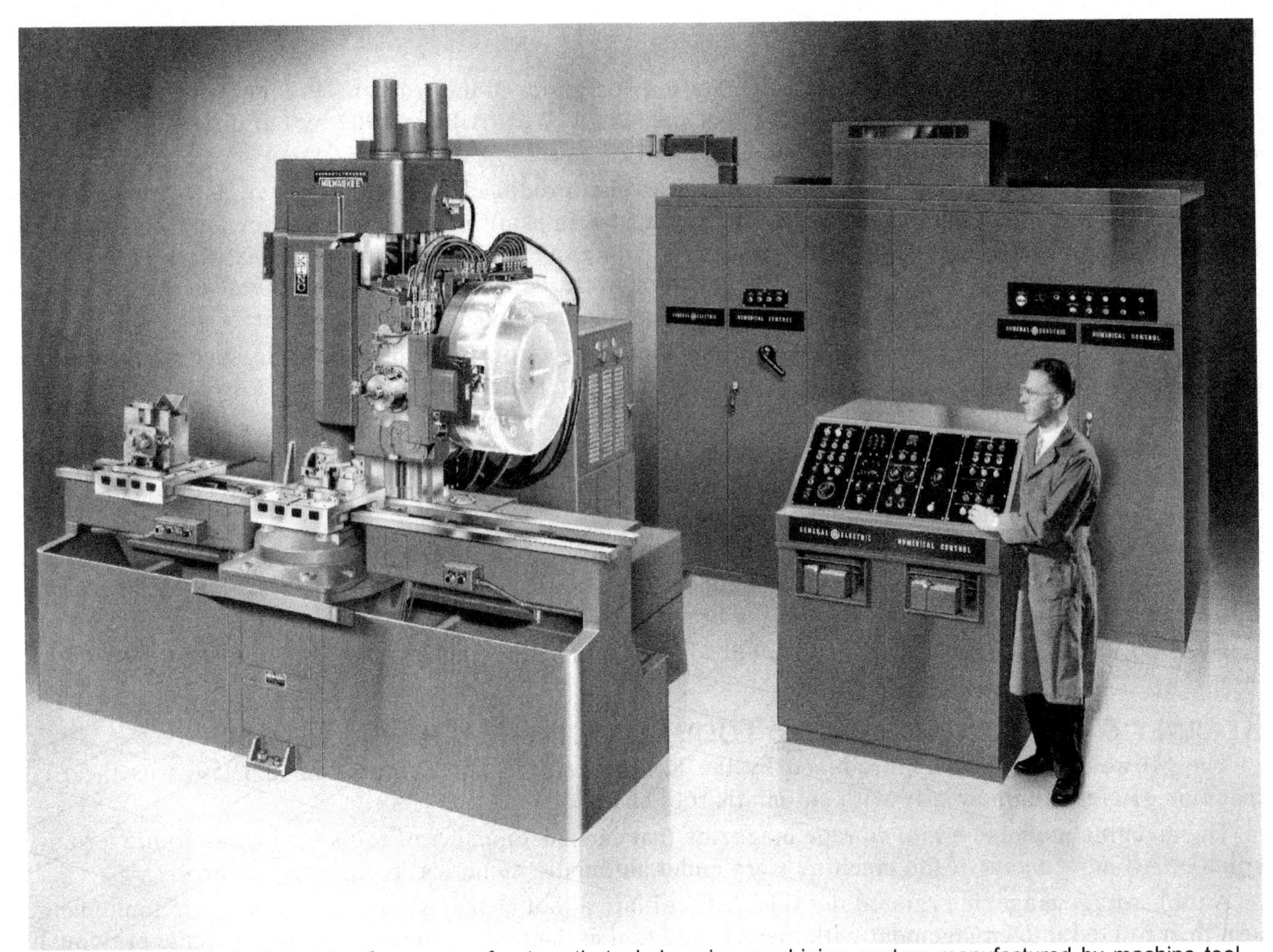

The Milwaukee-Matic II was the forerunner of automatic tool changing machining centers manufactured by machine tool builders all over the world and universally accepted as an outstanding improvement in machine tool construction. Photograph from ASME-Milwaukee files, courtesy of Kearney and Trecker.

The numerical control circuit designated the tool to be used in the next machining operation. All of the tool holders were provided with code rings that identified the tools. As the magazine rotated the tools in a circle, the code rings actuated a plurality of switches to identify the tools, and when the selected tool was identified, it moved it into the tool change station for transfer to the spindle by the tool change arm.

Automatic tool changing and numerical control rendered the machine tool fully automatic and constituted an outstanding advance in machine tool design. It greatly reduced the amount of human labor needed in machining operations. It also resulted in a substantial reduction of machining errors and thereby reduced scrap.

Because the machine was capable of performing a variety of machining operations on a workpiece, it decreased the requirement to move workpieces from one machine to another in completing its manufacture. This also resulted in reducing inventory inasmuch as it lessened the lead-time for parts manufactured in this manner.

REXNORD INCORPORATED

In 1970, Nordberg was acquired by Rex Chainbelt and by 1973, the merged companies adopted the name Rexnord. The resulting company, which continues to be headquartered in Milwaukee, is a leading worldwide industrial company comprised of two strategic platforms: process and motion control and water management. The company employs 7,700 individuals globally.

REXNORD HYDRAULICS

Rexnord hydraulic products specialists developed a new control system on a gigantic forging press at the Cleveland, Ohio plant of ALCOA in 1975. The twelve-story press, owned by the United States Air Force and used to make high strength structural components from lightweight alloys for commercial and military aircraft, has a forging capacity of 50,000 tons. The redesigned control system facilitates the fabrication of large pieces of titanium and aluminum metal alloys with fewer fasteners and less machining.

The unique control system offers an almost incredible multiplication of the efforts of its operator, as well as hairline precision. The press, which weighs 11,000 tons, responds instantly and precisely to the lone operator's hand on a lever just three inches long. The slightest touch governs 2,100 tons of moving weight, and can exert up to the full 50,000 tons of force.

Personnel of the Rexnord Hydraulic Products Division designed and manufactured the massive hydraulic system. Vital statistics on the system are staggering. There are twenty-three major control valves, some of them weighing as much as twelve tons. The system operates at a pressure of 4,500 pounds per square inch. Fluid flows range from 11,750 gpm at high pressure to 26,438 gpm at prefill pressure.

Eight main cylinders, all 60" in diameter, deliver the press's working force by advancing the press platen. Four filling check valves, weighing about twelve tons each and standing nine feet tall, prefill these main cylinders to bring the platen quickly into its first contact with the workpiece. Four pullback cylinders move the platen back through the return portion of its cycle. The majority of the control valves are driven by a single camshaft 30 feet long that synchronizes them through a series of rocker arms.

Two closed-loop electronic-hydraulic servo systems work simultaneously as the press operates. One of these systems controls the position of the main camshaft, which actuates through positive linkage most of the system's operating valves. A signal is transmitted from the operator's control lever, through the electronic-hydraulic servo control circuitry, resulting in positive mechanical positioning of all control valves, affecting a corresponding and proportionate speed and position of the forging platen of the press. Closed-loop feedback circuits maintain a precise response equal to the movement of the operator's control lever.

In addition to the main drive system, the press has a unique Tilt Control system. This was designed to compensate instantly for out-of-level conditions that can be caused during the power stroke by irregularly shaped forgings. The eccentric loads this creates can cause high stresses in localized portions of the press structure, weakening it and leading to premature failure unless the loads are counteracted immediately.

The digital analog leveling system uses four encoders to monitor and compares instantly the four-corner positioning of the moving platen. The actual position of the platen is shown on a video display screen at the operator's station. In the case of the tilt control system, these signals are also fed through a servo mechanism to two balancing valves, each controlled by a servo-controlled hydraulic cylinder, to make necessary corrections automatically by directing high-pressure fluid to the proper main cylinders so as to maintain the level of the platen.

The Rexnord Hydraulic Products Division has designed hydraulic systems for a multitude of other specialized and unique systems.

The Hydraulic Products Division of Rexnord developed the highly advanced electronic hydraulic control system capable of operating the massive 50,000-ton forging press. Photograph courtesy of Rexnord Corporation.

WAUKESHA MOTORS

Waukesha Motors was formed in 1906 by Harold LeVan Horning, Frederick Ahrens and Allan Stebbins at the Blue Front Garage, located at 139 East North Street in downtown Waukesha.

Stebbins, a retired farmer, was the major investor and was named president of the company. Horning, born in Wauwatosa in 1880, attended Carroll College in Waukesha and eventually became head of the mechanical engineering department of the Modern Steel Structural Company in Waukesha. While at Modern Steel, he met Frederick Ahrens. Ahrens was born in Germany about 1872 and came to the United States in 1883. The two men discovered that they shared an interest in internal combustion engines. In 1906, they purchased a building that was formerly the office building for Waukesha Woolen Mills and later served as a bicycle repair shop. Horning and Ahrens used the space to work on the repair of motors during their off hours. They erected a sign outside their business which read "The Blue Front Garage—Cars Washed, Stored, and Repaired—Morgan & Wright and Hartford Tires."[254]

When vehicles were brought in for repair, it is reported that Horning and Ahrens would often use the opportunity to disassemble and analyze the engine parts, before reassembling and tuning it up. They developed a reputation for excellent service—and figured out how to build a better engine.

They built their first engine from scratch, designing, casting and machining every part. The development cost them $12,000, which was an enormous sum for the small enterprise. It was designed as a boat motor and sold for $500. The first buyer quickly sold it for more than he paid for it. It quickly led to additional orders. In 1907, the design was standardized as the Waukesha Model 'A' engine. In that year the new company also introduced its Model 'H' engine. Orders were initially from boat owners, but soon Waukesha Motors was building engines for automobile and truck builders.[255]

Pictured is the Waukesha Model 'A' engine, the first produced by Waukesha Motors. Photograph used with the permission of the Waukesha Engine Historical Society Inc.

The company quickly exceeded the capacity of their small Blue Front building and acquired land on the west side of Waukesha along St. Paul Avenue, which had good railroad access and room for expansion. At the same time, it continued its research and development activities to produce additional engine designs and explore new applications. Between 1912 and 1913, Waukesha Motors released four new engine models. The company also significantly increased engine sales to truck manufacturers and began furnishing engines for Ingersoll Rand air compressors.

As war broke out in Europe there was a significant increase in demand for US agricultural products—resulting in an increased demand for farming equipment. At the same time, horses were being replaced by the internal combustion engine. All of this combined to create a large demand for Waukesha's engines. The City of Waukesha itself boasted three tractor manufacturers—Waite, Nilson and Paramount Farm Tractor—all of which used Waukesha engines. Waukesha Motors also began producing engines for military trucks for various European countries involved in the war. As the United States entered into the World War, President Woodrow Wilson established the War Industries Board. Harold Horning was appointed to a position in charge of standardizing the engine design for the military's Library truck program.

This photograph, taken on the grounds of the White House in 1917, shows Harold Horning of Waukesha Motors (center front without hat) talking to President Woodrow Wilson about the design of the Liberty Truck. Note the truck on the right of the photograph has its engine cowl open, to show the engine. Photograph courtesy of the Waukesha County Museum

Following the First World War, Waukesha experienced high demand for its engines from truck and tractor manufacturers. Most of the new vehicle manufacturers did not have the technological capability to build their own engines, which provided Waukesha Motors with a ready market.[iii] The company also entered into an agreement with Harry Ricardo of England to manufacture and market his new cylinder

[iii] Waukesha Motors ultimately produced engines for 132 different automotive and truck manufacturers, and 140 tractor manufacturers. Source, private correspondence with Dennis Tollefson, Waukesha Engine Historical Society, Inc.

head design, which used a unique design to promote fuel mixture during combustion, which permitted higher combustion rations while avoiding pinging, which resulted in higher efficiency and output.[iii]

In addition to their use on tractors and trucks, Waukesha engines were used on excavators, ditch diggers, oil derricks, irrigation pumps, concrete mixers, hoists and locomotives. The company also produced engines for stationary and portable electric power. Engine offerings increased and Waukesha was soon producing engines that ranged from six horsepower to 1,200 horsepower.[256]

Also in the 1920s, Waukesha developed the "cooperative fuel research engine" (see article below), which provided the industry with a standard test method to determine the octane rating of gasoline and similar fuels.

The depression of the 1930s hit Waukesha Motors hard. In order to increase business, the company continued to explore other markets for its engines, introducing its smallest four-cylinder engines to power refrigeration units for railway boxcars, air conditioning and power for railway passenger cars, as well as for industrial and commercial air-conditioning uses.

When the United States entered into the Second World War after the attack on Pearl Harbor, the demand for Waukesha engines increased dramatically. The plant ran around the clock to meet the high demand for military applications. The company was asked to produce engines for trucks, tanks and other vehicles, as well as to power pumps, blowers, generators and the like. Waukesha Motors greatly expanded its factory in order to produce the engines needed for the war effort, and the workforce increased to about 2,000—many of which were women that were trained to work as machinists. Barracks were built across the street from the plant to house workers.

Following the war, Waukesha engines found use in a number of significant markets—automotive, industrial, power generation, oilfield applications, logging and timber, agricultural and mobile refrigeration. In spite of the loss of military orders, sales increased.

During the 1950s, Waukesha Motors introduced its first turbocharged diesel and its first V-12 engine. It also opened a new research facility. In 1956, Waukesha Motors celebrated its 50th anniversary and marked the production of its millionth internal combustion engine.

[iii] Under the agreement, Ricardo assigned the United States patent rights to Waukesha Motors. US patent number 1,474,003

WAUKESHA 'CFR' KNOCK-TEST ENGINE

In the 1920's a committee was formed by gasoline refiners and engine builders to develop a means of measuring and defining gasoline combustion characteristics. The committee was named the Cooperative Fuel Research (CFR) Committee and consisted of a wide range of fuel producers and engine manufacturers, including Harold L. Horning, president of Waukesha Motor Company, and Arthur W. Pope, Jr., Chief Research Engineer of the company. In 1928, the committee reached the decision that a standardized single-cylinder test engine was needed as a first step in developing a gasoline knock-test method. In early December 1928, the CFR Committee accepted a design presented by Waukesha Motor Company, and detail drawings proceeded at once. Because of its genesis, the engine became known as the *CFR Engine.*

Arthur W. Pope of Waukesha Motors with one of the first CFR "Knock Test" engines. Photograph was taken June 29, 1931, courtesy of Waukesha Engine Historical Society. This engine was designated a historic mechanical engineering landmark by ASME in 1980.

The first engine was designed, built, tested, and delivered to Detroit in under forty-five days. It was put on display on January 14, 1929, at the Society of Automotive Engineer's Annual Meeting. The basic design represented the amalgamation of the ideas of Harry Horning, Arthur Pope, and Howard M. Wiles;

with the actual construction of numerous parts in very long days and nights by Pope and Wiles. The engine was assembled by them in the laboratory.

The CFR engine provided the engine and fuel industries with the first universally accepted standard test engine, which could be produced in sufficient quantity to meet industry needs. The initial plan was that Waukesha would build seventy-five of these engines to satisfy the entire needs of the industry. This was changed before production began and eventually well over 5,000 engines were built.

The means for varying the compression ratio quickly and accurately without affecting valve clearances or basic combustion chamber configuration was probably what caused this design to prevail over all other rivals. Moving the entire cylinder up and down with respect to the piston was far better than changing shims, moving a plug in the combustion chamber, or running with fixed compression as was done with some other early fuel research engines.

The engine improved the ability of the automotive and petroleum industries to tailor their products to perform better together, because it provided a recognized standard for defining fuel quality. It undoubtedly facilitated the rapid evolution of both fuels and engines. These CFR engines are still sold today for basic research in such new areas as exhaust emissions and alternate fuels suitability.

When folks today select regular or premium gasoline at the gas station pump, they are selecting a gasoline that was rated by the Waukesha CFR engine to meet the needs of their cars and trucks.

Diesel Engine for the Hemphill World's Diesel Speed Record

In 1935, Dave Evans drove a vehicle designed and built by students at the Hemphill Diesel School in Chicago, setting a record speed for a diesel-powered automobile. The official AAA record run was made on the sands of the Daytona Beach.

The Hemphill students used a Waukesha diesel engine—a model 6D140 known as the *Silver Comet Six*. The six-cylinder engine, as the name implies, had a five-inch bore and 5½ inch stroke, resulting in 648 cubic inches of displacement. The engine was rated at 140 horsepower at 2,100 RPM.

On the morning of February 15, 1935, on the Daytona sands, a vehicle designed and built by Hemphill Diesel school students using a stock Waukesha diesel engine, set a new world's speed record for a diesel-powered automobile. From author's collection.

The timed record speed was 125.69 miles per hour. While the engine was a stock Waukesha diesel, the students certainly took liberties in order to increase peak performance. It appears that the speed governor was removed from the injection pump, for example. A chain drive was used to drive the rear wheels, using a larger sprocket in front to provide an overdrive effect.

Cummins Diesel apparently didn't take kindly to seeing a bunch of students achieve the land-speed diesel record. A few months later, a vehicle powered by a Cummins diesel was able to beat the speed established by the Waukesha-powered Hemphill racer—achieving 133.023 miles per hour on March 1, 1935.

Dale Evans was a regular driver at the Indianapolis 500 race from 1927 through 1934. In 1931 he drove the first diesel-powered race car to compete at the Indianapolis 500—a vehicle powered by a Cummins diesel engine. It was the first race car to complete the race without stopping for fuel. It finished in thirteenth place.

The engine cowling is open in this view, showing the Waukesha *Silver Comet Six* diesel engine used by the Hemphill school team.

HAROLD LEVAN HORNING was born on March 23, 1880, in Wauwatosa, Wisconsin. He attended Carol College in Waukesha. In June 1899, he presented a science oration at the College that reviewed the inventions of the time and the achievements of modern science. Harry served for a time as assistant superintendent of the Milwaukee Gas Company and as head of the mechanical engineering department of the Modern Steel Structural Company in Waukesha.

In 1906, he formed Waukesha Motors along with Frederick Ahrens and investor Allan Stebbins. Serving as secretary-treasurer of the company, Horning's principal role was chief engineer and developer of internal combustion engines for the new company. He is regarded as one of the most innovative leaders during the formative decades of the automotive industry[257] and had at least seventeen patents to his name dealing with various aspects of engine design and construction. He also helped to develop the CFR engine, which has been referred to as the "Horning engine." This engine continues to serve as the standard test engine for determining the octane ratings of gasoline and other fuels.

In 1925, Horning was named president of the company and served in that role until his death in 1936.

During his career, Horning was a member and officer of several research and automobile societies, including serving as past president and life member of the Society of Automotive Engineers, the first president of the Internal Combustion Engine Institute, chairman of the automotive section of the War Industrial Board during First World War (serving as a "dollar a year man"), director of the Automotive Parts and Equipment Manufacturing Association. He also had membership in the English Automotive Engineering Society, the American Society for Testing Materials, the American Petroleum Institute and the National Association of Manufacturers.

Harold Horning. Photograph used with the permission of the Waukesha Engine Historical Society Inc.

In 1938, the Society of Automotive Engineers established the Harry L. Horning Memorial Award in his name to annually recognize the author(s) of the best paper(s) relating to the better mutual adaptation of fuels and internal combustion engines. The award is meant to "preserve the memory of the dedication of SAE's 1925 President, Harry L. Horning, to the pursuit of improved mutual adaptability of engines and fuels and serves as a motivation for others to follow in his footsteps.[258]"

Horning sponsored engineering courses at Carroll College and served as a trustee at the college as well as for Waukesha High School.[259]

CHAPTER 9: THE ONES THAT GOT AWAY

This chapter describes two important engineering innovations that were created soon after the innovators left Milwaukee. It is believed that a credible case can be made that the foundation for these creations was laid in Milwaukee.

As noted throughout this book, the century between 1860 and 1960 was an incredibly innovative one in Milwaukee. Most of the accomplishments were of large machines—hence the title of this book, *The Magnificent Machines of Milwaukee.* Milwaukee was known and celebrated as the *Machine Shop of the World* because of its prominence in the manufacture of grand-scale machines. In fact, it has been observed that during this era, the greater Milwaukee area contributed to the country's "age of largeness much as Silicon Valley, California, now pioneers in microelectronics, is presently contributing to the age of smallness."[260]

The two innovations described in this chapter, however, were at the other end of the scale—important in part because of their *smallness*. They were two extremely important achievements, resulting in devices that are used daily by billions of people on all corners of the planet—the O-ring and the microchip.

The O-ring is an interesting seal that is underappreciated. O-rings are hidden away in our home faucets, refrigerators, air conditioners, and automobiles. They are used in hydraulic systems everywhere, from airplanes to industry to spacecraft. They have been customized to seal a large range of fluids and gasses, and are used in some very difficult environments. They are truly ubiquitous.

The world without them would be quite different. Yet we only think about them when they fail—the bathroom faucet develops a stem leak, or an oil seal on our car leaks, or, in an extreme example, the space shuttle *Challenger* fails on take-off and takes the lives of its seven crewmembers. As a result, we have a love-hate relationship with the O-ring. But when properly designed and maintained, O-rings provide reliable, efficient and economic seals.

The creator of this marvelous device is almost unknown, but has deep Milwaukee roots. He left Milwaukee before he created the O-ring. However, his experiences in Milwaukee likely led him to continue his quest for a better seal.

The microchip has similar characteristics. It is truly ubiquitous in our modern environment—showing up in almost all electronic devices, as well as in controls for things that we wouldn't normally think of as electronic. They are also commonly hidden away from view, embedded behind protective plastics and screens of smartphones and computers, or in the internals of things like coffee makers. Microchips have changed the world and they continue to evolve into devices that are more sophisticated.

The creator of the microchip also had a deep Milwaukee experience that guided him and led him on the road to its development.

This chapter briefly explores both engineers, their creations, and the Milwaukee influences that shaped them.

NIELS CHRISTENSEN'S O-RING

Niels Christensen's engineering accomplishments while in Milwaukee were covered in Chapters 2, 6 and 7 of this book. As noted, he was very innovative—and more than a bit stubborn. Most of his innovations involved pneumatics—compressed air applied to various uses such as braking and starting internal combustion engines. He had several dozen patents. He also successfully fought a lengthy patent battle with the Westinghouse Air Brake Company.

It goes without saying that Christensen was often dealing with the difficult task of preventing leaks in the various pneumatic devices that he created and built. His streetcar braking system dealt with the potential for leaks in being "fail-safe." Loss of air pressure resulted in applying the brakes. The system had a large compressed air storage tank, so minor leaks could be accommodated. However, as Christensen's devices became more sophisticated, and were designed for environments such as aircraft where space and weight needed to be minimized, an efficient means of sealing the components became ever more important.

In 1926, Christensen left Milwaukee for Cleveland, where he accepted a position to work at Midland Steel Products to design various pneumatic and vacuum braking systems for automobiles and trucks. Most automobiles of that era had simple mechanical brakes, operated by levers and cables. Stopping distances were long and, as average speeds increased, a better braking system was needed. Christensen, working with Midland, attempted to fill that void.

In 1921, Fred Duesenberg introduced the first hydraulic braking system on a passenger car.[261] For a time the relative merits of the two systems were being debated. Duesenberg's hydraulic brakes had several advantages—principally their low-cost and relative ease of installation. Christensen's pneumatic system required the installation of an air compressor and storage tank, resulting in a more expensive solution. However, Christensen's system was considered safer, since the loss of air pressure would result in applying the brakes, whereas the loss of hydraulic fluid would cause the brakes to be entirely ineffective with potentially catastrophic results.

The Midland Steel Products Company of Cleveland was advertising Christensen's automotive air brake system in 1930.[262] However, in the meantime, several automobile manufacturers including Chrysler, Dodge, Desoto, and Plymouth had introduced standard hydraulic brakes. Ford was the last domestic holdout—it didn't introduce hydraulic brakes until 1939.

It appears that Christensen's system had more success for heavy-duty trucks, in part because of the ease of connecting truck trailers to the pneumatic braking system. Air brakes are standard equipment on most heavy-duty trucks to this day.

Niels Christensen applied for numerous patents during this time. It could be said that he saved his most important for last—his 1937 patent filing for a device that he referred to quite simply as a "packing."[263]

Working in his South Euclid basement, he developed an O-ring seal. Through trial and error, he discovered that a torus oval of rubber that was constrained in a properly sized groove would provide a reliable tight seal of a piston sliding in a cylinder. Christensen described his packing ring as, "perfectly circular in cross section prior to its assembly and insertion in a cylinder, and possesses a normal circular cross section"—a torus, or in lay terms a donut shaped ring.

The patent filing goes on to state, "When compressed within its groove, the packing ring flattens ... so that the portion of the ring contacting with the bottom of the groove is flat, as well as its outer periphery which contacts with the cylinder substantially over the area ... and at least one side of the ring flattens out.... If this groove is made fractionally wider preferably by a fraction of the diameter of the ring section than the width of the ring section when compressed, then one side of the ring is not as flat as its other side, which contacts with the side of the groove. In any event, it is advantageous that the packing ring be not compressed solid within the groove but that the dimensions of the latter be such that when the ring is assembled with the piston and cylinder it has slight freedom of movement to be worked or kneaded during the piston reciprocation in order to maintain the rubber in a live state and to prolong its serviceable life."

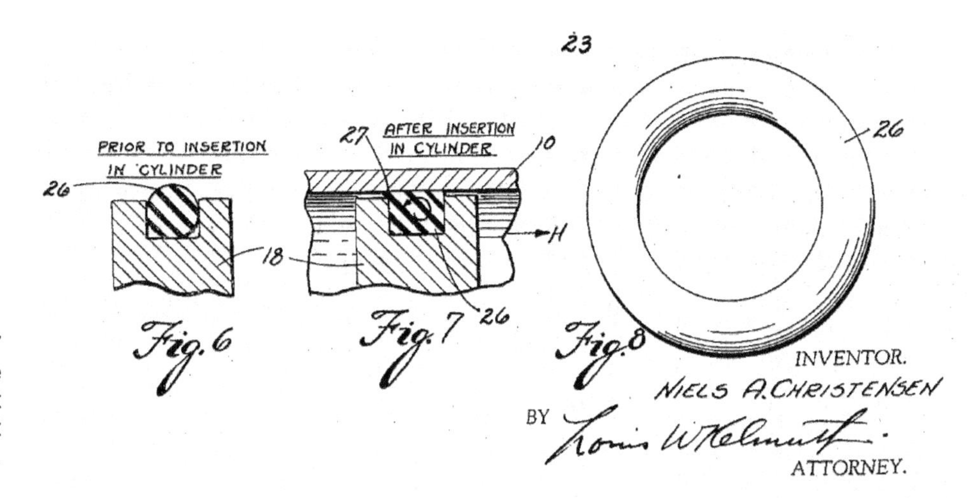

Figures from Christensen's 1937 US Patent filing, describing the packing ring and showing how it is constrained to provide a tight seal against a piston.

While there had been previous attempts to develop similar packing seals, none had been successful.[kkk]

Midland Steel wasn't impressed with Christensen's O-ring, apparently not believing his claims of reliability. In 1934, he was informed by Midland's president that his services were no longer required. As a result, Christensen's innovation largely escaped notice. An improved packing seal certainly didn't rise to the level of public attention. Nor did it excite industry. Christensen, however, was undaunted. He knew from his years of working with pneumatics and other sealing applications that he was on to something big—a small device with large implications. As demonstrated previously, Niels was a stubborn man—a characteristic that helped him to ultimately succeed.

After a few years of contacting potential users and not generating much interest, Christensen contacted the Army Air Corp and requested an interview. The United States was gearing up for the Second World War and would need millions of hydraulic seals. The Corp's procurement engineers agreed to a meeting. Christensen drove to Dayton's Wright Field to meet with Nicholas Bashark and Elsworth M. Polk, the two engineers at the Army Air Corps who were in charge of hydraulic seals. They installed Christensen's O-ring seals on a worn landing gear of a Northrop A-17A aircraft and tested it over eighty landings.[264] They then ran controlled laboratory tests. Impressed with both the field tests and the laboratory results they were convinced of the integrity of the O-ring for use in aircraft hydraulics, and the aeronautical industry finally adopted it into its standards in 1942. Christensen was on the brink of realizing the profits from his innovation. However, his timing wasn't good.

After the Japanese bombing of Pearl Harbor, the United States government quickly acquired the rights to many war-related patents and made them available to allied manufacturers royalty-free. Christensen's O-ring patent was determined to be critical to the war effort. Christensen was paid $75,000 for his patent rights during the war. It was a substantial sum, but not nearly what he might have realized from the innovation.

After the war ended the patent rights were eventually transferred back to Niels Christensen, but he only had four years left. Christensen litigated, apparently over the delays—his patent rights were not

[kkk] Early uses of round, resilient rings were not confined in a groove, allowing them to slide back and forth. These were not effective. Large cross-section India-rubber rings were used as gaskets in counter-bores for water-works piping in the mid-19th Century. Thomas Edison's 1882 light bulb patent shows a round rubber ring at the neck of the glass bulb to keep the mercury in and the air out. There is also a Swedish patent for an O-ring that is dated May 12, 1896. Better materials and Christensen's precise dimensional prescriptions were undoubtedly responsible for his success.

released until 1952. Again, he was successful in the courts, but not until nineteen years after his death. His heirs ultimately received a $100,000 settlement in 1971.

In the 1950s, O-rings came into widespread use for industrial hydraulics, plumbing, appliances, pumps, valves, and many other devices. Today, billions of O-rings are sealing every conceivable apparatus all over the world, in the air, on land, and sea, and in outer space. They are truly ubiquitous.

The O-ring is the most widely adopted seal in history because of its simplicity, low-cost, ease of installation, and small space requirements. It is suitable for dynamic or static seals <u>within the temperature limits of the elastomeric material used for the ring</u>.

This later, underscored, point is important. If the temperature becomes too low for the elastomer used for the seal, the O-ring can lose its elasticity. This was tragically shown with the *Challenger* disaster on January 28, 1986. The O-ring failure on that mission was predictable. Allan McDonald, an engineer from Morton Thiokol, had warned that the flight was at risk due to O-ring leakage, because the temperatures were too low on the morning of the flight.[265] Unfortunately, NASA mission control decided to proceed with the flight anyway, and the world would eventually hear much more about the nuances of O-rings. In an article entitled, "Seven Myths about the Challenger Shuttle Disaster, NBC News space analyst James Oberg concludes. "NASA managers made a bad call for the launch decision, and engineers who had qualms about the O-rings were bullied or bamboozled into acquiescence."[266]

NIELS ANTON CHRISTENSEN was born on August 16, 1865 at Tørring, a village on the east coast of the peninsula of Jutland, Denmark. After attending local public schools, he was apprenticed in Veile, a nearby city, at fourteen years of age. He became a journeyman when he turned eighteen while attending evening school to become proficient in the fundamental laws of physics and electricity, as well as in mechanical drawing. He entered the polytechnic institute in Copenhagen and also engaged in practical work in local machine shops. While there, he worked on the task of designing and installing the then largest lighthouse in the world, erected on the west coast of Jutland. He also acquired a good deal of experience with maritime construction.

In 1888, at the age of twenty-three, Christensen moved to England and was engaged in London for a time. In order to increase his proficiency in English, he took a job as a third engineer on an English merchantman ship engaged in trade in the Mediterranean and the Black Sea.

Returning to England, Christensen worked on general plans for a new water works in India, and designed the machinery and general plans for a Chilean nitrate works. Later on, he designed Corliss engines and textile and papermaking machinery.

He eventually came to the United States to take a position with the Fraser-Chalmers Company as designer of Corliss engines. While inspecting the new electric railway in Chicago known as the "Cicero and Proviso" system at Oak Park, an accident occurred resulting in deaths of two people and the injury of many others. Christensen found that electric streetcars did not have power brakes. He proceeded to remedy this situation, designing the Christensen scheme of streetcar air brakes for which he obtained patent protection. He successfully tested the system in Detroit in March 1893, but could not obtain financing.

Later that year, Christensen accepted a position at the E.P Allis Works in Milwaukee. He continued to work on air brakes and in 1894 developed a system for use on railroad cars that had fewer moving parts than competing braking systems. Leaving E.P Allis, in the early part of 1897, he interested investors in establishing the Christensen Engineering Company to produce the Christensen "triple valve" brake and related components.

In 1902, the company decided to broaden the scope of the business into general electric manufacture. Christensen, who was general superintendent but not a shareholder, objected to this diversification and decided to depart the company to concentrate on air compressors and other patented devices he designed. He entered into a relationship with Allis-Chalmers for a time, in which Allis manufactured Christensen's air brake systems, before going on to establish a new company devoted to providing compressed air starters for aircraft engines and other internal combustion engines.

Christensen eventually left Milwaukee for Cleveland Ohio, where he worked on air brakes for automobiles and trucks. Later in his career, while experimenting in his basement workroom, he developed a packing seal that has become known as the O-ring. The O-ring has become an extremely important sealing device and is used in numerous applications to provide a low-cost, compact, yet effective seal.

Christensen received the liberty medal of King Christian X for special service to his native country during the Second World War.

Niels Anton Christensen married Mathilda Thomessen Hagerup in 1894 and they had one child—Esther Marie (Young) born in 1895. Niels Christensen died on October 5, 1952, in Cleveland, Ohio.

JACK KILBY'S MICROCHIP

As noted in Chapter 6, Jack Kilby was employed by the Centralab Division of Milwaukee's Globe-Union between 1947 and 1958. He concentrated in the area of electric circuit miniaturization. In his memoirs Kilby related, "This was a fortunate choice for me, because they worked with hybrid circuits—an early form of miniaturization."

Centralab had developed a process that would become known as "thick-film hybrid circuits." In the process, silver paint was deposited on a ceramic base layer to form conductors. Carbon-based inks were also applied to the substrate to serve as resistors, and small capacitors were formed in the substrate.

Kilby also noted, "Centralab was ideal for me in another way. The group I worked in was small, so I saw the entire process, from engineering through sales and production. My initial duties included the design and product engineering work on hearing aid amplifiers and resistor-capacitor networks for television sets."

While employed at Centralab, Kilby took night classes toward his master's degree in electrical engineering from the University of Wisconsin's Extension Division. He received his master's degree from the University of Wisconsin in 1950. His thesis was "Design of a Printed Circuit for a Television IF Amplifier." His thesis complemented his work at Centralab, which introduced a product based on the technology in 1952.

Centralab also exposed Kilby to transistor design concepts, sending him away in 1952 to study the emerging technology via a course sponsored by Western Electric's Bell Laboratories in New Jersey. Upon returning from this course, Kilby was given the responsibility to design and build the tooling to make transistors—a reduction furnace to make germanium, a zone refiner to purify it, and a crystal puller to grow the germanium crystals. He had to also build or purchase related tooling, such as testing equipment and ultrasonic saws.[267]

Kilby also attended a lecture by Nobel laureate John Bardeen[III] sponsored by the Marquette University physics department on the research that led to the development of the transistor. During his talk in May 1957, Bardeen focused in the miniaturization of circuits made possible with transistors.[268]

Centralab initially applied Kilby's transistors to hearing aids, but soon other products followed. While at Centralab, Kilby was awarded ten patents for transistor improvements and circuit uses.

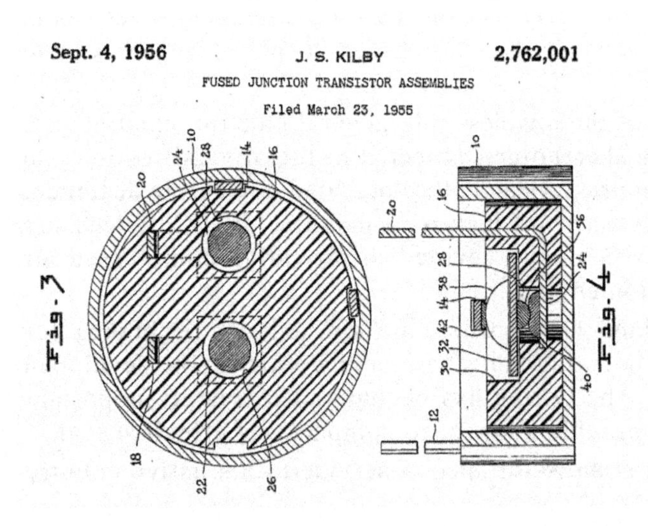

Drawing from a patent awarded to Jack S. Kilby in 1956 for fused junction transistor assemblies, which were "mechanically strong unusually small in which the parts are easily positioned and soldered using mass production methods."

[III] John Bardeen was born in Madison Wisconsin in 1908 and received his bachelor's and master's degrees in electrical engineering from the University of Wisconsin. He is the only person to have won the Nobel Prize in Physics twice: first in 1956 with William Shockley and Walter Brattain for the invention of the transistor, and again in 1972 with Leon N. Cooper and John Robert Schrieffer for a fundamental theory of conventional superconductivity known as the BCS theory.

In March 1958, Kilby received a bonus for doing outstanding work the preceding year. A letter accompanying the bonus from the president of Globe-Union congratulated him. However, the letter also noted the deteriorating business conditions the industry faced, and hinted that there may be difficult times ahead.[269]

Jack Kilby started to explore other positions—interviewing with Texas Instruments, IBM and Motorola. By May, he received offers from both Motorola and Texas Instruments, and decided to accept TI's offer, which stated, "We want you to work in our new micro-miniaturization department."[270] The position was obviously a good match for Kilby's experience, based on his decade of work at Centralab.

Kilby left Centralab on May 15, 1958, and traveled to his new job in Dallas, Texas. Soon after arriving, the company had its annual summer break during which most employees took a vacation. Kilby, as a new employee, did not yet have the right to take a vacation. As a result, he was left almost alone for the period. During that summer break, using borrowed and improvised equipment, Kilby designed and completed the first electronic circuit in which all components were fabricated on a single piece of semiconductor material. The circuit was half the size of a paper clip—a dramatic achievement in miniaturization.

Kilby recorded the various steps of his initial microchip development in his notebook. By July 24 of that year, when the vacationing employees returned, his notebook summarized his findings and showed the necessary steps for the production of the microchip. He outlined his concepts to his bosses at Texas Instruments. He demonstrated the individual electronic components satisfactorily and then sent the circuit design to the TI laboratory technicians for fabrication and testing.

The full circuit was successfully demonstrated on September 12, 1958. Kilby later invented the first integrated circuit based calculator, dubbed the *Pocketronic* at the time. It was initially a rather ornate affair that was later streamlined. By the 1970s, it became available for general consumer usage.

The microchip has been in existence for less than sixty years, but it would be fair to say that in that time it has changed the world. The technology has created completely new industries and provided the technology behind computers, smartphones and other devices that would have been impossible without Kilby's innovation.

Kilby returned to Milwaukee in 2001 to attend a ceremony held at Johnson Controls. The company recognized him for his innovations while at Centralab, as well as his contributions to the world of science and technology. Johnson Controls also used the opportunity to present a paper found in its archives to the Smithsonian's Natural Museum of American History. The fifteen-page paper, written by Kilby in 1951, was entitled, "Transistors—Their Manufacture and Use."

During the ceremony, Kilby noted that he got his first exposure to transistors while working at Centralab, which sparked his interest in the field.

At a ceremony held in 2001 at Johnson Controls, Jack Kilby (center) was greeted by dignitaries including (L-R) Jim Keyes, CEO of Johnson Controls, Thomas Ament, Milwaukee's County Executive, John Norquist, Mayor of Milwaukee, and Brian Stark, President of Johnson Controls' Building Efficiency Business Unit. Photograph courtesy of Johnson Controls

JACK ST. CLAIR KILBY was born on November 8, 1923, in Jefferson, City, Missouri, and raised in Kansas. An amateur radio buff, Kilby went on to serve in the military during the Second World War. He was stationed in India as a technician—before attending school in the United States on the G.I. Bill. Kilby received his degree in electrical engineering from the University of Illinois in 1947.

Following graduation, Kilby joined the staff of Globe-Union in Milwaukee, doing developmental work in miniaturized circuitry at the company. He left Globe-Union in 1958 for a position at Texas Instruments, where he was soon able to design an integrated circuit, combining previously isolated electronic elements to work together in a miniature device that would become known as the microchip.

Kilby's invention was revolutionary for the computing/technological world and marked the beginnings of devices that have become ubiquitous in electronic and other devices. Robert Noyce of Fairchild Semiconductor devised another type of integrated circuit shortly after Kilby's, and the men's respective companies eventually negotiated a cross-licensing agreement.

Kilby later invented the first integrated circuit based calculator, dubbed the *Pocketronic*. It was further streamlined by the 1970s for general consumer use, becoming the pocket calculator.

After receiving a director's position, Kilby took a leave from Texas Instruments in 1970, although he continued to work with the company even after his 1983 retirement. As an independent researcher he worked on solar power-based activities and joined the faculty at Texas A&M University.

Kilby was inducted into the National Inventors Hall of Fame and the Engineering and Science Hall of Fame. He also received a number of other accolades including the Nobel Prize in Physics, which he was awarded in the year 2000 along with Zhores I. Alferov and Herbert Kroemer. During the award ceremony, Kilby gratefully acknowledged the important contributions of Robert Noyce, who had passed away a decade earlier.

Over the course of his career, Kilby held dozens of patents and established the Jack Kilby International Awards Foundation, honoring peers and colleagues doing innovative, progressive work in science and medicine.

He married Barbara Annegers in 1948. The couple had two daughters and five granddaughters. Jack Kilby died on June 20, 2005, at the age of 81.

In this 1987 photograph, Jack Kilby is shown holding a handful of early microchips and the first Cal-Tech calculator. Photograph courtesy of DeGolyer Library, Southern Methodist University, Texas Instruments Records.

WHAT IF THEY HADN'T LEFT?

Perhaps it is fitting that the development of the O-ring and the microchip occurred elsewhere. Milwaukee has always been the city of *big* machines and innovations, and these innovations were at the other end of the size spectrum. There is no way to know with any certainty what may have occurred if Niels Christensen and Jack Kilby had spent the balance of their careers in Milwaukee. However, it is interesting to speculate what might have happened to the City's industrial environment if these two extremely significant innovations had occurred here.

CHRISTENSEN

From what is known about Niels Christensen, it is evident that he was not only stubborn, but also very patient. He took the time to see things through, and sought perfection. It seems likely that he would have tinkered with approaches to improving pneumatic seals in his Milwaukee basement, if he had not relocated to Cleveland. At that point in his career, he evidentially had the time and financial resources needed to test various sealing configurations, as well as to take the necessary legal steps to protect his innovations with patent applications.

The ultimate development of the O-ring in Cleveland seems to have had little direct impact on that city. The rights to use the O-ring were licensed to numerous manufacturers around the country, as well as overseas, during the Second World War. The patent rights to the device were lost soon after the war. As a result, the local effect seems to have been rather minor.

If the O-ring had been developed in Milwaukee, it seems unlikely that it would have had a significant impact.

KILBY

The impact of the development of the microchip is a more difficult matter to evaluate. Would Jack Kilby have had the freedom and resources to develop his microchip? He certainly had the background, having worked at Milwaukee's Centralab on miniaturization of electrical circuits. He had received an education on transistors, and had worked for some years on fabricating them, as well as their use.

The fact that he developed his successful microchip designs weeks after he left Milwaukee, and prior to any additional education or training by his new employer at Texas Instruments, make it clear that Kilby possessed the skills to design the microchip before he left Milwaukee.

Of course, other matters were important as well. Would he have had the motivation and resources, as well as the luxury of time to spend on this unsupervised project? We'll never really know the answers to these questions.

What may be a bit clearer, however, is the impact that Jack Kilby's innovation might have had on Milwaukee. Based on the impact that his innovations had at Texas Instruments, it seems likely that, had Kilby developed the microchip in Milwaukee, it could have had a significant impact on his employer, as well as the overall economic prosperity of the City.

CHAPTER 10: MILWAUKEE'S TRADE SCHOOL AND ENGINEERING ENVIRONMENT

It is interesting to note that almost all of the innovators highlighted in this book received their engineering education elsewhere. Many were immigrants, a few received their education out east, and a handful received their education from the University of Wisconsin in Madison. Milwaukee simply didn't have an engineering school until 1903.

An informal practice of apprenticeship was in place in Milwaukee, but it appears that this was haphazard at best—often left to the whims of individual tradesmen at the various area industrial companies. This was often limiting, because it was difficult for a tradesman to teach an apprentice more than he knew himself.[271]

Making matters worse, children as young as twelve often left school to take entry-level labor positions in industry. There was no way for them to continue their education and, without skills to draw upon—they had no realistic way to advance into other positions in industry. In 1909, were over 2,400 workers in Milwaukee that were under age sixteen. As students dropped out of school, classes began to be depleted before eighth-grade graduation, and then throughout high school.[272] While this provided industry with a steady source of entry-level, low-wage workers, it did little to advance the level of skills required in the workforce—especially as manufacturing, and the machines industry produced, became more complex.

TECHNICAL EDUCATION IN MILWAUKEE

Early Milwaukee industrialists eventually recognized the problems associated with young students entering the workforce at an early age without learning a trade, as well as the lack of opportunities for educational advancement once they entered the workforce. They began taking steps to alleviate the lack of education in the trades in 1906. Additionally, in 1911, the Wisconsin legislature introduced an innovative plan to provide continuing education for those that entered the workforce before completing school. These endeavors are discussed below.

THE MILWAUKEE SCHOOL OF TRADES

The Milwaukee School of Trades was established in early 1906 by the Milwaukee Merchants and Manufacturers Association, at the prompting of Frederick W. Sivyer, who was president of the Northwestern Malleable Iron Company. As a prominent member of the Association, Sivyer believed one of its duties should be to give young men in Milwaukee an opportunity to learn a trade of their choice from skilled, able teachers. A success from its start, the school eventually gained a national reputation. It was reported to be "one of a few of its kind in the west." Its courses included mechanical drawing, machine design, plumbing and wood and metal pattern making. When first opened the school was maintained by voluntary subscriptions, as well as by fees paid by students.[273]

The school initially operated out of the Pawling and Harnischfeger factory building in the Walker's Point area, while "a large modern trade school [was constructed] between Hanover and Greenbush Streets and Virginia Street and Colonial Place."[274] Most trade curriculums were completed in two years.[275] Classes were held fifty-two weeks a year and forty-four hours per week, with school hours from 8 to noon in the morning and 1 to 5 in the afternoon, except for Saturdays when school was dismissed at noon.

On July 1, 1907, the school was absorbed into the Milwaukee Public School System.

The Milwaukee School of Trades initially operated out of this Pawling and Harnischfeger factory, located at First and Oregon Streets in the Walker's Point neighborhood.

An article in the September 25, 1911, edition of the Milwaukee Journal reported, "The boys are all interested because they know they are doing something practical and there is hope ahead for the man who knows a skilled trade."[276]

The school was renamed the Milwaukee Boy's Technical High School in 1917. Girls were not admitted to the school until 1976, resulting in a name change to the Milwaukee Trade & Technical High School. In 2002, the school moved into a new building and its name was further changed to the Lynde and Harry Bradley Technology and Trade School, in recognition of a substantial donation from Jane Bradley Petitt.

The Wisconsin Idea

The lack of educational opportunities for those that entered the workforce at an early age was eventually addressed by the Wisconsin legislature. The background story is an interesting one. University of Wisconsin professor Charles Van Hise, with the support of Wisconsin's governor and fellow classmate Robert 'Fighting Bob' La Follette, was elected as the president of the University of Wisconsin in 1903. In one of his first speeches, Hise declared, "I shall never be content until the beneficent influence of the University reaches every home in the state."

Using his friendship with Governor La Follette, Van Hise helped to foster close ties between the University and state government. This collaboration eventually led to the establishment of the Wisconsin Legislative Reference Library, established during La Follette's tenure to provide policy research to help guide legislation. Charles M. McCarthy, who was a document clerk for the legislature and a member of the UW faculty, was selected as the head of the Reference Library. McCarthy turned out to be an excellent choice for the position. While an idealist who helped to incubate ideas, he was also a practical organizer. The Wisconsin legislature soon relied upon his analysis and assistance.

During the early 1900s, with McCarthy's assistance and the collaboration of university faculty experts, the Legislative Reference Library assisted legislators in drafting groundbreaking legislation, including the nation's first workers' compensation law, a groundbreaking law providing for the public regulation of utilities, and various educational initiatives. The progressive policies that were established during this era eventually became known as the *Wisconsin Idea.*[mmm]

The University of Wisconsin—Extension System

The first education initiative under the *Wisconsin Idea* was to formalize the University of Wisconsin's "extension" system. It was an effort to make Charles Van Hise's vision of reaching every home in the state a reality—providing access to the resources of the University throughout the state. In La Follette's first address to the Wisconsin Legislature in 1901, he stated that, "The state will not have discharged its duty to the University, nor the University fulfilled its mission to the people, until adequate means have been furnished to every young man and woman in the state to acquire an education at home in every department of learning."

In 1906, Charles McCarthy personally financed a State survey to determine the benefits to Wisconsin citizens of providing educational programs through the Extension System. UW President Charles Van Hise formally established the University Extension Division in 1907.[nnn] In the same year, the Wisconsin legislature appropriated $20,000 to the Extension to provide an extensive program of job training in Wisconsin factories. The measure, which was encouraged by the Milwaukee Merchant and Manufacturers Association, led to visitor-instructed courses in shop mathematics, blueprint reading, mechanical drawing and other subjects. The State's appropriation for the Extension program was substantially increased over the following decade.[277]

As the country mobilized for the First World War, the Extension system developed special correspondence courses for the armed forces, Red Cross nursing classes, and post-graduate medical refresher courses, along with information on food and fuel conservation, and the use of women in industry. It also provided for "emergency food agents" to encourage food production through *victory gardens* and improved crop production to meet wartime demand.

In 1923, the State Legislature appropriated funds for the construction of a University Extension building in Milwaukee to serve the growing number of students. The UW Extension program offered adult courses as well as a full-time freshman/sophomore university program—students could transfer to the University in Madison following their sophomore years. During the Depression years, the Extension Division offered employment rehabilitation programs to help the jobless to improve their skills and established in Milwaukee a Vocational Guidance Bureau.

During the Second World War, the Milwaukee Extension Center offered "swing shift" classes in military German, Japanese and Russian, as well as civilian pilot training, cryptography, and military mathematics. At the end of the war, enrollment surged as thousands of returning soldiers entered the Extension Center's two-year programs.

In 1955, the Milwaukee Extension Division merged with the Wisconsin State Teachers College in Milwaukee—forming the University of Wisconsin-Milwaukee (UWM). In 1963, the UW Board of Regents approved major university status for UWM.

[mmm] The term, *the Wisconsin Idea* was not applied to the progressive body of legislation of the early 1900s until Charles McCarthy of the Legislative Reference Library wrote a book on the subject by that name in 1912.

[nnn] The extension activities of the University were functioning before the University of Wisconsin-Extension was established as an institution. Wisconsin was one of the first states to institutionalize a university extension education.

Milwaukee Area Technical College

In 1909, several members of the Wisconsin legislature began discussing the forgotten groups in education—uneducated laborers, many of whom had dropped out of school at a young age and had no prospect for advancement. A committee was established to study the issues and report on its findings at the 1911 session of the legislature. McCarthy was appointed to the committee and traveled, at his own expense, to Europe to investigate solutions. His fact-finding tour took him through England, France, the Netherlands, Austria and Germany. McCarthy was impressed at some of the European practices and returned to Wisconsin full of ideas.

McCarthy made special note of the *fortbildungsschule,* or "continuing education school," in Munich, Germany. The school had been established by Dr. George Kierschensteiner in about 1880.[278] Based on the committee's recommendations, largely influenced by Charles McCarthy, the Wisconsin's Vocational Education Act passed in 1911 and launched a statewide network of "continuation" schools.

McCarthy discussed this legislation in a book that he wrote in 1912 entitled, *The Wisconsin Idea.*[279] Theodore Roosevelt was apparently so impressed by the book that he wrote the preface to the book. In it, McCarthy noted that the legislation required, "every child employed between the ages of fourteen and sixteen will have to attend school for five hours a week out of the time of the employer."[280] Industrial education was placed under the supervision of an industrial education board, which included three employers of labor and three skilled employees, as well as representatives of the State's educational system.

McCarthy stated that "Instead of concentrating upon a few costly trade schools, the plan is to build up a great system of industrial education for those actually in work; to do something, where nothing has been done to help all the workers; that is, the German continuation school in all its essentials has been incorporated into the school system of Wisconsin."

The law also required that all illiterates under twenty-one years of age must attend evening schools whenever these evening schools can be reached by them, unless excused because of lack of strength."[281]

In highlighting the advantages of the industrial education system, McCarthy stated, "The boy who is to become a bricklayer, while he is in apprenticeship will be taught not only the mere trade but also some essentials which will prepare him for life and a place in the civic body, giving him the opportunity to broaden his outlook that later he may be able to pass from the ranks of manual skill to the ranks of administrative ability. He will be taught not only bricklaying but architecture, buying and correlated subjects. And so in every industry the same general requirements, subject to approval by the state board of industrial education, must be met."[282]

The Milwaukee Continuation School opened in 1912 and began with part-time day continuation classes for boys and girls between 14 and 16 years of age who had left full-time school to go to word under a labor permit.[283] Classes were initially held at Mason and Water Street in the Manufacturers' Home Building. The school was headed up by Robert L. Cooley, who led the development of vocational education in Milwaukee.

The Manufacturers' Home Building, which still stands on the east bank of the Milwaukee River, was the first home of the Milwaukee Continuation School, which eventually evolved into today's Milwaukee Area Technical College. Photo courtesy of MATC.

In fall 1913, evening educational courses were offered for adults who could attend on a voluntary basis. At the same time, apprentice classes, formerly offered by the Extension Division, were transferred over to the Milwaukee Continuation School.

The Wisconsin legislature changed the law for the continuation group in 1915, raising the permit age to 17 and requiring eight hours of continuing education per week.

Cooley oversaw the construction of a new building for the Milwaukee Central Continuing School at Sixth and State Streets, the first phase of which opened in 1920. The name of the school was soon changed to "Milwaukee Vocational School" and eventually to the "Milwaukee Vocational and Adult School."

The first phase of the new building for the Milwaukee Central Continuing School included a power plant for heat and light. The power plant also served instructional purposes. Photograph courtesy of MATC.

The second construction phase was ready for occupancy in 1923, allowing all programs and courses to be offered under one roof. A third phase was eventually completed in 1927. It contained two gymnasiums, a swimming pool and a school cafeteria.

Nursing was offered beginning in 1923. In 1924, an evening technical engineering curriculum was offered. The technical engineering program was the result of an initiative by Dr. Roland, training director for the Milwaukee Electric Railway and Light Company, who saw the need to provide technical training for employees who would bridge the need between the skilled craftsman and the engineer—a profession now referred to as the "technician." Roland conferred with others from industry—collectively forming the Industrial Training Conference. The conference approached the Milwaukee Vocational and Adult School with a plan, which was quickly implemented.

During the Depression years, the need for continuing education for adults increased, resulting in the formation of an Adult High School as part of the Milwaukee Vocational and Adult School's curriculum. A Certificate Course was launched in 1934, which was accredited by the University of Wisconsin. This curriculum eventually became the vocational Junior College. It was placed in a new division known as the Milwaukee Institute of Technology.

In October 1951, at the urging of Milwaukee Mayor Frank Zeidler, plans were initiated for an educational television station. The following year the FCC allocated a station and an arrangement was made with station WITI-TV to make its broadcasting tower available to the vocational school for broadcasting its programs.[284] That effort eventually led to today's Milwaukee Public Television, which began broadcasting in 1957.

ENGINEERING EDUCATION IN MILWAUKEE

As noted, almost all of the innovators highlighted in this book received their engineering education elsewhere. Milwaukee didn't have an engineering school until 1903.

This section provides a brief history of the formation of engineering schools in the City of Milwaukee.

THE MILWAUKEE SCHOOL OF ENGINEERING

It appears that there may have been several early attempts to establish an engineering school in Milwaukee. However, the goal wasn't realized until Oscar Werwath formed his *School of Engineering.*

Oscar Werwath was born and educated in Germany and came to Milwaukee, where he obtained a position at the Mechanical Appliance Company. He quickly realized that skilled engineers were in high demand in Milwaukee. Apparently encouraged by Louis Allis, he approached the Rheude and Heine College Business College and Drafting School about holding night classes in practical electricity for young men. The Rheude school was operated out of the Germania Building at the time. Interest in Werwath's electricity course was so high that it quickly exceeded the capacity of the small school and Werwath was encouraged to open a separate engineering school. Louis Allis donated $500 toward the effort and Werwath opened the *School of Engineering* for training technicians and electrical engineers. Werwath was named its first president

The school started with an enrollment of one hundred students. From the beginning, the leaders of industry and business cooperated in its development. Early on Werwath realized that students needed financial assistance to attend the school. Werwath decided to meet this challenge by establishing a company where his students could work and make salable products. He established the Milwaukee Electric Construction Company, where students complimented their training by making electrical products to include lead-acid electric storage batteries. By 1911, the electric storage battery business was taking up too much room in the overcrowded facility and Werwath decided to sell the business to private investors. By the following year, the investor group moved battery production and established the Globe Electric Company.[285]

The School of Engineering pioneered cooperative engineering education in 1918, starting with local companies such as Allen Bradley. Under the cooperative program, students alternated periods between school and work.

In 1920, eager to adopt new radio technology, the School of Engineering established Milwaukee's first radio station. It went on the air in 1922. The station was acquired by the Hurst Corporation in 1928 and its "call letters" were changed to today's WISN.

The school's name was eventually changed to the *Milwaukee School of Engineering* (MSOE) in 1932 and reorganized as a non-profit school overseen by a board of regents. Werwath held the positon of president his death in 1948.

During the Second World War, many of the school's laboratories and classrooms were used to train war production workers for private industry. MSOE set up short courses to train women as welders. Following the war, returning veterans created an intense demand for MSOE's training programs.

Oscar Werwath is shown in front of his School of Engineering. Werwath enjoyed driving around in his electric powered vehicle.

Marquette University School of Engineering

Named after Jacques Marquette, a French Jesuit explorer and missionary, Marquette College was established in 1881 to provide an affordable Catholic education to the area's emerging German immigrant population. Between 1891 and 1906, the college employed only one full-time lay professor, with many classes being taught by Jesuit priests, part-time professors and master's students.

In 1906, Father A.J. Burrows, S.J. who was president of the young college, told the Free Press that, "The fact that Milwaukee is such a large manufacturing center, particularly in iron and steel, would make this an ideal point at which to maintain a school of engineering." He began exploring adding a college of engineering, as well as other professional colleges, and "making Marquette a real University."[286]

Father Burrows began by talking to a number of other schools about their engineering schools, including Cornell, Notre Dame and Cincinnati. He also consulted Bernard A. Behrend, chief engineer of the electrical department at Allis-Chalmers. Behrend, who had been educated in Germany, recommended a combination of classroom training and factory experience. With this advice and others, Marquette decided to proceed. Engineering classes opened in September 1908 with John C. Davis of Cornell University named as Dean of the new college.

Initially a two-year curriculum was offered. Tuition was set at one hundred dollars per year—forty more than charged arts and sciences students.[287] Nineteen students registered, who were referred to as the *Boiler House boys* since they took part of their coursework in the basement of Marquette's Johnston Hall. They shared the Liberal Arts College's chemistry and physics laboratories, and mechanical drawing was taught in the Academy a few blocks away on the State Street campus.[288]

As the number of students grew, the school faced a critical space shortage. To solve the program, Dean Davis proposed to allow the engineering students to build their own school. At the conclusion of the 1909-10 school year, as part of their training, Davis led a team of engineering students and their faculty in constructing a reinforced concrete school building with a brick facade. Before they got started, they also demolished two buildings on the site just south of Gesu Church—a Chinese 'hand' laundry and a duplex.

The new building was eventually expanded in 1916—again by the students overseen by the faculty. The completed structure was named *Bellarmine Hall.* Ultimately, the engineering building had five classrooms, four laboratories, three drawing rooms, a library and offices for the professors. A boiler room and the University's power plant were located in the basement.[289]

The early engineering students also participated in other school construction projects. They laid out a new baseball and athletic grounds and, in 1912, installed an automobile testing plant in *Bellarmine Hall.*[290]

This photograph from Marquette's 1916 Hilltop yearbook shows the initial phase of the engineering building (at the left) adjacent to a house that was retained for the college's offices. Later that year the house was torn down and the engineering building expanded. The former engineering building currently serves as the Parish Center for Gesu Catholic Church.

A four-year engineering curriculum was soon adopted and, in 1912, the initial nine graduates received degrees in civil and electrical engineering.

A Cooperative Education Program was established by the College in 1919, under Professor William D. Bliss. This program combined in-the-field work experience with the academic program. Students alternated monthly between factory assignments and academic work. While the curriculum was extended to five years to compensate, students earned salaries to offset tuition and gained valuable experience. As a result, the program was very popular and enrollment in the college expanded from fifty students in 1917 to four hundred in 1924. Eventually, the program was modified such that students alternated full semesters between work and academic assignments following their sophomore years.[291]

By the late 1930s, with over five hundred students in the engineering college, it was clear that a new building was needed to replace the one constructed by students in 1910 and 1916. A fund drive was held, and an advisory board made up of representatives from Nordberg Manufacturing, Falk Corporation, Allis-Chalmers, Wisconsin Electric, the Milwaukee Road, and the Wisconsin Highway Commission.[292] The effort raised just under $400,000—far less than the estimated cost of the new building. Rather than cancel or delay plans, an innovative approach was used to reduce the cost of the new building. While the exterior of the building conformed to the Tudor Gothic-Collegiate style prevalent on the campus at the time, the interior classrooms, laboratories and common spaces were left with cinder block walls and smooth-formed concrete floors and ceilings. Somehow, the appearance was appropriate for an engineering building and the building was completed at less than half the initial estimate.[293]

The timing was good. Had the effort started a few months later, it would likely have been delayed because of wartime production needs. As it was, the completion of the school allowed it to be used for training requirements during the Second World War.

A new engineering building was completed in 1941, located on Wisconsin Avenue between 15th and 16th Streets.

In the fall of 1943, the first female student enrolled in the chemical engineering program—changing the engineering school's men-only tradition.[ooo] In the fall of 1950, the College established an undergraduate evening program in engineering, which allowed students working full time to obtain engineering degrees.

[ooo] Marquette University was the first Catholic University in the world to go co-ed, admitting its first women students in 1909.

University of Wisconsin-Milwaukee

In 1866, the Wisconsin legislature established a normal school at Platteville—the first of eight teacher-training schools across the state. The term "normal school" originated in the early 16th century from the French *école normale*. Under the French concept, normal schools would provide teaching practices using model classrooms.[294] A normal school was finally established in Milwaukee in 1880.

The Milwaukee State Normal School opened for classes in 1885 at 18th and Wells Streets. It has an initial enrollment of forty-six. The school moved to Milwaukee's east side in 1909, when a new building was completed—now Mitchell Hall.

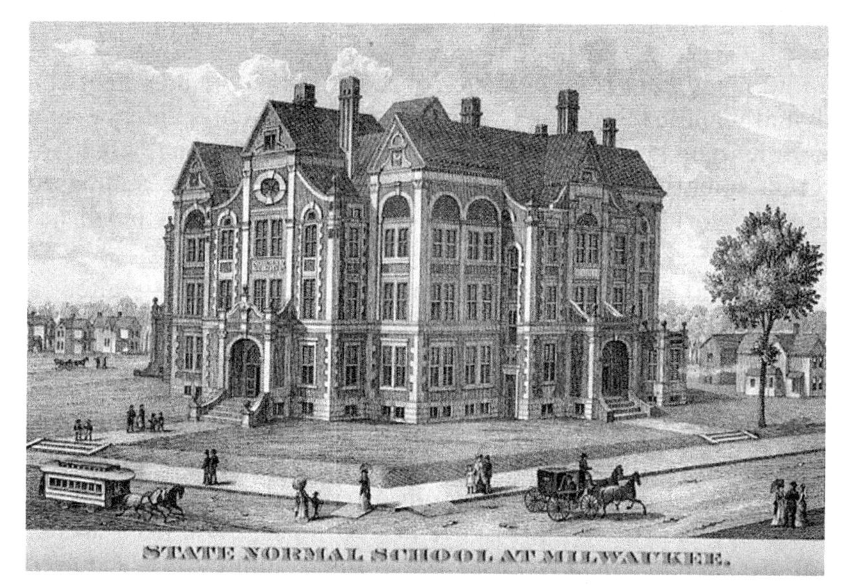

Illustration of the State Normal School at Milwaukee, published in the 1885 edition of the Wisconsin Blue Book.

In 1911, the legislature permitted state normal schools to offer two years of post-high school work in art, liberal arts and sciences, pre-law, and pre-medicine. While popular, the State Normal School Regents voted to discontinue college courses in 1922, in order to refocus on the education of teachers. By 1927, the school changed its name to the Wisconsin State Teacher's College and offered education-related four-year degrees. After the Second World War, the school added a graduate program in education.

The Wisconsin legislature reversed itself in 1951, empowering all state colleges to offer liberal arts programs. The Wisconsin State Teacher's College changed its name to the Wisconsin State College of Milwaukee. Five years later, it became part of the University of Wisconsin—formally known as the University of Wisconsin-Milwaukee. It began offering a two-year undergraduate engineering curriculum in 1964.

THE ENGINEERING PROFESSION

Milwaukee developed a strong engineering community during the period of innovation covered in this book. It is believed that this community contributed toward the advancements that occurred in the area. This chapter discusses some of the early engineering activities that helped to bring the innovators together.

The term "engineering" is derived from the Latin words *ingenium*, meaning *cleverness* and *ingeniare*, meaning, "*to contrive or devise.*" These words provide an apt foundation for the profession—engineers are individuals who invent, innovate, and design. The engineering profession is broader than that, however. It also includes individuals that maintain, research and improve machines, tools, systems, components, materials and processes.

There is also an important distinction to be made between the basic roles of engineers and scientists. Engineers typically apply scientific and mathematical principals, including empirical evidence and economic, social and practical knowledge. Scientists generally establish the scientific and mathematical principals that engineers rely upon. The great scientist-engineer Theodore von Karman once addressed this distinction by stating, "Scientists study the world as it is, engineers create the world that never has been."[295]

Putting this all together results in a definition of an engineer that applies to the individuals cited in this book: ***An Engineer is an individual who applies scientific and mathematical principals*** to invent and improve machines, as well as structures, tools, systems, components, materials, and processes. This is obviously a convoluted definition. I prefer to think of an engineer simply as a person trained with appropriate skills and knowledge to use the materials and forces of nature to create machines and other devices for the benefit of all. But it might be best to simply state, ***engineers are problem solvers***.

Engineering has evolved over the span of civilization. Each generation of engineers has built on the accomplishments of the generations preceding it. Historically the science has been passed on from master engineer to apprentice, or from professor to student. In any case, the trainee applies the acquired knowledge to new problems, acquiring new insights and advancing the science. At the same time, the tools and materials of the profession have continued to evolve and develop—all resulting in increasing levels of achievement. While the same might be said for several professions, it is clear that the engineers of today stand on the accomplishments of those that practiced the profession before them.

THE EARLY MILWAUKEE ENGINEER

While most of the individuals that developed the machines illustrated in this book are referred to as engineers, many of these innovators did not have engineering degrees. In the 1800s, in particular, there were two tracks for obtaining and demonstrating engineering expertise—the *shop culture* and the *school culture*.[296] Some of the most famous engineers, such as Edwin Reynolds who served for a time as the president of the American Society of Mechanical Engineers, received their training as apprentices working for other engineers, who may or may not have attended an engineering college.

The way engineers obtain the training and expertise in their profession has changed over time. Today, almost all practicing engineers hold one or more college degrees related to their profession. However, it has been observed that many of today's engineers graduate from college with a good theoretical understanding of their profession, but often lack the practical skills that can often only be obtained on the shop or laboratory floor, working with seasoned practitioners of the profession.

There does not appear to have been active engineering professional societies in Milwaukee until the late 1800s. The exchange of information or ideas must have been informal. That changed around the turn of the century, with several important events.

The Verien Deutcher Ingenieure

The *Verien Deutcher Ingenieure,* often referred to as simply the VDI, or the Association of German Engineers, began meeting in Milwaukee as early as 1889. This German engineering society provided a nucleus for organizing the professional engineers located in the area—many of which were of German heritage.[297] Given that seventy-two percent of the population of Milwaukee was of German birth or descent by 1900, it is not too surprising that the Association of German Engineers established gatherings here.

The VDI was founded in German in 1856 by fellow researchers from the Academic Society *Hütte.* Their first engineering journal was written the following year. At the time, engineering was not regarded as high as other scientific disciplines. VDI worked to promote the engineering discipline and in 1899 was able to reclassify German institutes of technology (Technische Hochschule) as universities.

ASCE's 1888 Annual Conference in Milwaukee

The American Society of Civil Engineers held its 1888 annual conference in Milwaukee[298]—which appears to have been the first national engineering society to hold a major conference here. The conference was held between June 28 and July 4. The Annual Address of the ASCE president, Thomas C. Keefer, was read. Keefer, a Canadian, used the opportunity to discuss the Canadian Pacific Railway and in particular the construction of the railway through the Canadian Rockies and the design and construction of snow sheds through the mountain passes. Milwaukee City engineer G.H. Benzenberg discussed several aspects of the City's municipal water system during the course of the meeting.

According to the Transactions of the society, one hundred and thirty-three engineers attended the conference. In addition to listening to a number of technical papers, the attendees were offered the opportunity to take excursions to "points of interest" in Milwaukee, Appleton, Kaukauna, Oconomowoc and Waukesha, as well to mining regions of Wisconsin, Michigan and the Sault Ste. Marie.

ASME's 1901 Spring Meeting in Milwaukee

The American Society of Mechanical Engineers (ASME) held its 1901 semi-annual meeting in Milwaukee.[299] It was a large gathering—one hundred and fifty-six ASME members were present from around the United States, along with numerous guests. A local committee was formed to organize the event,[300] which included Warren S. Johnson (who served as chairman), Bruno Nordberg, Irving H. Reynolds and Edwin Reynolds. They were all prominent engineers in Milwaukee, as noted elsewhere in this book. Warren S. Johnson had founded the company now known as Johnson Controls; Bruno Nordberg, who worked for a time at the E.P. Allis Company, founded Nordberg Manufacturing Company; and Edwin Reynolds was chief engineer for the E.P. Allis Company, which in 1901 became Allis-Chalmers. Numerous other Milwaukee engineers were listed as serving on the local committee including Irving Reynolds, Edwin's nephew and his eventual successor as chief engineer at Allis Chalmers.

LOCAL COMMITTEE.

Warren S. Johnson, *Chairman,*
Irving H. Reynolds,
Geo. M. Conway,
M. A. Beck,
Geo. P. Dravo,
E. P. Worden,
Edw. T. Adams,
W. G. Starkweather, *Secretary,*
Chas. P. Bossert,
W. P. Caine,
Wm. E. Dodds,
Jno. C. Finney,
Frank Kempsmith,
Bruno V. Nordberg,
G. J. Patitz,
J. F. Max Patitz,
Edgar Piercy,
Edwin Reynolds,
Jno. A. Bechtel,
Howard C. Slater,
Harry E. Smith,
James Tribe,
J. S. Unger,
Henry Weickel.

The Milwaukee "local committee for arrangements" for ASME's 1901 semi-annual meeting was included in the proceedings, as shown here.

It is evident that the engineers on the local planning committee had been meeting at least informally in Milwaukee well before the 1901 ASME meeting. The proceedings of that meeting state, "the members resident in Milwaukee and its business interests made common cause in urging upon the Council the selection of their city, and pursuant to this pressure the arrangement was completed." Since such meetings were typically organized a year or two in advance, the Milwaukee group likely started its campaign to attract the national meeting before the turn of the century. This meeting helped to serve as the nucleus to the formation of local engineering groups.

The proceedings of the 1901 ASME meeting provide some interesting insights. For example, it states, "The hotel headquarters and the auditorium for the reading of the reading of papers were located in the Plankinton House, Grand Avenue and West Water Street. The headquarters was particularly comfortable and convenient, by reason of its size and its location on the parlor floor. The auditorium was in a species of rotunda in a wing, splendidly lighted and airy. The opening session was set for the evening of Tuesday, May 18th and at half-past eight the meeting was called to order by Prof. Warren S. Johnson, chairman of the Local Committee of Arrangements, who in a few sentences welcomed the Society to Milwaukee and turned the business of the meeting over to President S. T. Wellman, who occupied the chair during all the sessions. The President, in brief reply, expressed the pleasure of the Society and their anticipations for a successful and profitable meeting."

The proceedings continue, discussing certain papers that were read and commented upon during the meeting, and then continues, "After announcements by the Secretary a recess was taken, and the members and their ladies became the guests of the Local Committee and their ladies for an informal social reunion in the hotel dining-room and parlors. The entertainment for the gentlemen took the form of a smoker. It was a noticeable peculiarity of the meeting, which manifested itself at this time that a large proportion of those in attendance were graduates of one or another of various institutions devoted to technical education."

Tours of local manufacturing concerns were offered, which provides some insight into what may have been considered prominent engineering-based companies at the time. As noted in the proceedings, "the afternoon of this day (Wednesday, May 29, 1901) was devoted to excursions to the manufacturing interest of Milwaukee. The party was divided along the lines of its individual preferences, so as to provide for visits to the E.P. Allis Company's works, the Filer & Stowell Manufacturing Company, the Nordberg Manufacturing Company, and the Christensen Engineering Company. Special cars of the electric system of Milwaukee were chartered and put at the service of the visitors, who made use of them under a well-ordered time table."

The Proceedings also summarize other local attractions attended by the *Ladies'* Committee, noting, "During the morning and afternoon of this day, the Ladies' Committee entertained their guests by a drive and a visit to the Public Museum and Library Building. In the evening the members and their ladies were the guests of the Local Committee at a reception and conversazione at the Deutscher Club (now known as the Wisconsin Club) on Grand Avenue. The officers of the Society and the chairman of the Local Committee received the guests on their entry into the large reception room and later in the evening a banquet was served in the dining-hall of the club. The club stands in attractive grounds, which were illuminated during the evening."

On the afternoon of May 30, "the Society [attendees] were again the guests of its Local Committee for a ride by trolley cars to the attractive surroundings of Waukesha. Carriages were awaiting the party at the terminus of the road, and after a visit to two or three of the springs, for which the locality is famous, and a drive through its principal streets, the party returned to the cars and were conveyed home. In the evening, a reunion of the members and guests was held in the dining hall of the hotel."

During technical sessions on May 31, "the ladies were again the guests of the hostesses of Milwaukee for a visit to the Layton Art Gallery. Small parties were also escorted in steam automobile carriages through the attractive parts of the city and parks. In the afternoon, carriages provided by the Local Committee conveyed the party to the Mechanical Massage Department of the Sacred Heart Sanitarium, where a most complete exhibit of the Zander machines for Swedish massage were exhibited and much amusement was caused by certain of the younger members who participated actively in the illustration of their completeness. The hydropathic and other departments of the sanitarium were also visited and admired. Leaving there, the drive took the visitors through the Soldiers' Home Park and other attractive park districts of the city, and so back to the hotel."

At the closing session, a lengthy resolution was passed, thanking the Local Committee for its efforts.

Reynolds Elected President of ASME

Edwin Reynolds of the Allis Chalmers Corporation obviously impressed the Society during its meeting in Milwaukee, because he was elected as ASME's president in 1902—one year following the meeting. It appears that his service during the year was interrupted because of health problems. The Proceedings of the year report that he was unable to attend the annual meeting in New York during his term in office because of "ill health."

It is likely that Reynolds' tenure as president aided Milwaukee's efforts to form a local Section of the Society. ASME-Milwaukee formed a local section two years after his term as President of the Society.

Another Milwaukee engineer, Frederick M. Prescott, was active in the national ASME organization in the early 1900s. Prescott was a Manager of ASME (equivalent to a vice president today) between 1906 and 1908. Prescott was president and owner of the Prescott Steam Pump Company, located at 60th and Greenfield Avenues. The company was acquired by the Worthington Pump & Machinery Corp. in about 1916.

A review of the Proceedings of ASME, as well as the *Centennial History of the Society*,[301] indicates that the Society wrestled with how to address the need for local section activity. Society's policies required that papers presented at all meetings of the Society be made available to all members. Furthermore, meetings were all summarized in the annual transactions. When monthly meetings were held in New York, these requirements were relatively easy to deal with. However, it was more difficult to document meetings held at remote locations.

Concerns were also expressed about the use of annual dues. The Society hadn't begun to consider how it might allocate membership dues toward local chapter activities. Gus Henning, an influential member of ASME and involved in its early Standards activities, was familiar with the practices of the *Verien Deutcher Ingenieure* of the German Polytechnic Society. He presented a thoughtful plan for the formation of local sections of the Society at the annual meeting in 1901. However, the plan did not gain traction.

Establishment of the Milwaukee Section of ASME

The Milwaukee Section of ASME was established in 1904 with a territory that initially encompassed a radius of 50 miles from the City of Milwaukee. Matthias Beck was named the president of this local chapter, and W.G. Starkweather the Secretary-Treasurer.[302]

The January 1905 edition of Power Magazine published a short article about the formation of the Milwaukee Section[303]

A branch of the American Society of Mechanical Engineers has been established at Milwaukee. Its affairs are managed by council, of which M. A. Beck is president; Geo. P. Drayo, M. L. Jenkins, S. L. G. Knox, E. P. Worden, members; W. G. Starkweather is secretary-treasurer.

Apparently this didn't sit well with ASME in New York. In 1905, the Society passed a set of rules of establishing local sections. In a contradictory action, however, it quashed the Milwaukee Section's formation, finding that it had taken action without awaiting the Council's permission.[304] The *American Machinist* reported that the Milwaukee Section dissolved as a result, forming a local engineering society.[305]

The rules the Society had established made the formation of local sections prohibitive. Local Sections were responsible for their financial affairs but had no control over technical papers, and the Society had the power to disband the local section on sixty days' notice.[306]

An April 1930 an article in Mechanical Engineering magazine explained the start of the 'Local Section movement.' It read, in part:

> *In the quarter of a century between the founding of the Society and the (1904) request from Milwaukee, the national organization had been holding stimulating meetings every year in cities scattered from the Atlantic to the Pacific coast, and had left behind them groups of men who had realized the healthy thrill that arises from the meeting together of technically trained men of the discussion of mutual professional problems.*
>
> *In the localities where Society meetings had been held were centered the industries and men who were in the vanguard of technical progress. The engineer was consciously contributing to the development of civilization. The mechanical engineer, in particular, had a vision of the potentialities of the machine. He was no longer merely the mechanic and looked after the machine. He was a man who was interested in scientific and mathematical principles because, in the pursuit of his daily work, he was using his head as well as his hands. He wanted to keep abreast of engineering progress because he was not only capable of it but responsible for it.*
>
> *The time was at hand when Society meetings in New York, with one, or at most two meetings each year in other places, were no longer sufficient.*

It took several years for local section issues to be addressed by the Society. A book on the Society's history, written at the time of its centennial in 1980, reports that [307] "In later years, the Society would list the Milwaukee section as its first, founded in 1904, but it was, in fact, another six years before the legitimacy of geographic divisions was even secured. The problem was common to the entire field of engineering professionalism. ... And in 1904, when engineering's institutional arrangements were still far from resolved, local sections seemed as much a threat as an asset." An American Machinist editorial in 1905 commented, "Throughout the profession, the matter of local sections was receiving a great deal of attention." It went on to describe the movement as perhaps the only way members distant from New York could effectively participate in a[n engineering] society.

ASME sanctioned sections were formed in St. Louis and Boston beginning in 1909. However, ASME did not fully resolve the matter of local sections until 1911, when it recommended that members in any city be allowed to organize local meetings and that the Society assume some of the costs of those meetings, using member dues. It also devised a formula for financial local activity that allowed section membership by those that were not ASME members—which had been a sore point by local engineers desiring to organize a section in Milwaukee.

Because of this action, the Milwaukee Section of ASME was officially re-established in 1911—seven years after it first formed and attempted to obtain the national society's recognition.

The Engineers and Scientists of Milwaukee

Between 1904 and 1911, the former officers of ASME-Milwaukee redirected their efforts by forming a local engineering society that was not directly affiliated with the national engineering societies—they formed the Engineers and Scientists of Milwaukee (ESM). ESM was officially incorporated in 1905 and held its first official meeting in the Colonial Room of the Plankinton House on December 13, 1905, to elect officers and nine directors. George P. Dravo was elected as their first President. Dravo was one of the founding members of the Milwaukee Section of ASME. ESM had an initial membership of fifty-nine engineers.

In addition to the group of engineers that had attempted to form a local chapter of ASME, ESM's early members included members of the American Institute of Electrical Engineers, the *Vereines Deutcher Ingenueure*, the American Chemical Society, American Society of Civil Engineers and other non-affiliated engineers and technical men in and around the City of Milwaukee. It is apparent that ESM's activities were intertwined with all of the local engineering professions in the early years. The small number of engineers present in the area made it inevitable that ESM, acting as a society of societies, would host joint meetings.

ESM's bylaws provided for the establishment of "affiliates." It created a council in which representatives of affiliate societies could participate in the affairs of ESM. It is reported that Matthias Beck traveled to a number of cities in the United States, attempting to set up similar local engineering societies and relationships. Apparently Beck's efforts were met with some success. By 1909, ESM was affiliated with local engineering societies in eight other cities and states and an Association of Engineering Societies was formed.

During the early 1900s, ESM held many joint meetings with the affiliates and other non-affiliated societies. It also was host to national presidents of these affiliates, who addressed joint meetings of the societies. Records show that fifteen national engineering society presidents came to Milwaukee for talks, including eight presidents of ASME.

Meanwhile, efforts to establish a local section of the American Society of Mechanical Engineers continued. The minutes of a Council meeting on August 30, 1911, quotes Matthias Beck as reporting that "not much progress had been made toward securing a branch of ASME" (in Milwaukee). However, Fred Dorner stated, "he had been east and had a talk with the Secretary of ASME and he thought that they were looking very favorably upon the proposition."

In 1912, ESM sponsored a joint meeting of all technical societies in Wisconsin. This meeting took place during the month of March in Milwaukee Auditorium. This was a large gathering—five hundred attended the morning session, six hundred and fifty at the afternoon session, and five hundred at the evening session. It is evident from this meeting that there was a large concentration of engineers living and working in the Milwaukee area.

In 1916, ASME established a "Committee on Sections" which visited four cities, Milwaukee included, to review Section activity and report back to the Society. The Committee members visited Milwaukee on October 23, 1916, and toured the plant of the Allis-Chalmers Company and later had dinner and discussed

Section matters with members of the Milwaukee Section. At the meeting, they reported, "a movement was started to develop the membership and scope of the Milwaukee Engineering Society with which the A.S.M.E. Section is affiliated."[308]

On August 1, 1944, the Engineers and Scientists of Milwaukee acquired a stately mansion at 3112 W. Highland Boulevard originally built by August Pabst for his son, Frederick. Remodeling was completed in 1946. With new headquarters building, the affiliated societies were able to expand their programs.

In 1948 William Monroe White, who was a former chairman of the ASME Milwaukee Section, presented the Engineers and Scientists of Milwaukee with much of his personal engineering library, along with funds to prepare a library room in the ESM headquarters.[309]

However, the organization was eventually forced to sell its headquarters building on West Highland Boulevard in order to settle a tax dispute with the Internal Revenue Service.

Professional and Social Cooperation

Not only did local engineers get together professionally to share information on the science of engineering—they also got together socially. In fact, the preamble to the Constitution of the Milwaukee Branch of the American Society of Mechanical Engineers stated:

> *The object of this Association is to* ***promote professional and social cooperation*** *among members of the American Society of Mechanical Engineers and others engaged in engineering work.*
>
> Preamble to the Constitution of the Milwaukee Branch of the American Society of Mechanical Engineers, 1904

In the first decade of the 1900s, there were only about seventy members of ASME in the entire State of Wisconsin and, of course, a somewhat lesser number in the Milwaukee area. But membership was growing rapidly. By 1920, the Milwaukee Section of ASME had 144 members. After the Engineers and Scientists acquired a headquarters building in 1944, the Section began holding its meetings and technical programs there and the Section's membership grew to 358 in 1947.

Along the way, the Milwaukee Section established a Ladies Auxiliary, which functioned for a number of years, holding its own meetings. The Ladies Auxiliary members were generally not engineers, but rather spouses of engineer members of ASME. Announcements of at least two of their meetings appeared in the local newspapers.

Trade Conventions

It is apparent that national trade conventions were periodically held in Milwaukee. For example, in 1918 the simultaneous gathering of four leading metal trade organizations was held here—along with a large exhibition. The trade convention brought in participants from across the country—Milwaukee's foundry businesses were lauded.

While a search of the local trade conventions held in Milwaukee was not conducted, references to several were found during the research for this book. Such trade conventions would certainly have fostered the interaction of manufacturing and labor professional and have resulted in reinforcing the environment for important manufacturing tradesmen.

OSCAR WERWATH was the founder and first president of the Milwaukee School of Engineering in Milwaukee.

Oscar Werwath was born in Stallupönen, Germany on May 3, 1880. His parents were Carl and Johanna Werwath; his father was a department store owner. Oscar received degrees in mechanical and electrical engineering from the Saxony Technical College and pursued graduate studies at the universities of Hanover and Darmstadt.

Oscar came to Milwaukee when he was twenty-three, initially taking a job as an electrician at the Mechanical Appliance Company, the predecessor of the Louis Allis Company. Soon, however, he was serving as an electrical engineering consultant in the area. He recognized that Milwaukee lacked an engineering school that would accommodate the needs of employers. He approached the president of Rheude's Business School and was given permission to establish a series of night classes for young men in practical electricity. Soon, enrollment in Werwath's courses exceeded the facilities available to Rheude's school. He was encouraged by Louis Allis to open an engineering school and, in 1905, with the donation of $500 from Louis Allis he opened the "School of Engineering." ("Milwaukee" was added to the school's name in 1932).

Oscar was president of the school until his death in 1948. Under his guidance, Milwaukee School of Engineering was one of the first engineering schools to adopt a plan of co-operative training. His program permitted students to take part-time jobs in industry while enrolled in school, thus partially financing their education while receiving practical work experience.

Werwath established the Milwaukee Electric Construction Company within the School of Engineering, to provide students with training by making electrical products while earning money to cover tuition. Among other products, the company manufactured lead-acid electric storage batteries. By 1911, the electric storage battery business was taking up too much room in the overcrowded facility and Werwath decided to sell the business to private investors. By the following year, the investors moved battery production and established the Globe Electric Company, which eventually became Globe-Union.

Oscar married Johanna Seelhorst, a daughter of Friedrich Seelhorst, in 1908. They had four children—Karl (1909-1979), Greta (1910-2003), Hannah (1913-1984), and Heinz (1916-1987). Karl was a graduate of MSOE in 1936 and became president after his father's death in 1948. Oscar's other children were also employed at the school from time-to-time.

Werwath was a member of the American Institute of Electrical Engineers (predecessor to IEEE), the Milwaukee Electric League and the Association of Commerce.

He is buried at Forest Home Cemetery.

CHAPTER 11: INSIGHTS

The previous chapters of this book illustrate the importance of engineering innovation. It clearly was a major driver of Milwaukee's economy, contributing to the "transformation of back yard shacks into great factories"—which William Bruce referred to as the as "the great romance in Milwaukee industry."[310]

This was a remarkable century of progress, and the innovation that occurred here has had a lasting impact on Milwaukee. The City's industries manufactured products that satisfied the needs of society and, in doing so, directly contributed to the area's prosperity.

Economist Robert M. Snow studied the connection between innovation and its impact on American prosperity.[311] His work was cited by the PEW Charitable Trust in 2015, concluding that:

> *Innovation is the engine of American prosperity. Economists now recognize that up to 80 percent of modern economic growth arises from technological innovation—the process of developing new knowledge and harnessing those discoveries to create better commercial products, more competitive industries, and well-paying jobs.*

This chapter attempts to review how technological innovation occurred during Milwaukee's industrial past and to provide insights that may help to develop a climate that will foster innovation in the future.

INNOVATION IS FOSTERED BY COMMUNITIES

Innovation does not occur in isolation; inspiration might, but without the support and foundation provided by others, inspiration will not lead to innovation. Innovation requires a foundation of knowledge and experience.

Innovation needs motivation, training, reinforcement, opportunities for trial-and-error, sharing of ideas, transfer of expertise, and a knowledge of the marketplace.

Innovation seems to prosper best in *communities of practice*—communities in which others are working toward similar, or somewhat allied efforts, using tools, backgrounds, and methods that are complementary. These communities were able to support, and indeed were necessary elements toward achieving the innovation that occurred in Milwaukee. Economists Gary Pisano and Willy Shih defined such a community as an *industrial commons.* They believe that the restoration of the *Commons* might be key to restoring American competitiveness.[312]

THE IMPORTANCE OF COLLISIONS

One of the current themes about generating innovation is the importance of "collisions"—the creation of environments where a diverse set of individuals can interact—as one key to generating new business ideas.[313] This book describes how many of the early Milwaukee firms responsible for engineering innovation were located in close proximity to one another in the City's early industrial neighborhoods. This was most evident in the Walker's Point neighborhood; it created numerous opportunities for employees to interact, which likely contributed to innovation and economic enterprise.

Walker's Point, which sits at the foot of Milwaukee's Menomonee Valley, was a truly remarkable community in the late 1800s. Perhaps more than any industrial neighborhood in the United States, Walker's Point personifies an industrial commons—a community of practitioners able to achieve remarkable innovation. In this neighborhood, companies such as E.P. Allis, Pawling & Harnischfeger, Filer & Stowell, C.J. Smith & Sons (predecessor of A.O. Smith), Kearney & Trecker, Nordberg Manufacturing, Obenberger Drop Forge, Louis Allis's Mechanical Equipment Company, Allen-Bradley,

George Meyer Manufacturing, Evinrude Outboard Motor and Christopher LeValley's Chain Belt Company, were all located within a few blocks of each another.

In some cases, the companies directly supported one another. For example, Pawling & Harnischfeger provided patterns for E.P. Allis's foundry, and Obenberger Drop Forge provided forgings to a number of local manufacturing firms. Pawling & Harnischfeger also provided shop space to allow the fledgling Nordberg and Chain Belt companies to become established, as well as space where the Milwaukee School of Trades initially operated.

However, there was another important way in which collisions occurred between employees. From the stories in the preceding chapters it is evident that the workforce was relatively mobile. Since the manufacturing companies were closely located, and employees lived in close proximity to the companies, employees could move from positions at the various companies as opportunities presented themselves.

In addition, employees from the respective companies interfaced with one another at the area's churches, bars and social clubs, and their children went to school together.

It is evident that this frequent interaction helped to foster the cross-pollination of talent and ideas in Walker's Point, as well as in other industrial neighborhoods of the Milwaukee area.

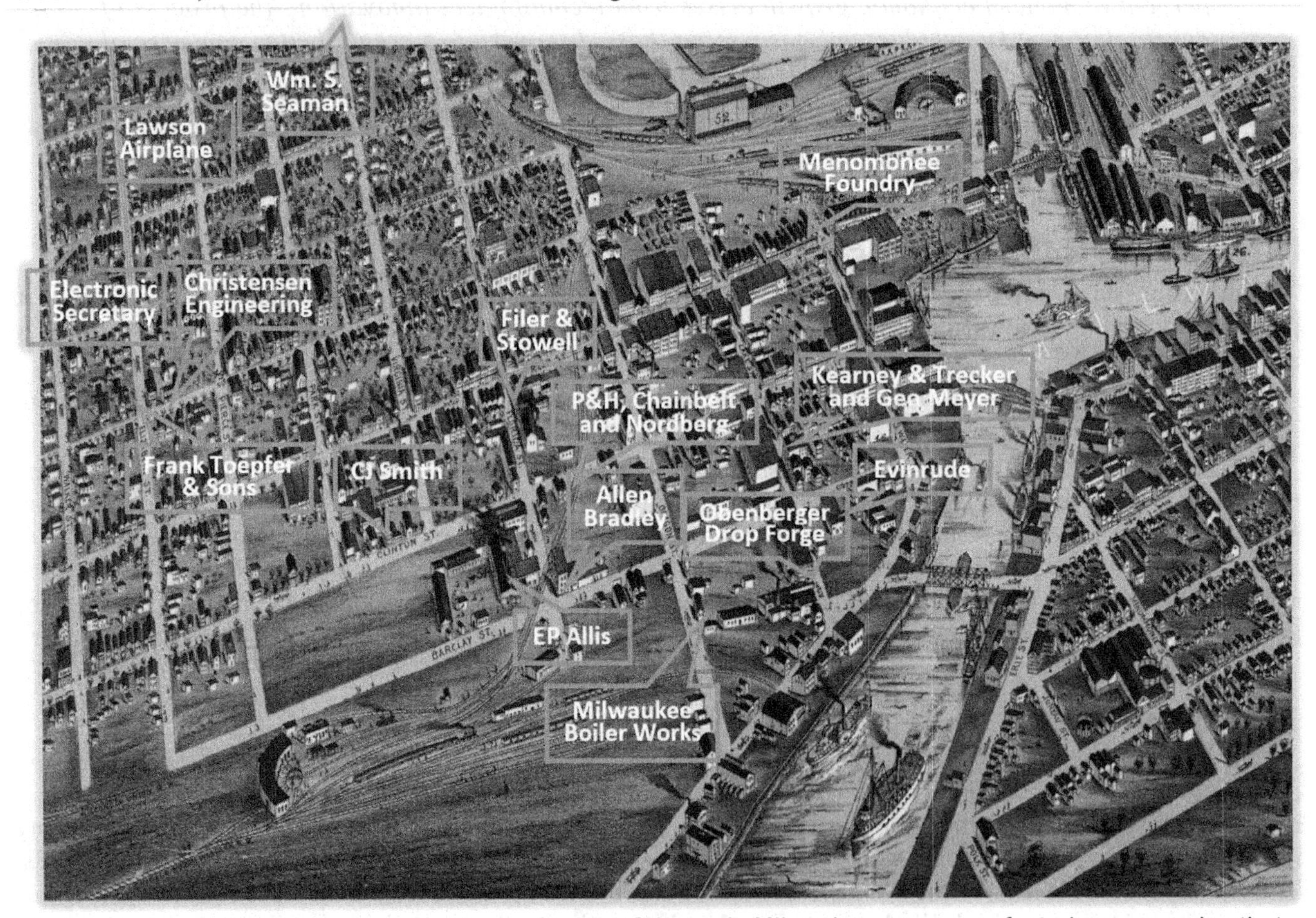

This map shows the approximate locations of some of the early Milwaukee-area manufacturing companies that existed in the Walker's Point neighborhood. It reveals how highly concentrated Milwaukee manufacturing was during this period. Interspersed with the above companies were also miscellaneous foundries, job shops and other manufacturing orientated companies. Note, not all of the above companies operated at the same time. From 1872 Illustrated map of Milwaukee.

The following diagram attempts to illustrate the relationship of companies that were in some way associated with E.P. Allis Company—companies that were formed by individuals that formerly worked there, or were influenced in some significant way by a relationship with Allis—as well as some of the other early interrelationships in the Milwaukee area.

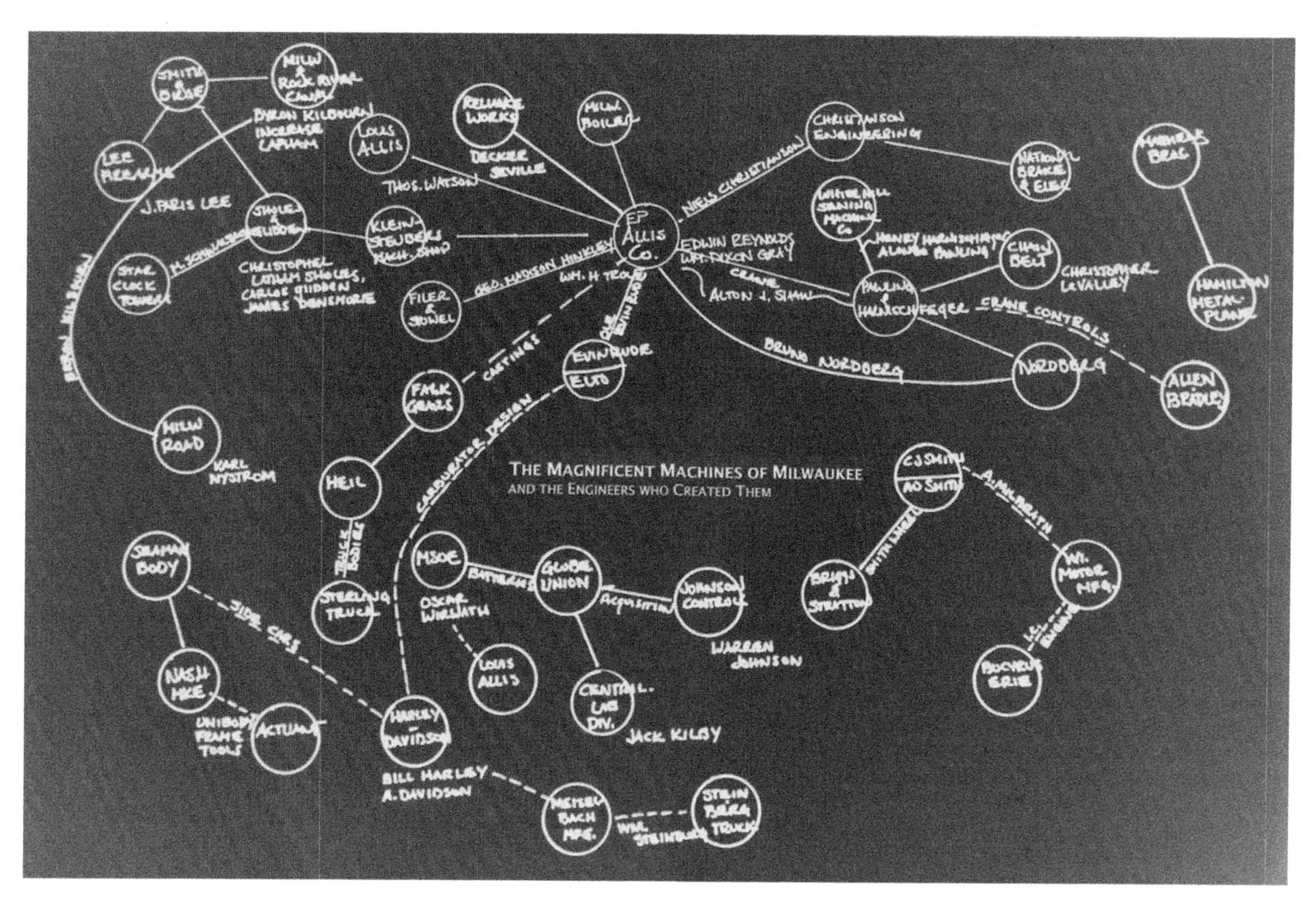

Currently Milwaukee has three consortiums designed to bring innovators together to collaborate in their respective fields—the Water Council, the Midwest Energy Research Consortium (M-WERC), and FaB-Wisconsin (the food and beverage industry cluster organization). It is hoped that these three organizations can help to bring back the industrial commons that once existed in Milwaukee.

The Importance of Collaboration

In one of the most remarkable accountings of the way innovation takes place, George Iles' *Leading American Inventors* tells the story of the development that led to the typewriter in Henry Smith's mill.[ppp] His account is based on an interview with Frederick Heath, who served as a member of the Milwaukee County Board for many years and also as president of the Milwaukee County Historical Society. As a young man, Heath was a friend of one of Christopher Latham Sholes' sons and was employed as a messenger by Sholes to deliver drawings and plans to those that were working on the development of his typewriter.

In describing the collaboration of Sholes, Soule and Glidden, on the design of a typewriter, Iles wrote:[314]

> *They began work at once in a small room on an upper floor of a mill owned by Henry Smith, an old friend. This two-and-a-half story building, in simple ashlar, stood on a narrow strip of land between the Milwaukee River and the Rock River Canal.*
>
> *Here, day by day, Sholes drew his plans with Soule's aid, and here their model gradually took form, proving to be a thorough success in a final test. On the same floor of the mill was the workshop of another tenant, Carlos Glidden, the well-to-do son of a retired ironmonger. Glidden was an inventor, too, and he was developing a spader which he believed would outdo the work of any plow on the market.*
>
> *Naturally, there arose many a colloquy betwixt the three inventors regarding their plans, with much debate of the weak points disclosed as their experiments followed one another. [...]*
>
> *In that grimy old mill on the Rock River Canal there were interludes to lighten and brighten the toil of experiment.*
>
> *All three partners were chess players of more than common skill, and they often turned from ratchets and pinions to moves with knights and pawns. Ever and anon a friend would drop in, and the talk would drift from writing by machinery to 'Reconstruction' in South Carolina, or to the quiet absorption by farms and mills of the brigades mustered out after Appomattox. Then, with zest renewed, the model was taken up once more, to be carried another stage toward completion. One morning it printed in capitals line after line both legibly and rapidly. Sholes, Soule, and Glidden were frankly delighted.*

The above description of the collaboration of Sholes, Soule, and Glidden provides insight into how they worked together—exchanging ideas, providing criticism and encouragement, and socializing together playing cards. While such collaboration wasn't evident with all of the innovators in this book, some of whom toiled in relative isolation, it certainly aided in the development of a number of important innovations.

Another example was the collusion that occurred between Ole Evinrude and Arthur Davidson. Evinrude eventually helped Davidson and his friend Bill Harley to design a carburetor that would provide the right fuel flow for their motorcycle engine, as well as provide other advice regarding their engine design—friends helping one another to achieve success.[qqq]

[ppp] Different accounts note the collaboration for the development of the typewriter by Sholes, Glidden and Soule that occurred at Kleinsteuber's Machine Shop. It is likely that the team used both facilities, from time-to-time, to meet their needs. Mattias Schwalbach listed Henry's Smith's Mill as his address and probably used the mill for his model work. We do know that Schwalbach and Densmore used the mill for the construction and assembly of the Milwaukee-built typewriters.

[qqq] According to some reports, Ole Evinrude provided more help with Harley-Davidson's motorcycle engine, citing the numerous similarities between Evinrude's outboard engine and Harley-Davidson's 1904 engine.

The collaboration between Arthur Davidson and Ole Evinrude is apparent from this photograph. Ole Evinrude (top) and Arthur Davidson (bottom) owned a pattern shop in 1902. Evinrude went on to share his expertise in building engines with Davidson as well as Bill Harley. The photo is included with the permission of Jean Davidson, author of "Growing Up Harley-Davidson."

Impact of Outsourcing

If the interaction and collisions that occur as part of communities are important to innovation, then it would seem that outsourcing innovation doesn't easily mix.

Two examples from Milwaukee's industrial history illustrate the problems with outsourcing and of separating engineering design and support from manufacturing. Both involve E. Remington & Sons. Founded in 1816 by Eliphalet Remington in Ilion, New York, the company had the skill and experience for manufacturing relatively intricate parts. Furthermore, their location near the eastern end of the Erie Canal would have made them familiar to many who would have passed through the area on their way to Milwaukee. By 1828, Remington had moved its forge and foundry to a site immediately adjacent to the canal. The Erie Canal provided a link between the East Coast and the Great Lakes, making transportation relatively convenient between the two areas—at least during the shipping season.

When James Paris Lee of Milwaukee needed assistance in boring the barrels for his slant breech-loading rifle in 1864, he turned to Remington. Accurately boring a rifle barrel required skills and tooling that were not available in the Milwaukee area. Remington specialized in such work and its barrels were recognized for their quality and reasonable price.[315] Unfortunately, a communication error resulted in the barrels being bored to the wrong caliber. As a result, they were rejected by the United States Army and did not see use during the Civil War.

It's a mystery what caused the rifle caliber error. Remington supplied the arms for the US Army during the Mexican-American War (1846-1848) and had considerable experience meeting military requirements. Perhaps the Army changed its specifications. It has been reported that James Lee sued the government for $15,000 and eventually settled for somewhat less than that amount.[316] In any case, the experience must have convinced James Lee that manufacturing firearms out of Milwaukee wasn't ideal. He soon relocated to New York State and spent most of the rest of his career in collaboration with Remington, which produced the majority of Lee's future firearm designs.

Lee went on to make significant contributions to firearms. The Lee Model 1879 rifle was a landmark rifle design, incorporating turn-bolt action and Lee's spring-loaded column-feed magazine system. It was adopted by the US Navy, and later models were used by the British military for decades. It is reported that virtually all existing bolt-action rifles were influenced by Lee's revolutionary design.

James Paris Lee is described as "one of the 19th Century's greatest gun designers."[317] It is unlikely that Lee would have been able to achieve this success had he remained in Milwaukee. He needed the frequent collaboration with Remington and other gun designers in the East in order to achieve his innovations.

The partners associated with the production of Milwaukee's Sholes & Glidden 'Type Writer' also outsourced manufacturing to E. Remington & Sons. The Sholes & Glidden group had initially set up manufacturing in Milwaukee, but the fifty or so typewriters manufactured were costly to produce and were inconsistent in quality. James Densmore took the design to Philo Remington, who agreed to manufacture the typewriter under contract. After manufacturing moved to Ilion, New York, additional refinements were made, resulting in a design that became standard in the industry.

The Remington version of the Sholes & Glidden 'Type Writer.' Photograph by the author taken at the Milwaukee Public Museum

Christopher Latham Sholes continued to work on typewriter design from Milwaukee. He eventually developed and patented a *visible* typewriter that would permit the typist to immediately see what was typed onto the page—a significant improvement over the Sholes & Glidden upstroke typewriter. However, by then others had achieved success using different approaches. As a result, Sholes's continued innovative efforts were too late and didn't have a significant impact on typewriter innovation.

Sholes died before his visible typewriter went into production. His sons eventually manufactured it under the C. Latham Typewriter Company name. While their venture was not commercially successful, they sold the rights to August D. Meiselbach of Milwaukee in 1900 who made the typewriter in Kenosha for several years.

In this stylized illustration from the early 1920s, Sholes is shown with one of his experimental typewriters. The artist idealized Sholes thinking about the impact of his invention upon women. Late in life, Sholes expressed pleasure that his invention had turned out to be "a blessing to mankind, and especially to womankind." The typewriter led to the entry of women into the office, a previously all male, inner sanctum of business. Illustration courtesy of the Herkimer County (NY) Historical Society.

Innovation Springs from a Foundation of Knowledge and Experience

Even the most creative and talented individuals could not simply sit at a drawing board and create innovative mechanisms. A review of the engineers described in this book illustrates that knowledge and experience are critical foundations to innovation.

As just one example, steam-engine builder Edwin Reynolds developed his improved Corliss valve mechanism only after working directly for the famed Corliss Engine Works. Prior to that, Reynolds worked for sixteen years at various jobs involving the machining and installation of steam engines. He had a foundation of knowledge upon which to build.

A metaphor of dwarfs standing on the shoulders of giants has been used to describe this foundation. The concept, attributed to Bernard Chartres in the 12th century, was expressed in English by Isaac Newton in 1676 as, "If I have seen further, it is by standing on the shoulders of giants." Reynolds obviously benefited greatly by "standing on the shoulders" of George Henry Corliss, which allowed him to improve and expand upon the Corliss valve design. Without this foundation of knowledge and experience, it is improbable that Reynolds would have succeeded as he did at E.P. Allis, becoming one of the country's foremost engine builders.

This picture, derived from Greek mythology, might be the original source of the term, *On the Shoulders of Giants*. It depicts the blind giant Orion carrying his servant Cedalion on his shoulders to act as the giant's eyes. Source: Wikipedia Commons

THE ROLE OF PERSEVERANCE

Perseverance was key for several of the inventors outlined in this book. For example, it is evident that the development of the typewriter would not have become a commercial reality without the skills of James Densmore, to both finance the project and serve as the catalyst and driver that motivated the team.

DON'T GIVE UP UNTIL YOU GET IT RIGHT

Christopher Latham Sholes and his fellow inventors seemed happy with the typing mechanism they initially patented. They enticed James Densmore to help them promote the product and finance its manufacture.

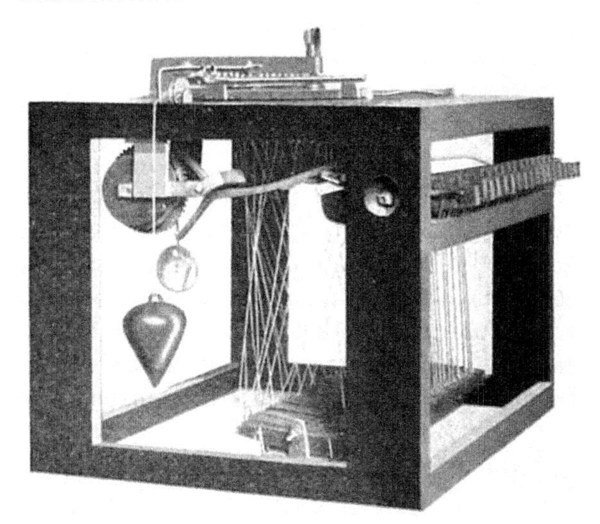

1868 Patent Model[318] of Sholes & Glidden 'Type Writer' From Charles Edward Weller, *The Early History of the Typewriter*, 1918, p. 81.

The mechanism was incredibly crude and only able to leave impressions on the bottom side of very thin paper, which was held flat on the top of the device. Fifteen of such typewriters were produced and observed in use—some of which were employed at a school for telegraphers in Chicago. It was clear to the team that this design was not yet marketable. Following these observations, Sholes and Densmore developed what they considered to be fundamental ideas that must be satisfied for the device to be successful: "the machine must be simple and not liable to get out of order," that "it must work easily and be susceptible of being worked rapidly," and "it be made with reasonable cheapness." Additionally, Densmore insisted that a successful typewriter must be capable of writing on paper of ordinary thickness—as opposed to the early designs by Sholes that only printed satisfactorily on paper that was tissue-thin.

In September 1869, Sholes declared that he had perfected all the necessary principles, writing to Densmore on the machine, "I am satisfied the machine is now done." However Densmore was not satisfied. He continued to press for additional improvements, much to the annoyance of Sholes. Somewhat reluctantly, Sholes continued to work on refinements. He next adopted a refined keyboard, devised by Schwalbach, which involved four rows of metal key levers and buttons set in ascending banks. At the urging of a customer who tried this design, a space bar was added underneath the four rows.

In the summer of 1871, the team began manufacturing their improved typewriters. However, they found that they were far from perfect. They had durability problems and the type bars didn't stay in alignment. Production was stopped and the team tried different approaches to keep the type bars in alignment. It wasn't until year-end 1872—three years after Sholes declared that he was satisfied that the machine was perfected—that a satisfactory design was developed.

By the time the Sholes & Glidden 'type writer' reached its final form in 1874, most of the team had dropped out. Even Sholes's financial interest in the device had been largely diluted. Densmore was the only one who became rich off the product—which might be appropriate since he was the one who had persevered to ensure that the device was a commercially successful product. He understood that you can't give up until you get it right!

While it took six years for Sholes, Glidden, Soule, Densmore and others on the team years to develop a workable, commercially salable typewriter, less than a century later it took Jack Kilby only a few weeks to develop the world's first microchip.

Of course, while Kilby's 1958 initial microchip design was functional, it wasn't ready for prime time. It needed the design and manufacturing muscle of Texas Instruments to ready it for the market. And the first microchips were very expensive. As a result, the use of the early microchips was relegated to critical military defense applications where size was critical. It wasn't until 1972 that Texas Instruments introduced its first calculator based on its integrated circuitry—priced at $149.99![rrr]

Thomas Edison summed up the need for perseverance with his famous quotation, "Genius is one percent inspiration and ninety-nine percent perspiration."

Keep Innovating!

By all accounts, Niels Christensen was a prolific inventor. Following his initial patent applications for electric streetcar braking systems, he continued to develop innovative machinery. Most of his mechanisms involved some form of compressed air systems, involving air compressors, control valves and storage systems. Piping leaks involved in the various components would have been a source of frustration. His braking systems were designed to be fail-safe upon loss of pressure. However, that meant that the brakes would be automatically applied upon loss of air. If the system was on a streetcar, or later a heavy-duty truck, an air leak could cause the vehicle to be immobile until air pressure was restored. Christensen knew that leaks had to be minimized.

He eventually went on to hydraulic systems, involving a hydraulic oil. Minor leaks around piston shafts resulted in leakage of hydraulic fluids. Again, Christensen realized that leak-free designs were critical.

Dealing with such systems for much of his life led Christensen to look for better sealing systems—reliable, yet less expensive and easier to connect. He eventually found his solution when he was in his seventies, patenting his innovative O-ring sealing system.

Christensen didn't give up; he kept innovating until he found a solution—and then had to work for years to convince others that his sealing system was reliable. Christensen's many years of working with the design of compressed air and hydraulic systems provided him with the necessary foundation to succeed.

[rrr] Texas Instrument supplied its integrated circuitry under contract to Canon and Monroe for pocket calculators as early as 1970.

THE IMPORTANCE OF BUILDING THE RIGHT TEAM

Edward P. Allis was able to transform a small manufacturing company producing mill wheels and various castings into one of the largest companies in the United States, if not the world. He was able to accomplish this feat by hiring the right talent and providing them with incentives to succeed. It's worth exploring in some detail how he was able to assemble the remarkable engineering talent and set them to work on designing innovative machinery to meet the needs of the market.

FINDING/HIRING THE BEST INNOVATIVE TALENT

After acquiring Dexter and Seville's Reliance Works, Edward P. Allis recognized the Milwaukee area marketplace needed better sawmills, milling equipment, and steam engines. He set about hiring the best talent available for these three technologies. In short order, he recruited George Madison Hinkley (sawmills), William Dixon Gray (milling equipment), and Edwin Reynolds (steam engines and rotating equipment). He recruited all of them from jobs elsewhere and, at least in the case of Edwin Reynolds, from better-paying positions.

In order to entice this troika of talent, Allis provided incentives that were remarkable for the time. He promised each that they would share in the ownership of any patented innovations they designed, and he honored them with naming rights on the associated machinery lines.[319] It appears that both of these concepts were unheard of at the time. Engineers often toiled in relative obscurity, unless they owned or established the company. Patents were almost always issued in the name of the owner of the company, and machines were likewise named. But by promising to share ownership of the patents, and then pay these engineers for their use, Allis must have created an irresistible incentive for these three men to join the fledgling company. All three gained prominence in their respective fields and the E.P. Allis Company became the largest area employer.

Edward Allis didn't stop his talent search with these three individuals. Over time he was able to recruit the likes of Alton J. Shaw (overhead cranes), William H. Trout (sawmill equipment), Irving Reynolds (steam engines/pumps), Bruno V Nordberg (mine hoists and steam engines), Niels Anton Christensen (streetcar air brakes), as well as Ray Ellsworth, Ole Evinrude, and Joseph Merkel who went on to fame after leaving the company.

Allis's incentive system wasn't without its flaws. William Trout wrote a remarkable memoir that reveals how he chafed under the system.[320] Trout was assigned to work for George Hinkley, who coveted patent coverage for any innovations by employees in his department. After seeing Hinkley benefit from his designs, Trout began doing his most innovative design work at home, and then seeking patent protection himself. In his book, Trout makes evident the resentment he felt about Hinkley taking credit for his innovations.

There is also a story of how E.P. Allis granted Bruno Nordberg the ability to sign his design drawings at E.P. Allis. This was an unusual practice at the time—drawings were typically signed by the chief engineer of the department. However, because of Nordberg's exemplary work and skills he was eventually allowed to sign his drawings. After a few years, Nordberg left E.P. Allis to form his own company. Shortly after leaving the firm, his upstart company bid on a steam-driven hoist for a mine in Michigan's Upper Peninsula. Since Nordberg Manufacturing Company was an unknown entity, the purchaser was reluctant to award him the contract. It was only after showing that his name was on the drawings for the previous mine hoist purchased by that company that they felt comfortable giving him the contract.

It is evident that employing engineers with the talent and skills to create new or better-designed products can result in innovation that will transform a company. If this impact were to be correctly recognized, one would expect to see recruitment of talent for industry similar to the recruitment of athletes for professional sports teams. Unfortunately, there is no equivalent of the National Football League's draft available for hiring engineers and other innovators.

INCENTIVIZE INNOVATION

The various inducements offered by Edward P. Allis to his key engineers provide excellent examples of the power of incentives. While Allis' incentive arrangements were not without flaws, including the negative influence the arrangements had on subordinate employees, the company was incredibly innovative during this period. Its principal engineers became well known and the company prospered. The incentives appear to have created a strong inducement that fostered innovation.

The evidence presented in this book suggests that innovation is fostered when innovators share in the profits associated with their inventions.

The design of an effective incentive program for innovation, however, is not straightforward. Providing high awards for good ideas can result in a flood of ideas. However, that can result in a chaotic situation in which the company has a difficult time processing numerous ideas looking for one or two gems that are truly merit-worthy.[321]

Currently, many tech companies provide employees with stock incentives to motive employees. A better incentive program to foster innovation might be to provide incentives directly linked to product successes, with significant awards to the innovators that provided the underlying insights and technology. Such an incentive system should require the innovators to work on final design and manufacturing phases through successful product launch, which would help ensure that the company's interests are properly aligned with its innovators.

WELCOME IMMIGRANTS

Many of the engineer-innovators covered in this book were immigrants. Over forty percent of the engineers highlighted with biographical information in this book were from other countries. Only twenty percent were born in the Milwaukee area. They were welcomed in the early Milwaukee factories, in some cases because of their education and training, but perhaps also because they were motivated and hardworking and may have been willing to work for less pay.

Today we have erected barriers to entry that make it more difficult to recruit and hire applicants from other countries, yet statistics seem to indicate that immigration fosters innovation. For example, in 2014 immigrants were nearly twice as likely as native-born Americans to start businesses. The Kauffman Foundation reports that more than 40 percent of the Fortune 500 companies in 2010 were founded by an immigrant or the child of an immigrant.[322] Similarly, a recent study by the National Foundation for American Policy, a non-partisan think tank based in Arlington, Virginia, indicates that immigrants were involved in more than half of current United States-based startups valued at $1 billion or more.[323]

CREATING AN ENTREPRENEURIAL CULTURE

Lately, significant emphasis is being placed on creating a corporate entrepreneurial culture. Companies have installed foosball tables in break rooms, provided free beverages, instituted casual dress policies, sponsored frequent parties, and on and on. There isn't much evidence that such policies were used by any of the Milwaukee-area companies during the Century of Progress studied in this book. However, other issues likely came into play, which might be much more important in the long run:

- Treating employees with respect and fairness
- Communicating well
- Establishing high expectations
- Rewarding employees appropriately
- Recognizing individual contributions
- Celebrating successes.

In addition to a corporation's desire to create an entrepreneurial culture, there are things that the community at large can do to encourage innovation and celebrate success. As a few thought-starters, such measures could include:

- Periodic news articles about innovative achievements
- Listing of patents in local newspapers
- Displays featuring past innovators at local museums
- Tweaking the educational curriculum to foster the next generation of inventors
- Celebrating past successes
- Naming streets, parks or schools after innovators.

One of the few Milwaukee-area streets named after a Milwaukee inventor is North Sholes Avenue, located just north of 66th Street and Center on Milwaukee's Northwest side. While there is an East Smith Street in the Bayview area, it is unlikely this street was named for a member of the A.O. Smith family. Milwaukee used to have streets named after Lee, Reynolds and Johnson, although it is not known if these streets were all named after the inventors. Waukesha has a middle school named after Harold Horning of Waukesha Motors and West Allis has a street named Theodore Trecker Way. Image courtesy of Google Maps.

DEVELOP THE SKILLS NEEDED TO INNOVATE

If knowledge and experience provide the foundation for innovation, how can they be acquired? The experiences of engineers Ole Evinrude and Jack Kilby provide contrasting examples of building the necessary foundation for innovation.

LEARNING THE TRADE

Ole Evinrude quit school after third grade, although he never stopped learning.[sss] He left to help his dad on the farm, as well as to take various other jobs, including work at local tobacco warehouses. He moved to Madison when he was fifteen and became an apprentice at the Fuller and Johnson farm machinery shop, and then moved on to similar positions at other Madison machine shops. He then relocated to Pittsburgh, Pennsylvania and worked in a steel-rolling mill, where he learned about metallurgy, and then moved to Chicago where he worked at a series of companies and learned machine tool making. By 1900, at age twenty-three, he moved to Milwaukee to run the pattern-making operations at the E.P. Allis Company. His curiosity and interest let him to read anything he could about mechanical science and internal combustion engines.

While working at Allis, Evinrude began building engines during his free hours, experimenting to develop efficient carburation. He also began testing his engines in a carriage. He formed a company with a machinist named Clark, but the partnership soon failed—in part because of the two men lacked marketing skills.[324] He soon established another partnership, forming a company called Clemick and Evinrude, to design and build custom engines for use by others. They received orders for engines from several automobile companies, as well as an order for fifty portable engines from the United States government. While successful for a time, the company eventually failed because of a disagreement between the principals. Evinrude tried again, forming a company he called the Motor Car Power Equipment Company. His partner was a retired furniture dealer. This company also failed, again partly because the partners lacked marketing skills and resources.[325] Somewhere along the way, Evinrude also formed a partnership with Arthur Davidson and the two ran a pattern shop in 1902. This shop was also unsuccessful.

After at least four unsuccessful ventures, Evinrude tried one more time. The story of how he designed his detachable outboard motor is discussed in Chapter 2. He eventually formed the Evinrude Detachable Outboard Motor Company with another partner—Chris Meyer. This enterprise was a winner. Sales were strong and the company expanded several times. Even then, however, Evinrude was not done. Ole Evinrude sold out to his partner, in part to care for his wife who was having health issues. Evinrude and his wife traveled extensively. Meanwhile his former company struggled. Meyer was reluctant to introduce innovations. Evinrude eventually approached his old company with an idea for a lightweight aluminum outboard motor. His former partner wasn't interested. Evinrude considered the idea too good to pass up and started a new venture—the ELTO Outboard Motor Company (ELTO was an acronym for "Evinrude Light Twin Outboard). The ELTO was successful and, in 1922, Evinrude acquired his former company, as well as competitor Johnson Motors of South Bend, Indiana.

Ole Evinrude's experience was not unlike that of Henry Ford, who experienced multiple failures before succeeding with his Ford Motor Company.

Evinrude acquired his knowledge largely through apprenticeships and self-study. His experience was garnered through employment at numerous companies and the successive failures of the companies that he started in partnership with others. It could be stated that Evinrude learned the trade through the 'school of hard knocks.' His failures set him up for his eventual success.

[sss] It is interesting that Ole Evinrude was able to quit school at such a young age, as even then Wisconsin law required at least twelve weeks of school annually until children reached sixteen.

Jack Kilby, on the other hand, had what is now considered a conventional education. He studied electrical engineering at the University of Illinois, graduating with a degree in electric engineering in 1947. He supplemented this training with post-graduate night school courses, attaining a master's degree in electrical engineering, and his employer sent him to technical courses on the design and manufacture of transistors.

The knowledge he obtained with this background was important—as was the experience that he gained in his early career. As a youth, he was an amateur radio hobbyist, which is likely why he was assigned work as a technician during his military service during the Second World War. After the war, working as a newly hired engineer at the Centralab Division of Globe-Union in Milwaukee, Kilby's work assignments included the design and manufacture of miniature circuits for products as diverse as hearing aids and military bomb fuses.

This combination of knowledge and experience provided the foundation that allowed Jack Kilby to create the world's first microchip shortly after he left Milwaukee.

Finding a Need Not Being Met

Innovators are often encouraged to find a customer need that is not being met, and then satisfying that need. However, this advice is more difficult to follow than it would seem.

First, solutions are not often obvious. Henry Ford, when commenting about the first successful automobile that he built, stated, "If I had asked my customers what they wanted, they would have said a faster horse."[ttt]

According to Dave Pollard in his 'How to Save the World' blog, "The market for products and services, though far from perfect, is reasonably efficient at identifying and satisfying needs. If you find an unmet need, there is almost surely a reason why that need isn't being met by some other enterprise. You need to find out what that reason is, and overcome it. And then you need to gather a team of people with the collective competencies to design, produce, market and distribute the product or service that meets that need, and the resources (physical, financial and intellectual) needed to do so effectively. Easier said than done.

"The key to doing this is in research, the difficult, time-consuming (but usually inexpensive) process of discovering the who, what, when, where, why and how of unmet needs."[326]

Becoming an Entrepreneur

There were thousands of patents filed by Milwaukee inventors during the *Century of Progress*. Most did not result in any significant impact. Many of the ideas were impractical; there was no market for others. However, one of the greatest problems was the lack of knowledge of what it takes to manufacture a new product. There is a considerable difference between the skills of patent inventors and the skills and resources needed to bring products to the marketplace.

Many of the inventors thought that their patent award would open the way to richness. But seldom does a patent award result in interested investors beating a path to the door. Many of the inventors featured in this book already had the resources of a manufacturing company behind them. Those that didn't often struggled to obtain the resources to bring their product to market.

Thomas Edison famously stated that he was more of a "sponge" than an inventor.

[ttt] Questions have been raised about whether Henry Ford actually uttered those words. His great-grandson William Clay 'Bill' Ford Jr., believed it to be appropriately attributed. Another interesting quote from Henry Ford is, "History is more or less bunk. It's tradition. We don't want tradition. We want to live in the present, and the only history that is worth a tinker's damn is the history that we make today." Chicago Tribune, 1916. However, in spite of that language, Ford spent considerable sums of money to establish Greenfield Village and the Henry Ford Museum, to preserve the history of American innovation.

I am more of a sponge than an inventor. I absorb ideas from every source. I take half-matured schemes for mechanical development and make them practical. I am a sort of a middleman between the long-haired and impractical inventor and the hard-headed business man who measures all things in terms of dollars and cents. My principal business is giving commercial value to the brilliant but misdirected ideas of others.

Edison's statement underlies the need of applying "hard-headed" business-sense to innovation. There is also the practical reality of finding the financial resources necessary to develop the product fully. Several of the innovators reviewed in this book struggled with this reality. As just two examples, Ole Evinrude's first four partnerships failed because of insufficient resources and lack of marketing skills. Niels Christensen found interested investors to bring his streetcar braking system to market, but failed to secure an ownership in the enterprise based on his inventions.

Today in the Milwaukee area, there are several immersive programs in enterprise creation for technology innovators. These programs typically help innovators to obtain the skills and resources necessary to succeed. While there is no guarantee that investors will ultimately fund the development of any given innovator's products, these programs often provide the bridge between innovation and enterprise creation.

AVOID COMPLACENCY

Several large Milwaukee manufacturing companies have continued to survive and prosper—some for well over a century. But many companies have dropped out along the way, some of which failed to remain competitive in evolving commercial markets. The successful companies have continued to evolve; their products have changed and they have entered into other markets. Some have relocated their manufacturing plants—presumably to reduce costs and remain competitive. Others have set up manufacturing activities in foreign countries in order to expand as commerce has become increasingly global.

ADVANTAGES OF RE-INVENTING THE COMPANY

The A.O. Smith Corporation might be the best example of a Milwaukee-area manufacturer that has frequently reinvented itself. The company, founded as C.J. Smith and Sons in 1874, first made buggy parts and miscellaneous hardware. Over the course of its history it made products as varied as bicycle parts (at one time being the largest manufacturer of bicycle components in the world), pressed-steel automotive frames, engine-drive motor wheels, welded pipe/tanks along with welding rods and equipment, and military materials (including aircraft propellers and landing gear, torpedo air flasks and welded casings for aerial bombs and shells), electric motors, glass-lined tanks, and eventually water heaters—as of this writing, the company has entered the water purification industry and is currently one of the world's largest suppliers of water heaters and related products.

It must have been difficult and at times painful as many of these adjustments were made over the years. However, without periodic adjustments, it is unlikely the company would have survived and prospered.

A.O. Smith history is evidence that the risk of bringing out new products is significant, but the reward can be lasting.

PREPARING FOR THE "TIDES"

A recurring theme underlies the histories of many of the companies chronicled in this book. Most did well during the First World War, suffered significantly during the Great Depression, and recovered and flourished during the Second World War. Those companies with an even longer history also suffered during the panic of 1857 and the recessions of 1873, but fared well in the recovery following the Civil War. The saying, "All ships rise in a rising tide," seems to have been evident by the increase in manufacturing resulting from United States wartime efforts. Likewise, "All ships fall in the falling tide" was equally true.

That said, it is also noted that some companies did far better than others in adjusting to the rising and falling tides of commerce. In response to wartime demands, many expanded their factories with new buildings and tooling. While these facilities often led to increased productivity following the wars, they also added significantly to the costs. The key for many companies was to increase demand for its goods by designing and manufacturing better products in order to keep their shops full and productive. But, of course, competing companies had a similar motivation.

Similarly, almost all Milwaukee manufacturing companies pared back significantly during the Depression. Not all survived. Those that did often struggled financially, as did their employees.

The ability of companies to weather these rising and falling tides has been a key to their survival and success. While these cyclical events (there have been as many as 47 recessions in the United States since 1790—thankfully there have been far fewer wars) affected almost all US manufacturing companies, there were other events that affected specific manufacturing sectors. For example, the price of metals and energy greatly affects mining activities, causing demand for mining equipment to rise and fall along with ore and energy prices. Mining equipment has been a prominent Milwaukee-area industry and these cycles have presented both obstacles and opportunities to area manufacturers. Similar statements can be made for agriculture and other products.

Preparing and adjusting for these rising and falling tides has to be a major struggle for manufacturers. A detailed review of the history of Milwaukee-area manufacturing companies could perhaps help guide companies for future periods of volatility. However, individual circumstances may be so unique as to make such planning difficult. Retaining a healthy financial balance sheet might be the only way to help weather future storms, since it provides companies with the resources and time to adjust to the changing conditions.

Impact of Globalization

During the century covered by this book, the reach of Milwaukee-area industry went from local, to regional, to national. Following the Second World War, the market became international. The expansion of the reach of the marketplace provided additional customers for Milwaukee's products. However, it also resulted in competition from other areas of the country (initially) and the world.

The impact of globalization is outside of the scope of this book, given that most of these impacts occurred after the period of focus. However, it is evident that as the market and source of products expanded, Milwaukee industry was faced with innovative products produced elsewhere, as well as by the economic pressures of competing with areas of the country and world with lower labor, energy and shipping costs.

Some Milwaukee companies have continued to prosper in this global environment. Harley-Davidson, Rockwell, Johnson Controls, A.O. Smith, Briggs & Stratton and a few others have all been able to prosper. An evaluation of how Milwaukee-area companies adapted to the global marketplace may be worthy of consideration in a future book.

APPENDIX: MILWAUKEE AREA LANDMARKS

The items presented in this book are meant to represent a cross-section of the significant mechanical engineering accomplishments from the Milwaukee area. While it is an extensive list, it likely only draws off the surface of a deep well of laudible engineering efforts.

In 1979, the History and Heritage Committee of the Milwaukee Section of the American Society of Mechanical Engineers started work on an inventory of the significant mechanical engineering accomplishments of the Section's area as a special Centennial project. It was felt that conducting this inventory would be an appropriate way to prepare for and to celebrate the Society's one-hundredth anniversary in 1980. The previous year (1979) marked the 75th anniversary of the Milwaukee Section, which added impetus to the project.

Well over one hundred letters, seeking nominations for inclusion in this inventory, were sent out to local industry. A News Release was prepared to publicize the inventory and was carried by several area newspapers, by Mechanical Engineering magazine, and by the Section's Newsletter. The response to the solicitation was most encouraging; nearly one hundred items were nominated for the inventory. The Committee met to review these nominations and initially selected fifty items that were felt to merit placement on the inventory. Since that time, the inventory has grown, to the point that now over 130 items are represented.

Items have been selected from our inventory and submitted to the National History and Heritage Committee for landmark consideration. The Milwaukee Section has been successful in this regard, with the following items receiving landmark status:

- The Vulcan Street Plant in Appleton, among the first hydroelectric central station power plants in the United States
- The Automated Automobile Frame Plant (also known as the 'Milwaukee Marvel') of A.O. Smith in Milwaukee, which in 1918 could produce 10,000 frames per day in almost complete automation.
- The East Wells Street (Oneida Street) Power Plant of Wisconsin Electric Power Company in Milwaukee, where successful experiments in pulverized coal firing were conducted, leading' to lower electric power production costs.
- The Port Washington Power Plant of Wisconsin Electric Power Company, which was the most efficient power plant in the world for many years and served as the model for many other power stations worldwide.
- The Evinrude Outboard—the world's first commercially and mechanically successful outboard motor. This development was the key to revolutionizing pleasure boating. It became the foundation for the "World's outboard industry." There are Evinrude singles built in 1910-11 still running today.
- Milwaukee River Flushing Station screw type pump. At 500,000,000 gallons per day capacity, it pumped a greater quantity of water than any machine in the world (1888). It was used for flushing the Milwaukee River and remains in partial operation today
- Quincy Mine Company No. 2 Mine Hoist. This thirty-foot diameter mine hoist was the world's largest steam-driven hoist. It was built by Milwaukee's Nordberg Manufacturing Company in 1919-20.
- The Development of Pneumatic HVAC Controls for Multi-Zone Applications by Johnson Controls. This patented innovation by Warren S. Johnson represented the first (1895) method of pneumatic temperature control for buildings. It remains in common use today.
- Cooperative Fuel Research Engine for determining the 'Knock Rating' of gasoline by Waukesha Motors. This test engine provided a recognized standard for defining fuel quality, which continues today.

- Sholes & Glidden 'Type Writer,' the first commercially successful typewriter. Patented by Christopher Latham Sholes, "This was the machine that finally succeeded on the market and established the modern idea of the typewriter." Sholes and his associates worked in the machine shop of Charles Kleinsteuber, as well as a mill owned by Henry Smith and Charles S. Birge.
- "Big Brutus" Coal Mining Shovel, built by Milwaukee's Bucyrus-Erie Corporation. When built in 1962, this shovel was the second largest in the world. It was used for the removal of overburden in the surface mining of thin coal seams. In its lifetime, it recovered nine million tons of bituminous coal.
- The City of Jacksonville's (Florida) Reynolds-Corliss Reciprocating Steam Engine and Water Pump. This engine-driven pump was designed and manufactured by Allis Chalmers. Following the expiration of the Corliss' patent, Edwin Reynolds designed a valve mechanism which had several distinct advantages over the releasing gears previously employed. The leverage of the releasing mechanism was constant so that the reaction on the governor was the same at all points of cut-off. The gear was quieter and could run at much higher speeds.
- Hiwassee Dam Unit 2 Reversible Pump-Turbine, built by Allis Chalmers. The Hiwassee pump-turbine was the first reversible pump-turbine built and installed in this country using wicket gates for control of turbine output power and improved pump efficiency. The rated turbine power output of 80,000 HP was over four times greater than the next largest pump-turbine in the world. Designed and manufactured by engineers at Allis-Chalmers to meet the requirements of the Tennessee Valley Authority.
- *SS Badger* Carferry, built in Manitowoc Wisconsin. The two 3,500-hp steeple compound Unaflow steam engines powering the S.S. Badger represent one of the last types of reciprocating marine steam engines. Built by the Skinner Engine Company, most Unaflow engines are single expansion. These feature tandem high- and low-pressure cylinders separated by a common head. The Badger's four Foster-Wheeler Type D marine boilers, which supply 470-psig steam to the engines, are among the last coal-fired marine boilers built. The car-ferry remains in service between Manitowoc and Ludington, Michigan.
- Montana Western #31 Gasoline-Electric Rail Car. This self-propelled railcar built by the Electro-Motive Company (EMC) for the Great Northern Railroad, is the oldest surviving equipment from the founding company of today's diesel-electric locomotive manufacturer: Electro-Motive Division of General Motors. This railcar is located in the railway museum in North Freedom Wisconsin.
- Fairbanks Morse Y-VA Diesel Engine, manufactured in Beloit in 1924 and installed on Useppa Island, Florida, as an outstanding example of an early high-compression diesel that did not require pre-combustion.

The Milwaukee Section of ASME is proud of the fact that seventeen of the Society's landmarks represent machines that were constructed in the greater Milwaukee area, which reflects well on the engineering innovations of Eastern Wisconsin.

REFERENCES

About This Book

[1] *Wisconsin's Engineering Gems*, Paul G. Hayes, Milwaukee Journal, January 8, 1981 in a review of *Mechanical Engineering: A Century of Progress.*

[2] *Technological Contributions of Milwaukee's Menomonee Valley Industries,* presented at the Annual Meeting of the Society for the History of Technology, Thomas H. Fehring, Milwaukee, WI, 1980.

[3] *The Mechanical Engineer in America, 1830-1910,* Monte A. Calvert, Johns Hopkins Press, 1967.

[4] *The Milwaukee Road: Its First Hundred Years,* August Derleth, Creative Age Press, New York, 1948, p. 115.

Introduction

[5] *Let There be Light: The Electric Utility Industry in Wisconsin,* Forrest McDonald, The American History Research Center, Madison, WI, 1957, p. 7.

[6] *History of Milwaukee,* William George Bruce, SJ Clarke Publishing Company, Milwaukee, 1922, p. 235.

Chapter 1

[7] Journal of the House of Representatives, First Session of the Third Legislative Assembly of Wisconsin, 1841, p. 121.

[8] *A Brief History – The Milwaukee Road*, published by the Chicago, Milwaukee, St. Paul and Pacific Railroad Company in 1968.

[9] *Industrial History of Milwaukee: the Commercial, Manufacturing and Railway Metropolis of the North-West*, E.E. Barton, Publisher, 1886, p. 22.

[10] *The Making of Milwaukee*, John Gurda, published by the Milwaukee County Historical Society, 1999, p. 46.

[11] *Industrial History of Milwaukee*, p. 22.

[12] *Leading American Inventors: Biographies of Leading Americans*, George Iles, Henry Holt and Company, New York, 1912.

[13] *James Paris Lee*, Wikipedia article, retrieved June 24, 2016.

[14] *The History of Remington Firearms,* Roy M. Marcot, the Lyons Press, 2005, P. 59.

[15] *The Sholes & Glidden 'Type Writer,* A Historic Mechanical Engineering Landmark, Thomas H. Fehring, American Society of Mechanical Engineers, 2011.

[16] *The Typewriter and the Men Who Made It*, Richard N. Current, Champaign: University of Illinois Press. (1954) ISBN 0911160884

[17] *The Original Typewriter Enterprise 1867–1873,* Richard N. Current, Wisconsin Magazine of History Madison: State Historical Society of Wisconsin. (June 1949).

[18] *Mastering the Dynamics of Innovation.* James M. Utterback, Boston: Harvard Business Press. (1999). ISBN 0875847404, pg. 4.

[19] *The Difficult Birth of the Typewriter,* Cynthia Monaco, American Heritage of Invention and Technology," Spring/Summer 1988, Vol. 4, Number 1, p. 20.

[20] *The Woman and the Typewriter: A Case Study in Technological Innovation and Social Change*, Donald Hoke Business and Economic History (1979) Milwaukee: Milwaukee Public Museum, Series 2.

[21] *Honoring the Inventor of the Typewriter,"* a lecture by Alan C. Reiley, broadcast from Marquette University radio station, Milwaukee, June 6, 1924.

[22] Information on *remontoire* time pieces from *My Time Machines* at http://www.my-time-machines.net/remontoire.htm, retrieved March 8, 2016.

Chapter 2

[23] US Patent 232,073 filed on April 30, 1880 and published September 7 of the same year.

[24] *Restoring American Competitiveness,* Gary Pisano and Willy Shih, July-August 2009 edition of Harvard Business Review

[25] *An Industrial Heritage*, Walter F. Peterson, published by the Milwaukee County Historical Society, 1978.

[26] Ibid, p. 19.

27 *American Lumberman*, December 23, 1905, p. 37, as cited in Peterson, *An Industrial Heritage.*
28 *Trout Family History,* William Henry Trout, Milwaukee, 1916
29 *The Milwaukee Road: Its First Hundred Years,* August Derleth, Creative Age Press, New York, 1948, p. 115.
30 Trout Family History, p. 229.
31 Ibid, p. 245.
32 *An Industrial Heritage*, Peterson, P. 28-39.
33 *A Quarter Century of Milling*, W. D. Gray, the Weekly Northwestern Miller, as cited by Peterson, *An Industrial Heritage.*
34 *Ninety-Eight Years of Thermal Power History: Allis-Chalmers Manufacturing Company*, M. C. Maloney, ~1970, unpublished, p. 11. Maloney was a former employee of Allis-Chalmers and wrote his accounts based on his recollections as well as his personal research into the company's history. Apparently some of his information was taken from a work by Mrs. Alberta Johnson Price entitled, *Mill Stones to Atom Smashers*, which also appears to have gone unpublished.
35 *Mechanical Engineers Handbook,* Lionel S. Marks, ed., McGraw-Hill, New York, 1st Ed., 1916, p. 973.
36 *Ninety-Eight Years of Thermal Power History: Allis Chalmers Manufacturing Company*, M. C. Maloney, ~1970, unpublished, p. 19-20.
37 *Niels Christensen,* from Wisconsin: Its Story and Biography, 1848-1913, Ellis Baker Usher, Lewis Publishing Co., Chicago, 1914.
38 *Ninety-Eight Years of Thermal Power History*: Allis Chalmers Manufacturing Company, M. C. Maloney, ~1970, Unpublished, p. 24.
39 The Milwaukee Journal, December 30, 1979.
40 *Ninety-Eight Years of Thermal Power History: Allis Chalmers Manufacturing Company*, M. C. Maloney, ~1970, unpublished, p. 62.
41 *An Industrial Heritage*, Peterson, P. 94
42 *Ninety-Eight Years of Thermal Power History: Allis Chalmers Manufacturing Company*, M. C. Maloney, ~1970, unpublished, p. 88-89.
43 Adapted from practicalmachinist.com articles on steam engine flyweights, retrieved May, 2015.
44 Both summaries are based on bios originally published in 1909 in *Memoirs of Milwaukee County.*
45 ASME Brochure for the Reynolds Corliss steam engine in Jacksonville, FL. Available on asme.org
46 *The Allis Chalmers Story,* C.H. Wendel, Crestline Publishing, Sarasota, FL, 1988, p. 258.
47 From the history of Pawling and Harnischfeger as published on: http://www.morriscranes.com/history.php, retrieved May, 2015.
48 *Ninety Eight Years of Thermal Power History*, M. C. Maloney, unpublished, ~1970.
49 *The Steam Engine at the End of the 19th Century,* Dr. Robert H. Thurston, past president of the American Society of Mechanical Engineers, as reported in the Transactions of ASME, Volume 21, 1900, p. 181.
50 Ibid, p. 214.
51 *History and Origin of the Nordberg Manufacturing Company,* H. W. Dow, Sr., unpublished, 1949, personal recollections of a former Nordberg employee.
52 *Mechanical Engineer's Handbook,* Lionel S. Marks, ed., McGraw-Hill, New York, 1st Ed., 1916, p. 962.
53 *Locomotives of the Milwaukee and Mississippi Railroad and the Milwaukee & Prairie du Chien Railway, 1850-1867*, Douglas L. Hays, Jr.
54 Ibid.
55 *Memoirs of Milwaukee County,* Vol. 2, Jerome J. Watrous, editor, Western Historical Association, 1909, p. 295.
56 *Milwaukee in the 1930s, a Federal Writer's Project City Guide,* Edited by John D. Buenker, Wisconsin Historical Society Press, 2016, p. 148.
57 *Oldest Automobile in the World Made in Milwaukee,* Milwaukee, A Magazine for Her Business Leaders, January, 1922 edition, p. 12.
58 *Milwaukee in the 1930s*, p. 148.
59 Source: A.O. Smith corporate website, retrieved May, 2015.
60 *Kearney Trecker New Plant Ready,* The Milwaukee Sentinel, February 18, 1940.
61 *Little Journeys to the Homes of Wisconsin Industries*, Number 17: Kearney & Trecker Co., The Milwaukee Journal, September 5, 1920.
62 *George Meyer Ill a Day, Dies,* The Milwaukee Journal, July 30, 1945.
63 *Meyer Ships Huge Bottling Machine,* The Milwaukee Sentinel, April 13, 1960.
64 *Milwaukee is Recalled as Shipbuilding Center*, The Milwaukee Journal, March 24, 1949.
65 *Wolf & Davidson's Ship-Yard,* History of Milwaukee, 1881, p. 468.
66 *Forging Ahead: A Centennial History of Ladish Co.,* John Gurda, 2005, p.3

[67] *Ring Master,* George Wise, Invention & Technology, Spring/Summer 1991, Volume 7. The article notes that the Christensen braking system differed from Westinghouse's braking system for railcars of the time which used mechanically driven compressors.
[68] *Niels Christensen,* from Wisconsin: Its Story and Biography, 1848-1913 by Ellis Baker Usher; Lewis Publishing Co., Chicago, 1914.
[69] Ibid.
[70] *Allen-Bradley: An American Story,* Harry Lynde Bradley with Norman Beasley, edited by Robert Smith, Rockwell Automation, 2003.
[71] *Time on His Side: Allen-Bradley Clock Patent Identifies Designer*, Amy Rabideau Silvers, The Milwaukee Journal Sentinel, August 10, 2006.
[72] *The Pictorial History of Outboard Motors,* W. J. Webb, Renaissance Editions, Inc., New York, 1967, p. 32.
[73] *Ole Evinrude and His Outboard Motor,* Bob Jacobson, Wisconsin Historical Society Press, 2009, p. 28.
[74] Ole Evinrude, from Wikipedia, retrieved June 17, 2016.
[75] *Built by Seaman: Four Generations of Family Enterprise*, John Gurda, 2000.
[76] Based on information about a Petrel being auctioned by Sotheby, retrieved from rmsothebys.com on October 21, 2015
[77] Some information on Seaman Body is from the CoachBuilt.com website, retrieved October 18, 2015.
[78] *The Rambler Automobile,* Wikipedia retrieved October 22, 2015.
[79] *History of AT&T,* Wikipedia, retrieved November 2, 2015.
[80] *Joseph J. Zimmermann Jr., 92, an Inventor, Is Dead,* obituary by The New York Times, Douglas Martin, April 11, 2004.
[81] *The Master Switch: The Rise and Fall of Information Empires*, Tim Wu, Vintage Books, 2011.

CHAPTER 3

[82] *Ninety-Eight Years of Thermal Power History: Allis-Chalmers Manufacturing Company*, M. C. Maloney, unpublished, written about 1970, p. 4.
[83] *The Milwaukee Road: Its First Hundred Years,* August Derleth, Creative Age Press, New York, 1948, p. 22.
[84] Ibid, Appendix B.
[85] Ibid, p. 237
[86] *High Ball for High Speed Railroading: Wisconsin Men Led the World Into Modern Era of Racing Streamliners*, Frank Sinclair, Milwaukee Journal, August 8, 1948.
[87] The Milwaukee Magazine, published by the Chicago, Milwaukee, St. Paul and Pacific Railway Company, July 1947, p. 10.
[88] Ibid, p. 52.
[89] *The Milwaukee Road: Its First Hundred Years,* August Derleth, Creative Age Press, New York, 1948, p. 261.
[90] *The Milwaukee Road: Its First Hundred Years,* August Derleth, Creative Age Press, New York, 1948, p. 239
[91] Ibid, p. 249
[92] The Illustrated Encyclopedia of North American Locomotives, Brian Hollingsworth, 1984, p. 113.
[93] *The Milwaukee Road: Its First Hundred Years,* August Derleth, Creative Age Press, New York, 1948, p. p. 8.
[94] The Milwaukee Magazine, published by the Chicago, Milwaukee, St. Paul and Pacific Railway, July 1947, p. 7-13.
[95] From United States patent description for a *Method of Repairing Railway Track Rails.* No. 593,953. Patented Nov. 16, 1897. Assigned to Herman Falk.
[96] *The Making of a "Good Name in Industry" – A History of the Falk Corporation, 1892-1992*, John Gurda, 1991, p.28.
[97] The Milwaukee Sentinel, January 1, 1926.
[98] *The Making of a "Good Name in Industry" – A History of the Falk Corporation, 1892-1992*, John Gurda, 1991, p. 54.
[99] *The Making of a "Good Name in Industry—A History of the Falk Corporation,* 1892-1992, John Gurda, 1991, p. 54.
[100] Ibid, p. 60.
[101] Ibid, p. 93.
[102] National Department of Historic Places Nomination Form for the *U.S.S. LST 325*, dated May 5, 2009. The Naval ship is maintained as a memorial in Evansville, Indiana by USS LST Ship Memorial, Inc.
[103] *The Making of a "Good Name in Industry" – A History of the Falk Corporation, 1892-1992*, John Gurda, 1991, p. 51.
[104] *Thomas L. Smith, from the History of Milwaukee, City and County*, Volume 2, William George Bruce, S. J. Clarke Publishing Co, Chicago, 1922. It is not known what became of Smith's attempt to organize the school—no further records of it could be located.
[105] *History of Milwaukee, City and County*, Volume 2, William George Bruce, S. J. Clarke Publishing Co, Chicago, 1922.
[106] Ibid.

107 *Milwaukee, A Half-Century of Progress*, 1846-1896, Consolidated Illuminating Company, Milwaukee.
108 *Koehring Built Large Concern: Visioned Auto Era and then Prepared for Good Roads*, The Milwaukee Journal, November 1, 1921.
109 *Wisconsin's Heavy Machinery is Helping to Change the World,* The Milwaukee Journal, Centennial Exposition Edition, August 9, 1948, p. 14
110 Ibid.
111 *An American Dream: A Commemorative History of Cutler-Hammer, Inc., 1892-1978.* Cutler-Hammer, Inc., p. 9-10.
112 Ibid., p. 11.
113 *Mechanical Engineering: A Century of Progress,* The American Society of Mechanical Engineers--Milwaukee Section, Thomas H. Fehring, ed., 1980, p. 19.
114 *The Hamilton Metalplane, the Vintage Airplane*, George Hardie, Jr., May 1976, p. 3-8.
115 *Thomas F. Hamilton,* Wikipedia, retrieved June 26, 2015.

CHAPTER 4

116 *City of Milwaukee,* Wikipedia article retrieved July 2015.
117 *The Bridge War,* Milwaukee Timeline, Milwaukee County Historical Society website, retrieved July 2015.
118 *Let There be Light: the Electric Utility Industry in Wisconsin,* Forest McDonald, The American History Research Center, Madison Wisconsin, 1956, p. 12.
119 Ibid, p. 13.
120 *Networks of Power: Electrification in Western Society*, 1880-1930, by Thomas Parke Hughes, p. 88.
121 *Let There be Light: the Electric Utility Industry in Wisconsin,* Forest McDonald, The American History Research Center, Madison Wisconsin, 1956, p. 16.
122 *Path of a Pioneer: A Centennial History of the Wisconsin Electric Power Company*, John Gurda, 1996, p. 10.
123 *Let There be Light: the Electric Utility Industry in Wisconsin,* Forest McDonald, The American History Research Center, Madison Wisconsin, 1956, p. 51.
124 *John I. Beggs*, Wikipedia, retrieved June 17, 2015.
125 *Let There be Light, the Electric Utility Industry in Wisconsin, 1881-1955,* Forest McDonald, The American History Research Center, Madison, WI, 1957, p. 211.
126 Average US price of electricity for all users. Sources: 1920-1925, U.S. Bureau of the Census, *Census of Electrical Industries;* 1926-1970, Edison Electric Institute, *Edison Electric Institute Statistical Bulletin,* New York, 1952 and 1970 issues.
127 *From Thermostats to Automobiles,* The Center for Automotive History, autohistorycenter.com, posted September 17, 2014 retrieved August 2015.
128 *The Coherer from an article about Édouard Branly,* Wikipedia, retrieved January 9, 2017.
129 News article in The Electrical World and Engineer, a weekly review of progress in electricity, Vol. 36.
130 *Dr. Lee de Forest, Professor Warren Johnson and the American Wireless Telegraph Company,* Glenn M. Trischan, Antique Wireless Association, 2001, Vol. 14, p. 148.
131 Diaries of Lee de Forest, Vols. 1-15, Jan. 1891-July 1903; Library of Congress Microfilm Accession Number 17,174; as reported by Glenn M. Trischan.
132 *Dr. Lee de Forest, Professor Warren Johnson and the American Wireless Telegraph Company*, Glenn M. Trischan, Antique Wireless Association, 2001, Vol. 14, p. 165.
133 Ibid, p. 166.
134 Lee de Forest, Wikipedia article, retrieved January 10, 2017.
135 *Historical sketches of the first quarter-century of the State Normal School at Whitewater, Wisconsin: 1868-1893,* Tracy, Gibbs & Co., 1893, p. 125.
136 *Into the Value Zone: Gaining and Sustaining Competitive Advantage*, Ron Wood, University Press of America, 2008, p. 142.
137 From description of US patent no. 950126 by Stephen F. Briggs; applied on February 2nd, 1909, granted February 1910.
138 *International Directory of Company Histories*, Vol. 27. St. James Press, 1999
139 Based upon private correspondence with Walter Smith of SmithPumps, descendent of Rueben Stanley Smith, on March 20, 2017, as well as information from the company website: smithpumps.com.
140 *The Small-Engine Handbook*, Peter Hunn, Motorbooks, June, 2005, and private communication with the company.
141 From the website of South Dakota State University at: http://state.sdstateconnect.org/stephen-briggs/, retrieved June 15, 2016.

[142] Based on DESCO corporate history from its website at: divedesco.com/divingequipment/CompanyHistory, retrieved January 5, 2016.
[143] Ibid, DESCO website.
[144] *The founders of DESCO, Maximilian Eugene Nohl*, The Journal of Diving History, Winter 2012 1 Volume 20, Issue 1, Number 70, p.23-30.
[145] From Wikipedia at: https://en.wikipedia.org/wiki/Phillips_Lord, retrieved September 7, 2015.
[146] Milwaukee Journal article dated February 11, 1935.
[147] *A Short History of Diving and Diving Medicine,* Eric P. Kindwall, Jefferson C. Davis and Alfred A. Bove, WB Saunders Company, p. 6-7.
[148] *Decompression Practice,* https://en.wikipedia.org/wiki/Decompression_practice, retrieved October 2015.
[149] Scuba technology born from one of Medical College of Wisconsin's earliest recorded discoveries, mdw.edu, retrieved October 2015.

CHAPTER 5

[150] *Memoirs of Milwaukee County,* Jerome Anthony Watrous, Western Historical Association, Chicago, 1909; Volume 1, p. 68-69
[151] *Cudahy, Wisconsin, Generations of Pride,* Joan Paul, Acadia Publications, 2002, p. 7.
[152] Wisconsin statute 66.0215
[153] *Oak Creek: Fifty Years of Progress,* Jim Cech, Arcadia Publications, Voices of America series, 2005
[154] *The Story of Bay View*, Bernhard C. Korn, Milwaukee County Historical Society, 1980, p. 52.
[155] *The Vilter Booster,* Harold Sloan, June 1950, p. 7-10.
[156] Ibid, p. 9.
[157] Ibid, p. 9.
[158] *Industrial History of Milwaukee,* 1886, E.E. Barton Publisher, p. 129.
[159] *Singer Vibrating Shuttle,* Wikipedia – retrieved February 24, 2016
[160] *Wisconsin Success Story No. 41, Nordberg Teems with Industry,* George Smedal, The Milwaukee Sentinel, August 5, 1956, p. 11.
[161] Wisconsin Industrial Hall of Fame.
[162] *History of Bucyrus International,* from the website Funding Universe, retrieved January 8, 2016.
[163] *The Wisconsin Motor Manufacturing Company*, Brian Wayne Wells, Belt Pulley Magazine, September/October 2004
[164] *Filer Stowell Output Jumps*, Milwaukee Journal, May 16, 1943, p. 3.
[165] *Forging Ahead: A Centennial History of Ladish Co.*, John Gurda, 2005, p. 18.
[166] Ibid, p. 49.
[167] Ibid, p. 58.
[168] Ibid, p 39.
[169] *Space Shuttle Challenger disaster*, Wikipedia article, retrieved February 2016.
[170] *Heil's Antics Jolt Wisconsin,* New York Times, June 4, 1939.
[171] *Milwaukee Shipyard One of Busiest on Great Lakes,* James L. Cabot, Ludington Daily News, September 2, 1995.

CHAPTER 6

[172] *The Milwaukee Road's Beer Line*, Art Harnack, Milwaukee Road Historical Association, 1970.
[173] *Riverwest: A railroad ran through it,* Carl Swanson, Milwaukee Notebook blog, retrieved July 6, 2016.
[174] *The Bigelow Defalcation,* from "The Romance and Tragedy of Banking," by Thomas P. Kane, The Bankers' Publishing Co., 1922, p. 260.
[175] *1914 Transactions,* Society of Automotive Engineers, Part 1, Volume 9, p. 264
[176] US Patent 1614452 A, filed August 1, 1924, granted January 18, 1927.
[177] *The Radio Products of the Globe Electric Company*, Glenn Trischan, The Antique Wireless Association Review, Vol. 24, 2011.
[178] *Jack St. Clair Kilby: A Man of Few Words*, a brief biography, Edwin G. Mills, 2008, p. 33.
[179] Institute of Electrical and Electronics Engineers, 1984 International Solid State Circuits Conference.
[180] *ASME Honors the Model T,* landmark brochure, American Society of Mechanical Engineers, May 20, 2005.
[181] *Original Model T Ford Prices by Model and Year*, the Model T Ford Club International, Inc., retrieved October 23, 2015. Prices for the 'runabout' model.
[182] *The Model T Ford, the Car that Changed the World,* Bruce McCalley, Krause Publications, 1994.

183 *Motown West: How Milwaukee once made claim at being automotive industry's heart,* Milwaukee Journal, John Gurda, July 4, 2009

184 *Nash Motors,* Wikipedia retrieved January 14, 2016.

185 *Nils Wahlberg and Nash - Salute To a Great Engineer and Unsung Automobiles,* retrieved from http://vwlarry.blogspot.com/2009/05/ive-been-wanting-to-do-piece-about-this.html on January 15, 2016

186 *Nash Offers New Style, Weather-Eye,* Milwaukee Sentinel, November 12, 1938.

187 *Weather Eye,* Wikipedia, retrieved January 15, 2016.

188 *Ford and Nash show first new cars,* Popular Science, August 1945, retrieved January 11, 2016.

189 *Damage Uncruncher for Unibody Cars,"* Popular Science, August 1982, retrieved January 11, 2016.

190 NHTSA Crash Test Scores

191 *The Economic Growth Engine: How Energy and Work Drive Material Prosperity*, Robert U. Ayres and Benjamin Warr, Edgar Elgar Publishing, 2009, p. 132.

192 Wikipedia, retrieved October 16, 2015.

193 *Stoking Furnace as a Farm Boy Turned Him to Boilermaking,* Doyle K. Ketter, Milwaukee Journal, July 3, 1959.

194 *Aqua-Chem Leads in Salt Water Conversion,* Mervin C. Nelson, Milwaukee Sentinel, October 31, 1966.

CHAPTER 7

195 *American Bicycle Company,* Wikipedia, retrieved October 12, 2015.

196 *Automobile Review,* Vol. 4, p 679, Technical Press of America, Chicago, May 7, 1904.

197 *Crown Worm-Drive Trucks,* Automotive Trade Journal, Vol. 18, Issue 10 p. 243, April 1914

198 From the website www.theflyingmerkel.com

199 Ibid.

200 *Joseph Merkel: Founder of the Merkel Motorcycles, engineer.* From 1998 induction into the AMA Motorcycle Hall of Fame

201 *Growing Up Harley Davidson, Memoirs of a Motorcycle Dynasty*, Jean Davidson, Veloce Publishing Ltd., 2001, p. x.

202 *Growing Up Harley-Davidson*, p. 39.

203 *Biography of William Sylvester Harley, Inventor of the World Famous Harley-Davidson Motorcycle, Co-founder and First Chief Engineer of the Harley-Davidson Motor Company, Milwaukee, Wisconsin,* Herbert Wagner, originally appeared in American Rider Magazine, July 1998.

204 *Growing Up Harley-Davidson*, p. 51.

205 *Biography of William Sylvester Harley, Inventor of the World Famous Harley-Davidson Motorcycle, Co-founder and First Chief Engineer of the Harley-Davidson Motor Company, Milwaukee, Wisconsin,* Herbert Wagner, originally appeared in American Rider Magazine, July 1998.

206 *Classic Harley-Davidson, 1903-1941*, Herbert Wagner, MotorBooks International, ISBN 978-0-7603-0557-7, 1999, p. 13.

207 Motorcycle Illustrated, September 21, 1911

208 Website sidecar.com/mbbs22/forums, retrieved February 1, 2016.

209 *The Harley-Davidson Motor Company, Little Journeys to the Home of Wisconsin Industries*, Milwaukee Journal, July 4, 1920.

210 *Harley-Davidson WLA,* Wikipedia article, retrieved February 2016.

211 *Biography of William Sylvester Harley*, Herbert Wagner, an updated and expanded version of his article in the July 1998 edition of American Rider magazine's "Special Harley-Davidson 95th Anniversary" issue.

212 Ibid.

213 Ibid.

214 Both racers are cited in the AMA Motorcycle Hall of Fame. Information retrieved from its website, motorcyclemuseum.org in February 2016.

215 *Biography of William Sylvester Harley*, Herbert Wagner, an updated and expanded version of his article in the July 1998 edition of American Rider magazine's "Special Harley-Davidson 95th Anniversary" issue.

216 *Miracle of World War II, How American Industry Made Victory Possible*, Francis Walton, MacMillan Co., New York, p. 396.

217 *Then & Now: The Mechanical Marvel*, Austin Weber, Assembly Magazine, Feb.1, 2002.

218 "*A Truly Amazing Legacy-The Smith Family,*" Walter Smith (grandson of Reuben Stanley Smith), from the website of Smith Precision Products, retrieved October 2015.

219 Apparatus for forming metal elements, US Patent 1628751 A, by Reuben Stanley Smith assigned to A.O. Smith; filed November 15, 1921 and published May 17, 1927

[220] *Welding, A Journey to Explore Its Past*, Andre A. Odermatt, Hobart Institute of Welding Technology, Troy, Ohio. Swedish born Oscar Kjellberg appears to have received the first patent (German Imperial patent number 231733) for the coated welding electrode in June 1908 – one year before Stohmenger. Stohmenger's use of asbestos exposed welders to asbestos fumes – obviously dangerous to their lungs.
[221] *Miracle of World War II*, p. 396.
[222] *How a newspaper "sparked" one of the biggest developments in A.O. Smith history*, published by A.O. Smith to celebrate its 140th anniversary, 2014. Orrin Andrus was related to the Smith family. He was the first cousin of Reuben Stanley Smith on his mother's side of the family.
[223] *Pioneering Advancements in Welding*, Walter Smith, from the website of Smith Precision Products, retrieved October 2015.
[224] *Miracle of World War II*, p. 396.
[225] *Handbook of Structural Welding: Processes, Materials and Methods*, John Lancaster, Abington Publishing, 1992, p. 52.
[226] *Encyclopedia of Chemical Processing and Design*, John J. McKetta, Chapter 67, water and wastewater treatment, protective coatings and zeolites. 1999, Marcel Dekker, Inc. New York
[227] *Miracle of World War II*, p. 397.
[228] *Freedom's Forge: How American Business Produced Victory in World War II*, Arthur Herman, Random House, 2012, p. 99.
[229] *Liberty Ship: Problems,* Wikipedia, retrieved December 2015.
[230] *Wisconsin's Heavy Machinery is Helping to Change the World,* The Milwaukee Journal, Centennial Exposition Edition, August 9, 1948, p. 14
[231] *Largest Mixer is Shipped Out,* The Milwaukee Journal, January 1, 1928.
[232] *Former Wells Resident with Milwaukee Firm,* The Escanaba Daily Press, May 3, 1949. It was reported that this machine had a guaranteed capacity of 6.6 cubic yards. An article in the Milwaukee Journal on May 11, 1949 simply reported that the machine weighed more than 20 tons.
[233] *Brake Patent Brings Award,* Milwaukee Journal, Feb. 27, 1930, p.
[234] *International Directory of Company Histories*, Vol. 45. St. James Press, 2002.
[235] Based on Master Lock's website at: http://www.masterlock.eu/about-us, retrieved September 1, 2016.
[236] Biography relied extensively on *The Harry Soref Story—a legacy of mind and heart,* which was prepared based on recollections of his daughter, Ruth Sklar Coleman and published in Milwaukee, in 2003.

CHAPTER 8

[237] *The Allis-Chalmers Story,* Charles H. Wendel, Crestline Publishing Co., 1988, p. 14.
[238] Ibid, p. 169.
[239] *The Tesla Bladeless Turbine & Related Turbomachinery*, Nikola Tesla, C.R. Possell, compiled by Gary Peterson, Twenty-First Century Books.
[240] Ibid.
[241] *A Century of Milwaukee Water: An Historical Account of the Origin and Development of the Milwaukee Water Works*, Elmer W. Becker, Milwaukee Water Works, 1974.
[242] *The Allis Chalmers Story,* C.H. Wendel, Crestline Publishing, Sarasota, FL, 1988, p. 258.
[243] *The Riverside Pumping Station*, brochure prepared by Milwaukee Water Works Department,
[244] *World War II, Milwaukee,* Meg Jones, The History Press, 2015, p. 69.
[245] From http://www.u-s-history.com/pages/h1652.html, retrieved October 26, 2016.
[246] Wikipedia at https://en.wikipedia.org/wiki/K-25, retrieved October 25, 2016.
[247] *Hiwassee Dam Unit 2 Reversible Pump-Turbine,* Landmark Brochure, The American Society of Mechanical Engineers, 1981.
[248] *An Important Development in the Art of Welding,* Milwaukee Journal, August 2, 1936.
[249] *A. F. Milbrath, Designer of Wisconsin Engine,* Motor Age, May 25, 1916, p. 27. It is not known what engineering school was cited in the article, since the Milwaukee School of Engineering was not established until 1903, and Marquette didn't start an engineering college until 1908.
[250] *Arthur F. Milbrath, Obituary,* The Milwaukee Sentinel, February 17, 1955.
[251] Information on the history of the Sternberg Manufacturing Company was adopted from, *The History of Sterling Trucks,* Ernest R. Sternberg, published by the Society of Automotive Engineers, SP-941, February 1993.
[252] *The History of Sterling Trucks,* Ernest R. Sternberg, published by the Society of Automotive Engineers, SP-941, February 1993, p. 36.
[253] *Number 17: Kearney & Trecker Co.,* The Milwaukee Journal, September 5, 1920.

254 *Powering Waukesha and the World: The Motor Works,* John Schoenknecht writing for The Freeman, Waukesha County's newspaper. Published in ten parts during 2014.
255 Ibid, June 2014.
256 *Powering Waukesha and the World: The Motor Works*, John Schoenknecht writing for The Freeman
257 Website of the Waukesha Engine Historical Society Inc., retrieved December 2015.
258 *The Harry L. Horning Award,* SAE International's website, retrieved December 2015.
259 *Harry Horning, Engineer, Dies,* The Milwaukee Journal, January 4, 1936.

CHAPTER 9

260 *Wisconsin's Engineering Gems*, Paul G. Hayes, Milwaukee Journal, January 8, 1981 in a review of *Mechanical Engineering: A Century of Progress*, Thomas H. Fehring, ASME-Milwaukee, 1980.
261 *10 Best Engineering Breakthroughs,* Csaba Csere, Car and Driver, January 1988, 33, p. 61.
262 Passenger Cars Soon Will Have Air Brakes, The Tuscaloosa News, 5-Oct-1930.
263 United States Patent Number 2180795, filed 2-Oct-1937 and granted 21-Nov-1939.
264 *Ring Master,* George Wise, Invention & Technology, Spring/Summer 1991, Volume 7, Issue 1.
265 *Truth, Lies, and O-Rings: Inside the Space Shuttle Challenger Disaster*, Allan J. McDonald and James Hansen, 2012.
266 *Seven Myths about the Challenger Shuttle Disaster*, James Oberg, NBC News, 25-Jan-2011, at: nbcnews.com retrieved 26-Mar-16
267 *Jack St. Clair Kilby, a Man of Few Words*, a brief biography, Ed Mills, 2008, p. 40.
268 *Two Inch Box Could Work TV, Bardeen Claims*, The Milwaukee Journal, 16-May-1957.
269 Ibid, p. 42.
270 Ibid, p. 42.

CHAPTER 10

271 *A History of the Milwaukee Vocational and Adult Schools,* Robert W. Tarbell, MVAS Press, Milwaukee, 1958, p. 5.
272 Average Number of Wage Earners in Manufacturing Industry, Table 23, 1909 United States Census.
273 *New School of Trades Opened,* Spokane Daily Chronicle, January 2, 1906.
274 *How Boys Are Taught at School of Trades,* The Milwaukee Journal, September 25, 1911.
275 The plumbing curriculum was the notable exception; it was a one-year course.
276 *How Boys Are Taught at School of Trades,* The Milwaukee Journal, September 25, 1911.
277 The History of UW-Extension from uwex.edu/about/uw-extension-history.html, retrieved April 28, 2016.
278 *A History of the Milwaukee Vocational and Adult Schools,* Robert W. Tarbell, MVAS PRESS, Milwaukee, 1958, p. 17.
279 *The Wisconsin Idea*, by Charles McCarthy, MacMillan Company, Norwood Press, 1912.
280 Chapter 660, laws of 1911, chapter 505, Wisconsin Laws of 1911
281 Chapter 522, Wisconsin Laws of 1911
282 *The Wisconsin Idea*, by Charles McCarthy, MacMillan Company, Norwood Press, 1912
283 *A History of the Milwaukee Vocational and Adult Schools,* Robert W. Tarbell, MVAS Press, Milwaukee, 1958, p. 52-53.
284 Ibid, p. 237.
285 *The Radio Products of the Globe Electric Company*, Glenn Trischan, The Antique Wireless Association Review, Vol. 24, 2011.
286 *The story of Marquette University*, Rev. Raphael Hamilton, S.J., Marquette University Press, 1953
287 *Milwaukee's Jesuit University: Marquette, 1881-1991,* Thomas J. Jablonsky, Marquette University Press, 2007, p. 75.
288 *The story of Marquette University*, p. 87-91.
289 *Marquette Is Part of Community,* Milwaukee Journal, 2-Mar-1961, p. 22.
290 Milwaukee's Jesuit University: Marquette, 1881-1991, p. 76.
291 *The story of Marquette University*, p. 227-229.
292 *Milwaukee's Jesuit University: Marquette, 1881-1991*, p. 168.
293 *The story of Marquette University*, p. 294-297.
294 *Normal School,* Wikipedia, retrieved May 23, 2016.
295 *Engineering Education,* American Society for Engineering Education, 1970, p. 467.
296 *The Mechanical Engineer in America, 1830-1910,* Monte A. Calvert, Johns Hopkins Press, 1967.
297 *Milwaukee Masters Manufacturing through Engineering,* Erwin G. Spewachek, Assistant Professor, Department of Engineering at Marquette University, presented to the 1954 Annual Meeting of ASME, held in Milwaukee that year.
298 ASCE Annual Conference, Transactions of the American Society of Civil Engineers, Volume XIX, 1888.

[299] *A Brief History of ESM: Serving Engineers for Seventy-Five Years*, prepared by the Engineers and Scientists of Milwaukee in 1979.
[300] *Transactions of the American Society of Mechanical Engineers*, Volume 22, 1901.
[301] *A Centennial History of the American Society of Mechanical Engineers, 1880-1980*, Bruce Sinclair, published by University of Toronto Press for ASME, 1980.
[302] Bibliographies of the founding officers, along with those of early Section Chairmen, is included in this report.
[303] Power Magazine, Volume 25, Published January, 1905. Article occurs on page 122.
[304] *A Centennial History of the American Society of Mechanical Engineers*, Bruce Sinclair, ASME, 1980, p. 79-80.
[305] *American Machinist*, 1905 XXVII 2 813
[306] A Centennial History of ASME, p. 80.
[307] Ibid. p. 80.
[308] From the Journal of the ASME, Volume 39 (January 1917).
[309] *A Brief History of ESM: Serving Engineers for Seventy-Five Years*, prepared by the Engineers and Scientists of Milwaukee, 1979.

CHAPTER 11

[310] *History of Milwaukee,* William George Bruce, SJ Clarke Publishing Company, Milwaukee, 1922, p. 235.
[311] *A Contribution to the Theory of Economic Growth,* Robert Merton Solow, Quarterly Journal of Economics 70 (1):65-94. 1956. (Solow received the 1987 Nobel Prize in Economics for his contributions in this area.)
[312] *Restoring American Competitiveness,* Gary Pisano and Willy Shih, July-August 2009 edition of Harvard Business Review
[313] *The Business Model Innovation Factory*, Saul Kaplan, John Wiley & Sons, Hoboken, NJ, 2012.
[314] *Leading American Inventors,* George Iles, Henry Holt and Company, New York, 1912.
[315] *E. Remington and Sons,* Wikipedia article, retrieved June 24, 2016.
[316] *James Paris Lee,* from forgottenweapons.com/biographies/james-paris-lee/, retrieved June 24, 2016.
[317] *The History of Remington Firearms,* Roy M. Marcot, the Lyons Press, 2005, p. 59.
[318] United States Patent 79265A for "Improvement in type-writing machines," by C. Latham Sholes, Carlos Glidden and Samuel W. Soule, dated June 23, 1868.
[319] *An Industrial Heritage*, Walter F. Peterson, published by the Milwaukee County Historical Society, 1978.
[320] *Trout Family History,* William Henry Trout, Milwaukee, 1916.
[321] *Don't Offer Employees Big Rewards for Innovation,* Oliver Baumann and Nils Stieglitz, Harvard Business Review, June 12, 2014, located at hbr.org.
[322] *The Economic Case for Welcoming Immigrant Entrepreneurs,* Dane Stangler and Jason Wiens, Kauffman Foundation, 2014.
[323] *Immigrants and Billion Dollar Startups,* Stuart Anderson, National Foundation for American Policy, NFAP Policy Brief, March 2016.
[324] *Ole Evinrude and His Outboard Motor,* Bob Jacobson, Wisconsin Historical Society Press, 2009, p. 14.
[325] Ibid, p. 15.
[326] *Natural Enterprise: Filling an Unmet Need,* Dave Pollard, posted on September 3, 2004, retrieved at howtosavetheworld.ca

INDEX

ABOUT THE AUTHOR

I am an engineer by training and practice, having worked for thirty-five years at "keeping the lights on" at Wisconsin Energy Corporation and its various subsidiaries. Earlier in my career, I also worked for a time at Falk Corporation, Briggs & Stratton and Ford Motor Company.

I was born in Milwaukee and grew up in the shadows of the A.O. Smith factory near 35th and Capitol Drive. I attended Marquette University and attained bachelor and master's degrees in Mechanical Engineering. Along the way, I married and my wife Suzan and I raised three children.

I consider myself an accidental historian. I didn't really enjoy history in school because of the rote memorization of events and dates that was generally required. But early in my engineering career one of my supervisors helped to "plant the seed" that led to my fascination with industrial history. He encouraged me to give back to my profession by volunteering. Taking to the call, I attended a local meeting of the American Society of Mechanical Engineers (ASME) and offered my services. Never a group to turn away free help, they assigned me to an open position as chair of their history and heritage committee. I found the assignment to be very rewarding. Forty years later, industrial archeology continues to be an important part of my life.

My foray into history continued after my retirement when I was appointed to ASME's history and heritage committee—which I currently chair. I have also been involved with community history, serving on the Whitefish Bay Historic Preservation Commission. I have published two books on the Village's history.

I first published a book on Milwaukee's industrial heritage in 1980. It was a modest effort, but was surprisingly well received. Since then I have continued to gather information and materials about the early companies of Milwaukee and the incredible machines that were built here over the years. My hope was to publish a more thorough book on the subject eventually. This effort is my attempt to fulfill this aspiration.

It has given me great pleasure to explore Milwaukee's amazing history and I am grateful for the chance to share it with you—just don't ask me to remember all of the dates!

ABOUT THE AUTHOR

CPSIA information can be obtained
at www.ICGtesting.com
Printed in the USA
LVOW04s1150190717
541880LV00019B/554/P